FRACTAL MECHANICS OF POLYMERS

Chemistry and Physics of Complex Polymeric Materials

FRACTAL MECHANICS OF POLYMERS

OF POLYMERS

Chemistry and Physics of
Complex Polymeric Materials

G. V. Kozlov, DSc, and Yu. G. Yanovskii, DSc

Gennady E. Zaikov, DSc, and A. K. Haghi, PhD
Reviewers and Advisory Board Members

Apple Academic Press
TORONTO   NEW JERSEY

Apple Academic Press Inc. | Apple Academic Press Inc.
3333 Mistwell Crescent | 9 Spinnaker Way
Oakville, ON L6L 0A2 | Waretown, NJ 08758
Canada | USA

©2015 by Apple Academic Press, Inc.

First issued in paperback 2021

Exclusive worldwide distribution by CRC Press, a member of Taylor & Francis Group
No claim to original U.S. Government works

ISBN 13: 978-1-77463-357-1 (pbk)
ISBN 13: 978-1-77188-041-1 (hbk)

Library of Congress Control Number: 2014952078

Library and Archives Canada Cataloguing in Publication

Kozlov, G. V., author
Fractal mechanics of polymers: chemistry and physics of complex polymeric materials/G.V. Kozlov, DSc, and Yu. G. Yanovskii, DSc; Gennady E. Zaikov, DSc, and A.K. Haghi, PhD, Reviewers and Advisory Board Members.

Includes bibliographical references and index.
ISBN 978-1-77188-041-1 (bound)
1. Polymers--Fracture. 2. Fracture mechanics. 3. Fractal analysis. I. Yanovskii, Yurii Grigorevich, author II. Title.

TA455.P58K69 2014   620.1'920426   C2014-906643-0

Apple Academic Press also publishes its books in a variety of electronic formats. Some content that appears in print may not be available in electronic format. For information about Apple Academic Press products, visit our website at **www.appleacademicpress.com** and the CRC Press website at **www.crc-press.com**

ABOUT THE AUTHORS

G. V. Kozlov, DSc

G. V. Kozlov, DSc, is a Senior Scientist at UNIID of Kabardino-Balkarian State University in Nal'chik, Russia, during 1981–1994 and from 1997 until now. His scientific interests include the structural grounds of properties of polymeric materials of all classes and states, including the physics of polymers, polymer solutions and melts, and composites and nanocomposites. He has proposed to consider polymers as natural nanocomposites. He is the author of more than 1500 scientific publications, including 30 books, published in Russia, Ukraine, Great Britain, German Federal Republic, Holland, and USA.

Yu. G. Yanovskii, DSc

Yurii Grigor'evich Yanovskii, DSc, is Director of the Institute of Applied Mechanics, Russian Academy of Sciences, Moscow, Russia. He is a specialist in the field of physics and chemistry of polymers, composites, and polymer materials. The majority of his publications are concerned with the mechanical properties of polymers. He is contributor of 1000 original papers as well as many reviews. He published more than 100 books.

REVIEWERS AND ADVISORY BOARD MEMBERS

Gennady E. Zaikov, DSc

Gennady E. Zaikov, DSc, is Head of the Polymer Division at the N. M. Emanuel Institute of Biochemical Physics, Russian Academy of Sciences, Moscow, Russia, and Professor at Moscow State Academy of Fine Chemical Technology, Russia, as well as Professor at Kazan National Research Technological University, Kazan, Russia. He is also a prolific author, researcher, and lecturer. He has received several awards for his work, including the Russian Federation Scholarship for Outstanding Scientists. He has been a member of many professional organizations and on the editorial boards of many international science journals.

A. K. Haghi, PhD

A. K. Haghi, PhD, holds a BSc in urban and environmental engineering from University of North Carolina (USA); a MSc in mechanical engineering from North Carolina A&T State University (USA); a DEA in applied mechanics, acoustics and materials from Université de Technologie de Compiègne (France); and a PhD in engineering sciences from Université de Franche-Comté (France). He is the author and editor of 65 books as well as 1000 published papers in various journals and conference proceedings. Dr. Haghi has received several grants, consulted for a number of major corporations, and is a frequent speaker to national and international audiences. Since 1983, he served as a professor at several universities. He is currently Editor-in-Chief of the *International Journal of Chemoinformatics and Chemical Engineering* and *Polymers Research Journal* and on the editorial boards of many international journals. He is a member of the Canadian Research and Development Center of Sciences and Cultures (CRDCSC), Montreal, Quebec, Canada.

CONTENTS

LIST OF ABBREVIATIONS

BPE	Branched Polyethylenes
CSC	Crystallites with Stretched Chains
DLA	Diffusion-Limited Aggregation
DS	Dissipative Structures
EP	Epoxy Polymer
HDPE	High Density Polyethylene
IP	Inclined Plates
PAASO	Polyarylate Arylene Sulfide Oxide
PAr	Polyarylate
PASF	PolyarylateSulfone
PC	Polycarbonate
PMMA	Poly(Methyl Methacrylate)
PP	Polypropylene
PPO	Polyphenyleneoxide
PS	Polystyrene
REP	Cross-Linked Epoxy Polymer
SIA	Slit Islands Analysis
UHMPE	Ultra-High-Molecular Polyethylene
ZD	Deformation Zones
CSC	Crystallites with Stretched Chains

PREFACE

As it is known, the construction of relationships, which gives intercommunication between structure and mechanical properties of medium, is the main goal of polymers mechanics. The main grounds of fractal analysis methods usage expediency for general purposes are based on the known experimental data number. As it is well-known at present, the polymer medium structure represents itself as a fractal object within the range of linear scales from several Ångströms up to several tens of Ångströms. This fact was confirmed experimentally more than once. The correct description of a similar fractal object is easily attained only within the frameworks of the fractal analysis, since the usage for these purposes of Euclidean geometry notions can give only relatively approximate description. The latter is explained by the fact that Euclidean objects with whole numerical dimension possess translational symmetry and fractal ones—dilatational one. The most simple confirmation of the above is the usage of the equations, obtained strictly enough within the frameworks of high-elasticity theory concepts, for glassy polymers behavior description. These equations give at best an approximate description, since high-elastic materials structure, for example, rubbers, is simulated as an Euclidean object, but the introduction into the real (fractal) dimension of the structure allows obtaining a precise description of glassy polymers behavior. However it is necessary to note that fractal analysis often gives more a precise prediction of rubber composites behavior in comparison with continuous models.

The important argument in favor of fractal approach application is the usage of two order parameter values, which are necessary for correct description of polymer mediums structure and properties features. As it is known, solid phase polymers are thermodynamically nonequilibrium mediums, for which Prigogine-Defay criterion is not fulfilled, and therefore, two order parameters are required, as a minimum, for their structure description. In its turn, one order parameter is required for Euclidean object characterization (its Euclidean dimension d). In general case three parameters (dimensions) are necessary for fractal object correct description: dimension of Euclidean space d, fractal (Hausdorff) object dimension d_f and its spectral (fraction) dimension d_s. Hence, actually at the value d (e.g., $d = 3$) assignment condition two parameters of order (dimensions, d_f and d_s) are required for fractal object description.

Let us note that the fractal dimension is represented by general characteristic, which do not take into account material structure specific features. As a matter of fact, the value d_f describes object density reduction gradient as its size growth. Therefore, various classes' materials (e.g., natural mineral and polymer) can have the same fractal dimension of structure. The indicated feature of fractal dimension may have both positives and negative points. Polymers possess complex multilevel structure (molecular, topological, supramolecular levels). These levels are characterized by parameters of various kinds that may even complicate analytical relationships among them. The usage of physically uniform parameters (fractal dimensions) for structural levels allows them to obtain easily and strictly their relationships. It should be noted that the cluster model of amorphous polymeric medium structure, based on structural model. However, in this book it will be shown that polymeric medium structure fractality is due to local order regions availability.

The purpose of the present book is the consideration of the relationships connecting structural and basic mechanical properties of polymeric mediums within the frameworks of fractal analysis with cluster model representations attraction. Incidentally the choice of any structural model of medium or their combinations is defined by expediency and further usage convenience only.

This new book:

- highlights some important areas of current interest in polymer products and chemical processes,
- focuses on topics with more advanced methods,
- emphasizes precise mathematical development and actual experimental details,
- analyzes theories to formulate and prove the physicochemical principles, and
- provides an up-to-date and thorough exposition of the present state-of-the-art of complex polymeric materials.

CHAPTER 1

THE INTERCOMMUNICATION OF FRACTAL ANALYSIS AND POLYMERIC CLUSTER MEDIUM MODEL

CONTENTS

During the last 25 years, fractal analysis methods obtained wide-spread attention in both theoretical physics [1] and material science [2], in particular, in physics-chemistry of polymers [3–8]. This tendency can be explained by fractal objects wide distribution in nature.

There are two main physical reasons, which define intercommunication of fractal essence and local order for solid-phase polymers: the thermodynamical nonequilibrium and dimensional periodicity of their structure. In Ref. [9], the simple relationship was obtained between thermodynamical nonequilibrium characteristic – Gibbs function change at self-assembly (cluster structure formation of polymers $\Delta \tilde{G}^{im}$ – and clusters relative fraction φ_{cl} in the form:

$$\Delta \tilde{G}^{im} \sim \phi_{cl} \qquad (1.1)$$

This relationship graphic interpretation for amorphous glassy polymers – polycarbonate (PC) and polyarylate (PAr) – is adduced in Fig. 1.1. Since at $T = T_g (T_m)$ (where T, T_g and T_m are testing, glass transition and melting temperatures, accordingly) $\Delta \tilde{G}^{im} = 0$ [10, 11], then from the Eq. (1.1) it follows, that at the indicated temperatures cluster structure full decay ($\varphi_{cl} = 0$) should be occurred or transition to thermodynamically equilibrium structure.

As for the intercommunication of parameters, characterizing structure fractality and medium thermodynamical nonequilibrium, it should exist indisputably, since precise nonequilibrium processes formed fractal structures. Solid bodies' fracture surfaces analysis gives evidence of such rule fulfillment—a large number of experimental papers shows their fractal structure, irrespective of the analyzed material thermodynamical state [12]. Such phenomenon is due to the fact that the fracture process is a thermodynamically nonequilibrium one [13]. Polymers structure fractality is due to the same circumstance. The experimental confirmation can be found in Refs. [14–16]. As for each real (physical) fractal, polymers' structure fractal properties are limited by the defined linear scales. So, in Refs. [14, 17] these scales were determined within the range of several Ångströms (from below) up to several tens Ångströms (from above). The lower limit is connected with medium structural elements finite size and the upper one – with structure fractal dimension d_f limiting values [18]. The indicated above scale limits correspond well to cluster nanostructure specific boundary sizes: the lower – with statistical segment length l_{st}, the upper – with distance between clusters R_{cl} [19].

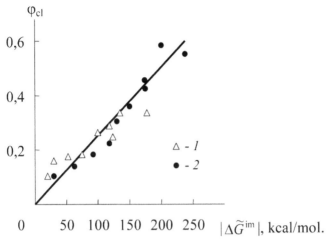

FIGURE 1.1 The dependence of clusters relative fraction j_{cl} on absolute value of specific Gibbs function of nonequilibrium phase transition $|\Delta \tilde{G}^{im}|$ for amorphous glassy polymers – polycarbonate (1) and polyarylate (2) [3].

Polymeric medium's structure fractality within the indicated above scale limits assumes the dependence of their density ρ on dimensional parameter L (see Fig. 1.2) as follows [1]:

$$\rho \sim L^{d_f - d}, \tag{1.2}$$

where d is dimension of Euclidean space, into which a fractal is introduced.

In Fig. 1.3 amorphous polymers nanostructure cluster model is presented. As one can see, within the limits of the indicated above dimensional periodicity scales Fig. 1.2 and 1.3 correspond each other, that is, the cluster model assumes ρ reduction as far as possible from the cluster center. Let us note that well-known Flory "felt" model [20] does not satisfy this criterion, since for it $\rho \approx$ const. Since, as it was noted above, polymeric mediums structure fractality was confirmed experimentally repeatedly [14–16], then it is obvious, that cluster model reflects real solid-phase polymers structure quite plausibly, whereas "felt" model is far from reality. It is also obvious, that opposite intercommunication is true – for density ρ finite values change of the latter within the definite limits means obligatory availability of structure periodicity.

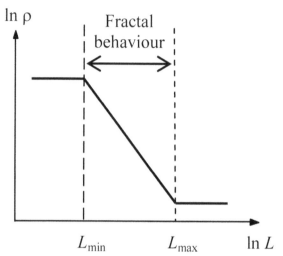

FIGURE 1.2 The schematic dependence of density ρ on structure linear scale L for solid body in logarithmic coordinates. The fractal structures are formed within the range $L_{min} - L_{max}$ [18].

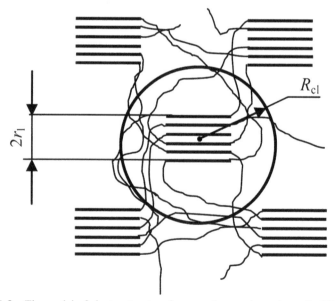

FIGURE 1.3 The model of cluster structure for amorphous polymeric media [7].

Meanwhile, we must not over look the fact that fractal analysis gives only polymers structure general mathematical description only, that is, does not identify those concrete structural units (elements), which any real polymer consists. At the same time, the physical description of polymers thermodynamically nonequilibrium structure within the frameworks of local order notions is given by the cluster model of amorphous polymers structure gives, which is capable of its elements quantitative identification [21].

The well described in literature experimental studies was served as premise for the cluster model development. As it is known (see, for example, [22, 23]), high-molecular amorphous polymeric mediums, which are in glassy state, at their deformation above glass transition temperature can display high-elastic behavior (when high-elasticity plateau is formed), which can be described by the high-elasticity theory rules. Personally, Langievene [24, 25] and Gauss [26] equations can be used for polymers behavior description large strains. In the latter it is assumed that at deformation in plateau region polymeric chain is not stretched fully and the relationship between true stress σ^{tr} and drawing ratio λ at uniaxial tension is written as follows:

$$\sigma^{tr} = G_p(\Lambda^2 - \Lambda^{-1}), \tag{1.3}$$

where G_p is the so-called strain hardening modulus.

The value G_p knowledge allows formally to calculate the density v_e of macromolecular entanglements network in polymeric medium according to the well-known expressions of high-elasticity theory [23]:

$$M_e = \frac{\rho RT}{G_p}, \tag{1.4}$$

$$v_e = \frac{\rho N_A}{M_e}, \tag{1.5}$$

where ρ is polymer density, R is universal gas constant, T is testing temperature, M_e is molecular weight of chain part between entanglements, N_A is Avogadro number.

However, the attempts to estimate the value M_e (or v_e) according to the determined from Eq. (1.3) G_p values result to unlikely low calculated values M_e (or unreal high values v_e), contradicting to Gaussian statistics requirements, which assume availability on chain part between entanglements no less than ~ 13 monomer links [27].

As alternative the authors of Ref. [28] assumed, that besides the indicated above binary hooking network in polymers glassy state another entanglements type was available, nodes of which by their structure were similar to crystallites with stretched chains (CSC). Such entanglements node possesses large enough functionality F (under node functionality emerging from it chains number is assumed [29] and it was called cluster. Cluster consists of different macromolecules segments and each such segment length is postulated equal to statistical segment length l_{st} ("stiffness part" of chain [30]). In that case the effective (real) molecular weight of chain part between clusters M_{cl}^{ef} can be calculated as follows [29]:

$$M_{cl}^{ef} = \frac{M_{cl}F}{2},$$

(1.6)

where M_{cl} is molecular weight of the chain part between clusters, calculated according to the Eq. (1.4).

It is obvious, that at large enough F the reasonable values M_{cl}^{ef}, satisfied to Gaussian statistics requirements, can be obtained. Further on for parameters of entanglements cluster network and macromolecular binary hooking's network distinction we will used the indices "cl" and "e," accordingly. Therefore, the offered in Ref. [28] model assumes, that amorphous polymer structure represents itself local order regions (domains), consisting of different macromolecules collinear densely packed segments (clusters), immersed in loosely packed matrix. Simultaneously clusters play the role of physical entanglements network multifunctional nodes. The value F (within the frameworks of high-elasticity theory) can be estimated as follows [31]:

$$F = \frac{2G_{\infty}}{kT\nu_{cl}} + 2,$$

(1.7)

where G_{∞} is equilibrium shear modulus, k is Boltzmann constant, ν_{cl} is cluster network density.

In Fig. 1.4, the dependences $\nu_{cl}(T)$ for polycarbonate (PC) and polyarylate (PAr) are adduced. These dependences show ν_{cl} reduction at T growth, that assumes local order regions (clusters) thermofluctuational nature. Besides, on the indicated dependences two characteristic temperatures are found easily. The first from them, glass transition temperature T_g, defines clusters full decay (see also Fig. 1.1), the second T_g' corresponds to the fold on curves $\nu_{cl}(T)$ and settles down on about 50 K lower T_g.

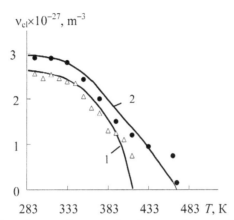

FIGURE 1.4 The dependences of macromolecular entanglements cluster network density n_{cl} on testing temperature T for PC(1) and PAr (2) [28].

Earlier within the frameworks of local order concepts it has been shown that temperature T_g is associated with segmental mobility releasing in polymer loosely packed regions. This means, that within the frameworks of cluster model T_g can be associated with loosely packed matrix devitrification. The dependences $F(T)$ for the same polymers have a similar form (Fig. 1.5).

Two main models, describing local order structure in polymers (with folded [34] and stretched [28] chains) allows to make common conclusion: local order regions in polymer matrices play role of macromolecular entanglements physical network nodes [28, 34]. However, their reaction on mechanical deformation should be distinguished essentially, it at large strains regions with folded chains ("bundles") are capable to unfolding of folds in straightened conformations, then clusters have no such possibility and polymer matrix deformation can occur by "tie" chains (connecting only clusters) straightening, that is, by their orientation in the applied stress direction.

Turning to analogy about mediums with crystalline morphology of polymer matrices, let us note, that semicrystalline polymers (e.g., high-molecular polyethylenes) large strains, making $1000 \div 2000\%$, are realized owing to crystallites folds unfolding [35].

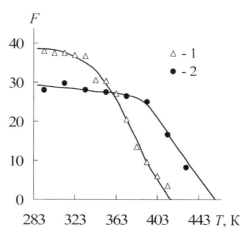

FIGURE 1.5 The dependence of clusters functionality F on testing temperature T for PC (1) and PAr (2) [28].

Proceeding from the said above and analyzing values of polymers limiting strains, one can obtain the information about local order regions type in amorphous and semicrystalline polymers. The fulfilled by the authors of Ref. [36] calculations have show that the most probable type of local nanostructures in amorphous polymer matrix is an analog of crystallite with stretched chains, that is, cluster.

Let us note, completing this topic, that for the local order availability substantiation in amorphous polymer matrix (irrespective to concrete structural model of medium) strict mathematical proofs of the most common character exist. For example, according to the proved in numbers theory Ramsey s theorem any large enough quantity $N_i > R(i, j)$ of numbers, points or objects (in the considered case – statistical segments) contains without fail high-ordered system from N_j $R(i, j)$ such segments. Therefore, absolute disordering of large systems (structures) is impossible [37, 48].

As it is known [39], structures, which behave themselves as fractal ones on small length scales and as homogeneous ones – on large ones, are named homogeneous fractals. Percolation clusters near percolation threshold are such fractals [1]. As it will be shown lower, cluster structure is a percolation system and in virtue of the said above – homogeneous fractal. In other words, local order availability in polymers condensed state testifies to there structure fractality [21].

The percolation system fractal dimension d_f can be expressed as follows [39]

$$d_f = d - \frac{\beta_p}{v_p},$$ (1.8)

where β_p and v_p are critical exponents (indices) in percolation theory.

Hence, the condition $\beta_p \neq 0$, which follows from the data of Table 1.1, also testifies to also polymer matrix structure fractality. It is obvious, that polymer medium structure fractality can be determined by the dependence of φ_{cl} on temperature, that is, using analysis of local order thermofluctuational effect (see Fig. 1.4). Let us note that structure fractality and "freezing" below T_g local order (φ_{cl} = const) from the physical point of view are interexepting notions.

TABLE 1.1   A Percolation Clusters Characteristics [5]

Parameter	Experimentally determined magnitudes		Theoretical "geometrical" magnitudes
	EP-1*	EP-2*	
b_p	0.54	0.58	0.40
n	1.20	1.15	0.88
b_p/n_p	0.45	0.50	0.46
d_f	2.55	2.50	2.55

*EP-1 and EP-2 are epoxy polymers on the basis of resin ED-22.

The considered above principles allow to connect by analytical relationship the fractal dimension d_f and local order characteristics φ_{cl} (or v_{cl}). According to the Ref. [18], the value d_f is connected with Poisson's ratio v as follows:

$$d_f = (d - 1)(1 + v),$$ (1.9)

and the value v is connected, in its turn, with v_{cl} according to the relationship [21]:

$$v = 0,5 - 3,22 \times 10^{-10} \left(l_0 v_{cl} \right)^{1/2},$$ (1.10)

where the following intercommunication between φ_{cl} exists:

$$\varphi_{cl} = S l_0 C_\infty \varphi_{cl},$$ (1.11)

where S is macromolecule cross-sectional area, l_0 is main chain skeletal length, C_∞ is characteristic ratio.

From the Eqs. (1.9) ÷ (1.11) the dependence of d_f on ϕ_{cl} or v_{cl} can be obtained for the most often used case $d = 3$ [21]:

$$d_f = 3 - 6,44 \times 10^{-10} \left(\frac{\phi_{cl}}{SC_\infty} \right)^{1/2} \qquad (1.12)$$

or

$$d_f = 3 - 6,44 \times 10^{-10} \left(l_0 v_{cl} \right)^{1/2} \qquad (1.13)$$

Let us note, returning to the Eq. (1.1), that its substation in (1.12) allows to obtain the following relationship:

$$\Delta \tilde{G}^{im} \sim C_\infty S \left(3 - d_f \right)^2 \qquad (1.14)$$

The value $\Delta \tilde{G}^{im}$ (Gibbs specific function of local, for example, supramolecular structures formation) is given for nonequilibrium phase transition "supercooled liquid → solid body" [11]. From the Eq. (1.14) it follows, that the condition $\Delta \tilde{G}^{im} = 0$ is achieved at $d_f = 3$, that is, at $d_f = d$ and at transition to Euclidean behavior. In other words, a fractal structures are formed only in nonequilibrium processes course, which is noted earlier [12].

Above general reasoning's about possibility of existence in polymeric media of local order regions, based on the Ramsey theorem, were presented in a simplified enough form. It can be shown similarly, that any structure, consisting of N elements, at $N > B_N(j)$ represents it self totality of finite number $k \leq j$ of put into each other self-similar structures, Hausdorff dimension of which in the general case can be different one. Therefore, any structure irrespective of its physical nature, consisting of a large enough elements number, can be represented as multifractal (in partial case as monofractal) and described by spectrum of Renyi dimensions d_q, $q = -\infty \div +\infty$ [18]. In Ref. [40] it has been shown, that the condensed systems attainment to self-organization in scale – invariant multifractal forms is the result of key principles of open systems thermodynamics and d_q is defined by competition of short- and long-range interatomic correlations, determining volume compressibility and shear stiffness of solid bodies, accordingly.

It can be noted as a brief resume to this chapter, that close intercommunication exists between notions of local order and fractality in a glassy polymers case, having key physical grounds and expressed by the simple analytical relationship (the Eqs. (1.12) and (1.13)). It has also been shown, that the combined usage of these complementing one another concepts allows to broaden possibilities of polymeric mediums structure and properties analytical description. The indicated expressions will be applied repeatedly in further interpretation.

KEYWORDS

- cluster model
- fractal analysis
- local order
- percolation model
- polymer medium
- Ramsey's theorem

REFERENCES

1. Feder, E. (1989). Fractals. New York, Plenum Press, 248 p.
2. Ivanova, V. S., Balankin, A. S., Bunin, I. Zh., & Oksogoev, A. A. (1994). Synergetics and Fractals in Material Science. Moscow, Nauka, 383 p.
3. Novikov, V. U., & Kozlov, G. V. (2000). A Macromolecules Fractal Analysis. Uspekhi Khimii, *69(4)*, 378–399.
4. Novikov, V. U., & Kozlov, G. V. (2000). Structure and Properties of Polymers within the Frameworks of Fractal Approach. Uspekhi Khimii, *69(6)*, 572–599.
5. Kozlov, G. V., & Novikov, V. U. (1998). Synergetics and Fractal Analysis of Cross-Linked Polymers. Moscow, Klassika, 112 p.
6. Kozlov, G. V., Yanovskii, Yu. G., & Zaikov, G. E. (2011). Synergetics and Fractal Analysis of Polymer Composites Filled with short Fibers. New York, Nova Science Publishers Inc., 223 p.
7. Kozlov, G. V., Yanovskii, Yu. G., & Zaikov, G. E. (2010). Structure and Properties of Particulate-Filled Polymer composites: The Fractal Analysis. New York, Nova Science Publishers Inc., 282 p.
8. Shogenov, V. N., & Kozlov, G. V. (2002). A Fractal Clusters in Physics-Chemistry of Polymers. Nal'chik, Poligrafservis IT, 268 p.
9. Kozlov, G. V., & Zaikov, G. E. (2001). The Generalized Description of Local Order in Polymers. In: Fractal and Local Order in Polymeric Materials. Kozlov, G. V., Zaikov, G. E. Ed., New York, Nova Science Publishers Inc., 55–64.

10. Gladyshev, G. P., & Gladyshev, D. P. (1993). About Physical-Chemical Theory of Biological Evolution (Preprint). Moscow, Olimp, 24 p.

11. Gladyshev, G. P., & Gladyshev, D. P. (1994). The Approximative Thermodynamical Equation for Nonequilibrium Phase Transitions. Zhurnal Fizicheskoi Khimii, *68(5)*, 790–792.

12. Hornbogen, E. (1989). Fractals in Microstructure of Metals. Int. Mater. Rev., *34(6)*, 277–296.

13. Bessendorf, M. H. (1987). Stochastic and Fractal Analysis of Fracture Trajectories. Int. J. Engng. Sci., *25(6)*, 667–672.

14. Zemlyanov, M. G., Malinovskii, V. K., Novikov, V. N., Parshin, P., & Sokolov, A. P. (1992). The Study of Fractions in Polymers. Zhurnal Eksperiment. I Teoretich. Fiziki, *101(1)*, 284–293.

15. Kozlov, G. V., Ozden, S., Krysov, V. A., & Zaikov, G. E. (2001). The Experimental Determination of a Fractal Dimension of the Structure of Amorphous Glassy Polymers. In: Fractals and Local Order in Polymeric Material. Kozlov, G. V., Zaikov, G. E., Ed., New York, Nova Science Publishers Inc., 83–88.

16. Kozlov, G. V., Ozden, S., & Dolbin, I. V. (2002). Small Angle X-Ray Studies of the Amorphous Polymers Fractal Structure. Russian Polymer News, *7(2)*, 35–38.

17. Bagryanskii, V. A., Malinovskii, V. K., Novikov, V. N., Pushchaeva, L. M., & Sokolov, A. P. (1988). Inelastic Light Diffusion on Fractal Oscillation Modes in Polymers. Fizika Tverdogo Tela, *30(8)*, 2360–2366.

18. Balankin, A. S. (1991). Synergetics of Deformable Body. Moscow, Publishers of Defence Ministry of SSSR, 404p.

19. Kozlov, G. V., Belousov, V. N., & Mikitaev, A. K. (1998). Description of Solid Polymers as Quasitwophase Bodies. Fizika I Tekhnika Vysokikh Davlenii, *8(1)*, 101–107.

20. Flory, J. (1976). Spatial Configuration of Macromolecular Chains. Brit. Polymer J. *8(1)*, 1–10.

21. Kozlov, G. V., Ovcharenko, E. N., & Mikitaev, A. K. (2009). Structure of the Polymers Amorphous State. Moscow, Publishers of the D.I. Mendeleev RkhTU, 392 p.

22. Haward, R. N. (1995). The Application of a Gauss-Eyring Model to Predict the Behavior of Thermoplastics in Tensile Experiment. J. Polymer Sci.: Part B: Polymer Phys, *33(8)*, 1481–1494.

23. Haward, R. N. (1987). The Application of a Simplified Model for the Stress-Strain Curves of Polymers. Polymer, *28(8)*, 1485–1488.

24. Boyce, M. C., Parks, D. M., & Argon, A. S. (1988). Large Inelastic Deformation in Glassy Polymers. Part I. Rate Dependent Constitutive Model. Mech. Mater, *7(1)*, 15–33.

25. Boyce, M. C., Parks, D. M., & Argon, A. S. (1988). Large Inelastic Deformation in Glassy Polymers. Part II. Numerical Simulation of Hydrostatic Extrusion. Mech. Mater, *7(1)*, 35–47.

26. Haward, R. N. (1993). Strain Hardening of Thermoplastics. Macromolecules, *26(22)*, 5860–5869.

27. Bartenev, G. M., & Frenkel, S. Ya. (1990). Physics of Polymers. Leningrad, Khimiya, 432 p.

28. Belousov, V. N., Kozlov, G. V., Mikitaev, A. K., & Lipatov, Yu. S. (1990). Entanglements in Glassy State of Linear Amorphous Polymers. Doklady AN SSSR, *313(3)*, 630–633.

29. Flory, J. (1985). Molecular Theory of Rubber Elasticity. Polymer J., *17(1)*, 1–12.

30. Bernstein, V. A., & Egorov, V. M. (1990). Differential Scanning Calorimetry in Physics-Chemistry of the Polymers. Leningrad, Khimiya, 256 p.
31. Graessley, W. W. (1980). Linear Viscoelasticity in Gaussian Networks. Macromolecules, *13(2)*, 372–376.
32. Perepechko, I. I., & Startsev, O. V. (1973). Multiplet Temperature Transitions in Amorphous Polymers in Main Relaxation Region. Vysokomolek. Soed. B, *15(5)*, 321–322.
33. Belousov, V. N., Kotsev, B. Kh., & Mikitaev, A. K. (1983). Two-step of Amorphous Polymers Glass Transition Doklady. AN SSSR, *270(5)*, 1145–1147.
34. Arzahakov, S. A., Bakeev, N. F., & Kabanov, V. A. (1973). Supramolecular Structure of Amorphous Polymers. Vysokomolek. Soed. A, *15(5)*, 1145–1147.
35. Marisawa, Y. (1987). The Strength of Polymeric Materials. Moscow, Khimya, 400 p.
36. Kozlov, G. V., Sanditov, D. S., & Serdyuk, V. D. (1993). About Suprasegmental Formations in Polymers Amorphous State. Vysokomolek. Soed. B, *35(12)*, 2067–2069.
37. Mikitaev, A. K., & Kozlov, G. V. (2008). The Fractal Mechanics of Polymeric Materials. Nal'chik, Publishers KBSU, 312 p.
38. Balankin, A. S., Bugrimov, A. L., Kozlov, G. V., Mikitaev, A. K., & Sanditov, D. S. (1992). The Fractal Structure and Physical Mechanical Properties of Amorphous Glassy Polymers. Doklady AN, *326(3)*, 463–466.
39. Sokolov, I. M. (1986). Dimensions and other Geometrical Critical Exponents in Percolation Theory. Uspekhi Fizicheshikh Nauk, *150(2)*, 221–256.
40. Balankin, A. S. (1991). Fractal Dynamics of Deformable Mediums. Pis'ma v ZhTF, *17(6)*, 84–89.

CHAPTER 2

MOLECULAR MOBILITY

CONTENTS

Sufficient attention is always paid to problems of molecular mobility in polymeric mediums (see, e.g., [1–3]). The reasons for it are obvious: polymers represent thermodynamically nonequilibrium solids, and their physical properties are defined by molecular relaxation processes proceeding in them. In its turn, molecular relaxation (mobility) processes depends on macromolecular chains constitution features and structural organization of polymeric medium. However, there is no common point of view on molecular mechanisms, it is assumed [1], that fast relaxation processes are defined by free chains mobility, located between structure densely packed regions, which simultaneously are network nodes of physical entanglements, formed by macromolecules. Such treatment displays full conformity to the basis statements of the cluster model of amorphous polymers structure [4, 5], one of the advantages of which is the possibility of quantitative description of structure elements. Spreading of these notions to amorphous glassy polymers means "freezing" of densely packed regions, that is, a sharp increase of their lifetime. This means that in glassy state the main factor defining molecular mobility level in polymeric medium is the mobility of free chains, that is, macromolecules parts, located between clusters [6]. The authors of Ref. [7] study this problem on example of a series of diblock-copolymers of oligoformal 2,2-di-(4-oxyphenyl)-propane-oligosulfone phenolphthalein (SP-OPD).

The mentioned above common character of fractal description application allows to use it for polymer structure prediction beginning at synthesis stage, since both macromolecular coils in solution and structure of polymer in condensed state are fractals with dimensions D and d_f accordingly [8]. The value D with appreciation of excluded volume interactions can be calculated as follows:

$$D = \frac{d_s(d+2)}{d_s + 2},\qquad (2.1)$$

where d_s is spectral (fracton) dimension of macromolecular coil, characterizing its connectivity degree [9], d is dimension of Euclidean space, in which a fractal is considered (in is obvious, that in our case $d=3$).

Since all the considered in Ref. [7] diblock-copolymers are linear polymers, then for them it follows to assume $d_s = 1$ [9]. The transition to gelation or to condensed state in polymerization process of polymeric medium is characterized by macromolecular coil environment change, that is, instead of solvent low-molecular molecules in the first case it is surrounded by similar coils in the second case. This results to fractal dimension change and for condensed state system its value d_f is determined as follows [8]:

$$d_f = \frac{d_s(d+2)}{2}. \qquad (2.2)$$

The Eqs. (2.1) and (2.2) combination at the indicated above conditions d_s = 1 and d = 3 gives for linear polymers [10–12]:

$$d_f = 1.5\ D. \qquad (2.3)$$

The Eq. (2.3) demonstrates clearly "genetic" intercommunication between reaction product (macromolecular coil in solution) and polymer condensed state structures. A methods number exists for the value D estimation in case of various combinations polymer – solvent [13–16].

Within the frameworks of model [4, 5] polymers amorphous state structure represents itself local order regions (clusters), immersed in loosely packed matrix (consisting of free chains according to the terminology [1]). In general case loosely packed regions fraction increase stimulates molecular mobility intensification. Therefore, it is to be expected that such structural factor as relative fraction of polymer matrix loosely packed part $\varphi_{l.m.}$ increase results to enhancement of dielectric losses angle tangent tgδ. The following expression is a general relationship for $\varphi_{l.m.}$ estimation [17]:

$$\varphi_{l.m.} = 1 - \varphi_{cl}, \qquad (2.4)$$

where φ_{cl} is relative fraction of local order regions (clusters).

The value $\varphi_{l.m.}$ estimation can be fulfilled, proceeding from magnitude d_f and polymer molecular characteristics using the Eq. (1.12) or Eq. (1.13). The indicated relationships are the example of general parameter d_f concretization with appreciation of molecular characteristics of polymer given type C, l_0 and S. The dependence tg$\delta(\varphi_{l.m.})$ plotting for copolymers SP – OPD shows [7], that, as it was to be expected, the loosely packed component fraction increase, characterized by $\varphi_{l.m.}$ growth, results to molecular mobility intensification: tgδ increases. The linear dependence tg$\delta(\varphi_{l.m.})$ extrapolation to $\varphi_{l.m.}$ = 1.0 gives the value tgδ = 0.062, that is typical for polymers considered type at glass transition temperature T_g. Such result has been foreseen, since within the frameworks of cluster model polymer s devitrification is identified as the decay of "frozen" local order, that is, $\varphi_{l.m.}$ = 0 or $\varphi_{l.m.}$ = 1.0 [19].

The dependence tg$\delta(\varphi_{l.m.})$ extrapolation to the value tgδ = 0, in course of which finite nonzero value $\varphi_{l.m.}$ 0.56 was obtained, is of greater interest (it can be expected, that molecular mobility suppression should occur at φ_{cl} =

1.0 or $\varphi_{l.m.} = 0$). The finite value $\varphi_{l.m.}$ existence possibility at tg$\delta = 0$ has the key physical significance, that will be discussed below. In the considered context the common dependence tg$\delta(\varphi_{l.m.})$ absence is important, since the data for copolymers with formal blocks content $C_{form} = 0$ located in left-hand side from the common dependence and with $C_{form} = 70$ mol.% − in right-hand side. This means that one more structural characteristic exists, characterizing molecular mobility and, hence, the value tgδ unequivocally [6, 7].

The fractal dimension of the chain part between clusters D_{ch} was proposed to use as such characteristic in Refs. [6, 7] that is due to the following reasons. Firstly, the mentioned chain part possesses self-similarity property and has the dimension, differing from its topological dimension, that is, it is fractal by definition [20]. Secondly, it has been shown [21], that the value D_{ch} within the range of values $1.0 < D_{ch} \leq 2.0$ characterizes exactly molecular mobility of a chain part (free chain according to the terminology [1]) in loosely packed matrix. The condition D_{ch} means this chain part straightening, transition from fractal behavior to Euclidean one and mobility full loss, that is, tg$\delta = 0$. The condition $D_{ch} = 2.0$ means this chain part maximum mobility, corresponding to its viscoelastic state. The value D_{ch} can be determined according to the equation [20]:

$$D_{ch} = \frac{\ln n_{st}}{\ln\left(4-d_f\right) - \ln\left(3-d_f\right)}, \tag{2.5}$$

where n_{st} is statistical segments number on chain part between entanglements (clusters).

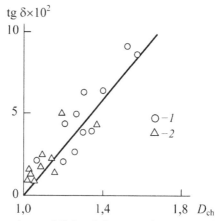

FIGURE 2.1   The dependence of dielectric losses angle tangent tgd on fractal dimension of chain part between clusters D_{ch} for copolymers SP − OPD at frequencies 1 (1) and 10 kHz [7].

The dependence tgδ (D_{ch}), obtained for the considered copolymers at two measurement frequencies (1 and 10 kHz) is adduced in Fig. 2.1. This dependence exists within the clearly determined limits. At D_{ch} = 1.0 it tries to attain tgδ = 0, that was to be expected from the made above assumptions. At D_{ch} = 2.0 the value tgδ is approximately equal to its maximum value. Therefore, the linear dependence tgδ(D_{ch}) can be used for this parameter prediction in different D_{ch} ranges.

Similar linear dependences for SP – OPD with various C_{form} were obtained in Ref. [7] and they testify to molecular mobility level reduction at d_f decrease and extrapolate to various (nonintegral) d_f values at D_{ch} = 1.0. The comparison of these data with the Eq. (1.5) appreciation shows, that D_{ch} reduction is due to local order level enhancement and the condition D_{ch} = 1.0 is realized at d_f values, differing from 2.0 (as it was supposed earlier in Ref. [23]). This is defined by polymers structure quasiequilibrium state achievement, which can be described as follows [24]. Actually, tendency of thermodynamically nonequilibrium solid body, which is a glassy polymer, to equilibrium state is classified within the frameworks of cluster model as local order level enhancement or φ_{cl} increase [24–26]. However, this tendency is balanced by entropic essence straightening and tauting effect of polymeric medium macromolecules, that makes impossible the condition φ_{cl} = 1.0 attainment. At fully tauted macromolecular chains (D_{ch} = 1.0) φ_{cl} increase is ceased and polymer structure achieves its quasiequilibrium state at d_f various values depending on copolymer type, that is defined by their macromolecules different flexibility, characterized by parameter C_∞.

Let us consider within the frameworks of irreversible aggregation model one more possible interpretation of polymers structure features – the so called diffusion-limited aggregation (DLA) [27, 28].

According to the Ref. [29] for DLA the value $d_f \approx 2.5$. The spatial distribution of DLA elements (in our case – statistical segments) can be characterized with the aid of two dimensions: d_f and chemical dimension d_l, which are defined as follows:

$$N \sim r^{d_f}, \tag{2.6}$$

$$N \sim l^{d_l}, \tag{2.7}$$

where N is particles number in cluster between two arbitrary points of cluster, r and l are distances between these points [30]. The value r is the piece length, connecting these points and l is defined as the shortest way over cluster particles between these points. In common case $l \geq r$ and equality sign is achieved in that case, when the indicated chain piece is stretched fully between points, that is, it has dimension $D_{ch} = 1.0$. The condition $l = r$ assumes the equality $d_f = d_l$, that follows from the Eqs. (2.6) and (2.7) and according to Ref. [31] achieves for DLA only. Therefore, polymers structure quasiequilibrium state realization at $D_{ch} = 1.0$ within the frameworks of DLA model assumes also polymer chains straightening also ($l = r$) and impossibility of φ_{cl} further enhancement (d_f reduction).

For semicrystalline polymers it has been shown [32, 33] that their molecular mobility is realized in noncrystalline regions. For polyethylenes at relatively small temperatures (> 240 K [34]) noncrystalline regions are devitrificated. In Refs. [35, 36] the hypothesis was proposed, that the indicated regions have peculiar conformational structure, which cannot be described by interpenetrating chaotically tangled macromolecular coils model ("felt" model [37]). Therefore, the question arises about influence of one or another structural organization on their chains molecular mobility. Proceeding from this authors of Ref. [38] studied the influence of molecular and structural characteristics of high density polyethylene (HDPE) noncrystalline regions on the fractal dimension D_{ch} value and, hence, on molecular mobility level in these regions.

Apart from the Eq. (2.5), at present there are a number of methodics, which allow to estimate the value D_{ch}. One from them was proposed in Ref. [20]:

$$\cos \theta = 1 - 2^{2/(D_{ch}-1)}, \qquad (2.8)$$

where θ is an angle between chain neighboring segments (Fig. 2.2).

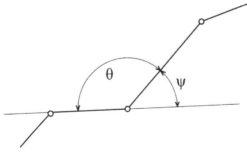

FIGURE 2.2 The schematic picture of macromolecule fragment and angles, used for D_{ch} calculation according to the Eq. (2.8) [20].

From the adduced in Fig 2.2 scheme one can see, that angle θ and angle between segment direction and its projection on neighboring segments direction ψ are connected by the relationship [38]:

$$\theta = 180° - \psi. \tag{2.9}$$

In its turn, the angle ψ is a function of C_∞ value, which is characteristic of chain statistical flexibility. For model with fixed valent angles this relationship has the following form [18]:

$$C_\infty = \frac{1 + \cos\psi}{1 - \cos\psi}, \tag{2.10}$$

and for model with braked internal rotation the following relationship is true [18]:

$$C_\infty = \frac{(1 + \cos\psi)}{(1 - \cos\psi)} \cdot \frac{(1 + \cos\overline{\varphi})}{(1 - \cos\overline{\varphi})}, \tag{2.11}$$

where $\overline{\varphi}$ is mean angle of internal rotation [18].

The other method of D_{ch} determination is based on the Richardson known equation usage [22, 39]:

$$\frac{L_{ch}}{l_{st}} = \left(\frac{R_{ch}}{l_{st}}\right)^{D_{ch}}, \tag{2.12}$$

where L_{ch} is length of chain part between its fixation points, R_{ch} is distance between these points, l_{st} is statistical segment length.

The value L_{ch} can be determined with the aid of the equation [22]:

$$L_{ch} = \frac{2}{v_e}, \tag{2.13}$$

where v_e is macromolecular entanglements network density, S is macromolecule cross-sectional area [22]. For polyethylenes $v_e \approx 4.3 \times 10^{26}$ m^{-3} [40] and $S = 14.7$ Å^2 [41]. The value R_{ch} is estimated approximately as follows [42]:

$$R_{ch} \approx 10\, C_\infty, \tag{2.14}$$

and l_{st} is determined according to the equation [40]:

$$l_{st} = l_0 C_\infty \qquad (2.15)$$

where l_0 is the length of main chain skeletal bond, which is equal to 1.54Å for polyethelenes [40, 42].

In Ref. [7] the following equation was offered, which also allows to estimate the dimension D_{ch}:

$$\frac{2}{\phi_{cl}} = C_\infty^{D_{ch}} . \qquad (2.16)$$

And at last one more method is based on application of the Eq. (2.5), where the value n_{st} was calculated according to the equation [22]:

$$n_{st} = \frac{L_{ch}}{l_{st}} = \frac{2}{\phi_{cl}} . \qquad (2.17)$$

Let us note, that the chain flexibility, characterized by parameter C_∞, in one form or another is included in all relationships, adduced above for D_{ch} estimation. Hence, for D_{ch} correct determination it is necessary to define the temperature dependence C_∞ [18], which can be estimated with the help of the equation [43]:

$$C_\infty = \left(1 - \sqrt{1 - \frac{3T_m}{4eT}}\right) \Big/ const , \qquad (2.18)$$

where T_m is melting temperature, which is equal to 408 K for HDPE [34], T is testing temperature and the constant in the Eq. (2.18) is determined from the condition $C_\infty = 6.8$ [42] at $T = 293$ K.

In Fig. 2.3, the temperature dependences of D_{ch}, calculated according to the Eqs. (2.5), (2.8), (2.12) and (2.16), are adduced. The attention is paid to various character of the dependences $D_{ch}(T)$, calculated according to the Eqs. (2.8) and (2.12) on the one hand and the Eqs. (2.5) and (2.16) on the other hand. If the two first ones show very weak temperature variation of D_{ch}, which is due to C_∞ temperature variation only, then the two last predict strong influence of temperature on D_{ch}. In addition, if in the first case the absolute values D_{ch} are differed strongly, then in the second case the data of calculation according to the Eqs. (2.16) and (2.5) are approximated practically by one curve.

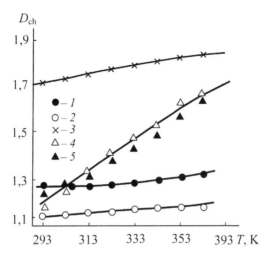

FIGURE 2.3   The dependence of fractal dimension D_{ch} of chain part between entanglements nodes on testing temperature T for HDPE. The calculation of D_{ch} according to the equations: 1 – (2.8), the model of fixed valent angles, 2 – (2.8), the model with braked internal rotation, 3 – (2.12), 4 – (2.16), 5 – (2.5) [38].

For the observed distinctions explanation it is necessary to point out, that the Eqs. (2.8) and (2.12) take into consideration only molecular characteristic, namely, macromolecule flexibility, characterized by the value C_∞. Although the Eq. (2.12) takes into account additionally topological factor (traditional macromolecular binary hooking network density v_e), but this factor is also a function of C_∞ [40, 42]. The Eqs. (2.16) and (2.5) take into account, besides C_∞, the structural organization of HDPE noncrystalline regions within the frameworks of cluster model of polymers amorphous state structure [5] or fractal analysis with the aid of the value d_f [22]. Hence, HDPE noncrystalline regions structure appreciation changes sharply the dependence $D_{ch}(T)$.

Let us also note that the chain model choice influences on the value D_{ch}. As it follows from the data of Fig. 2.3, the model with braked internal rotation ($\overline{\varphi} = 0$ [18]) gives smaller values D_{ch}, than the model with fixed valent angles. Besides, the first from the indicated models gives better correspondence to calculation according to the Eqs. (2.5) and (2.16), from which it follows, that it better describes real polymer chains behavior.

The calculation of D_{ch} according to the Eqs. (2.8) and (2.12) as a matter of fact gives D_{ch} lower and upper boundaries, accordingly, as one can see from the plots of Fig. 2.3. The small absolute D_{ch} values, received with the aid of the Eq. (2.8), are due to relatively small n_{st} values for HDPE. As it was shown in Ref. [20], the value D_{ch}, received by computer simulation, was small up to $n_{st} \approx 40$ (~ 1.33) and approximately constant. This corresponds to the adduced in Fig. 2.3 results, since for HDPE $n_{st} \leq 32$.

As it is known [35], the main chain mobility "freezing" occurs at certain temperature T^*, that supposes the condition $D_{ch} = 1.0$ achievement. For polyethylenes the value $T^* \approx 150$ K [35]. At $T_m = 408$ K polyethylenes turn into rubber-like state [44] and therefore, for them it is excepted $D_{ch} = 2.0$ [22]. In Table 2.1 the values D_{ch} at the indicated limiting temperatures, obtained according to the four used equations, are adduced. As one can see, the full range of D_{ch} change is obtained for the Eqs. (2.5) and (2.16) only, that confirms the necessity of structural parameters using at molecular mobility level estimation.

TABLE 2.1 The Limiting Values of Fractal Dimension D_{ch} of Chain Part Between Entanglements For HDPE [38]

An equation, using for D_{ch} calculation	*Dch*	
	$T^* = 150$ K	$T_m = 408$ K
the Eq. (2.18)	1.14	1.35
the Eq. (2.12)	1.31	1.89
the Eq. (2.16)	0.82	1.91
the Eq. (2.5)	0.78	2.25

As it is known [1–3], at present tangent of angle of mechanical (dielectric) losses tgδ is used most often for the experimental estimation of molecular mobility level. In Fig. 2.4, the temperature dependences of tgδ for HDPE according to the data of Ref. [45] and D_{ch}, calculated according to the Eqs. (2.8) and (2.16), are adduced. As it follows from the indicated plots comparison, full similarity of the dependences $D_{ch}(T)$ (or, more precisely, $(D_{ch} - 1)$ (T)), calculated according to the Eq. (2.16), and tgδ(T) is observed. And, on the contrary, such similarity is absent for the dependence $D_{ch}(T)$, calculated according to the Eq. (2.8).

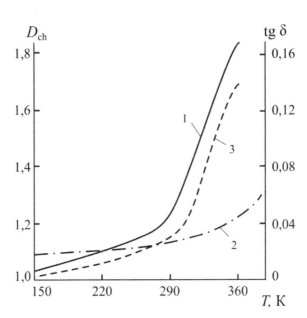

FIGURE 2.4 The schematic dependences mechanical losses angle tangent tgd (1) and fractal dimension of chain part between entanglements nodes D_{ch} (2, 3) on temperature T for HDPE. The calculation of D_{ch} according to the equations: 2 – (2.8), model of fixed valent angles, 3 – (2.16) [38].

Hence, the adduced above results shown that the main factor, influencing on molecular mobility level in HDPE noncrystalline regions, is these regions structure, characterized by fractal dimension d_f or relative fraction of local order regions (clusters) φ_{cl}. Definite influence is exercised by molecular characteristics, especially if to take into account, that between d_f and φ_{cl}, on the one hand, and S and C_∞, on the other hand, the close intercommunication exists (see, for the example, the Eqs. (1.11) and (1.12)). As consequence, the equations using, taking into account their structural state, will be correct for polymers dimension D_{ch} estimation [38].

Glassy state structural components distinction defines their behavior distinctions in both deformation [46] and relaxation [47] processes. It is known [48, 49], that in its turn polymers glassy state includes a substates number, differing by mechanical properties temperature dependences. A breaking (bend) on the corresponding parameter dependence, for example, of yield stress on temperature, is a typical indication of transition from one substate to another. At present unequivocal structural identification of these states is absent. The cluster model [4, 5] notions attraction allows to consider possible mechanisms of glassy polymers structural relaxation and realize the indicated identification.

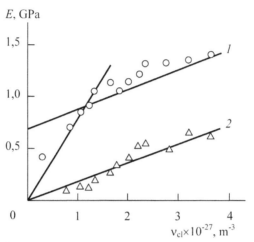

FIGURE 2.5   The dependences of elasticity modulus E (1) and equilibrium modulus E_∞ (2) on macromolecular entanglements cluster network density n_{cl} for PC [47].

The plotted according to the experimental data dependencies of elasticity modulus E on macromolecular entanglements cluster network density v_{cl} (Figs. 2.5 and 2.6) break down into two linear parts, the boundary of which serves loosely packed matrix glass transition temperature T'_g, which is lower on about 50K of polymer glass transition temperature T_g [50]. Below T_g the value E is defined by the total contribution of both clusters and loosely packed matrix and above T_g – only by clusters contribution. It becomes clear, if to taken into consideration, that above T_g the elasticity modulus value of devitrificated loosely packed matrix has the order of 1 MPa [51], that is, negligible small. It is an extremely interesting the observation, that loosely packed matrix in the value E, determined by the plot E (v_{cl}) extrapolation to $v_{cl} = 0$, is independent on temperature. Such situation is not occasional and deserves individual consideration.

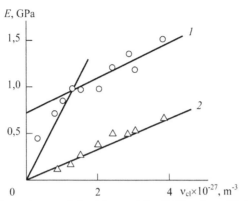

FIGURE 2.6   The dependences of elasticity modulus E (1) and equilibrium modulus E_∞ (2) on macromolecular entanglements cluster network density v_{cl} for PAr [47].

The value of polymers fluctuation free volume f_g can be calculated according to the equation [52]:

$$f_g = 0{,}012\left(\frac{1+v}{1-2v}\right),\qquad (2.19)$$

where v is Poisson's ration, determined according to the results of mechanical tests with the aid of the relationship [52]:

$$\frac{\sigma_Y}{E} = \frac{1-2v}{6(1+v)},\qquad (2.20)$$

where σ_Y is yield stress.

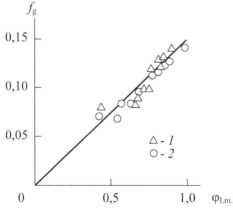

FIGURE 2.7   The dependence of relative fluctuation free volume f_g on relative fraction of loosely packed matrix $\varphi_{l.m.}$ for PC (1) and PAr (2) [47].

Within the frameworks of cluster model it is postulated, that fluctuation free volume is concentrated in loosely packed matrix of polymer structure, the relative fraction of which can be estimated according to the Eq. (2.4). In Fig. 2.7, the correlation $f_g(\varphi_{l.m.})$ is adduced, which has an expected character. The straight line on this Figure is drawn according to the following consideration: $f_g = 0$ at $\varphi_{l.m.} = 0$ and $f_g = 0.104 \times 0.159$ at $\varphi_{l.m.} = 1.0$. These boundary conditions are obvious: at densely packed structure $f_g = 0$ and clusters full decay ($\varphi_{l.m.} = 1.0$ or $\varphi_{l.m.} = 0$) means polymer devitrification and its free volume achieves incidentally the indicated value [53]. From the data of Fig. 2.7, a very important key conclusion follows: the value f_g in loosely packed matrix is always equal to the value f_g at glass transition temperature and integral value f_g counting per volume of the entire polymer is defined by the value $\varphi_{l.m.}$ [54]. Nevertheless, loosely packed matrix at $T < T_g$ is in glassy state, that is clearly demonstrated by the data of Figs. 2.5 and 2.6 and its devitrification is due to small (instable) clusters [46]. It is natural, that the condition $f_g = $ const defines constancy of loosely packed matrix contribution in value E as well.

The dependences $E_\infty(\nu_{cl})$ for PC and PAr, also adduced in Fig. 2.5 and 2.6 differ from the dependences $E(\nu_{cl})$ at $T < T_g$ by loosely packed matrix zero contribution only. As a matter of fact, the plot $E(\nu_{cl})$ parallel transference is observed and this circumstance indicates, that relaxation processes in amorphous polymers at $T < T_g$ are realized completely in loosely packed matrix.

In Fig. 2.8, the dependences $E(\nu_{cl})$ and $E_\infty(\nu_{cl})$ for HDPE, received in quasistatic tests, are adduced. The principally different picture in comparison with PC and PAr behavior at $T < T_g$ (Figs. 2.5 and 2.6) is observed. Both adduced in Fig. 2.8 plots are linear and extrapolate to zero at $\nu_{cl} = 0$ and relaxation processes occurrence is expressed by the plot $E_\infty(\nu_{cl})$ slope change in comparison with $E(\nu_{cl})$. If equality to zero of values E and E_∞ at $\nu_{cl} = 0$ is defined by HDPE devitrificated loosely packed matrix small elasticity modulus, then the slope decrease indicates on structural relaxation mechanism change, which is now controlled by structure second component-clusters [47].

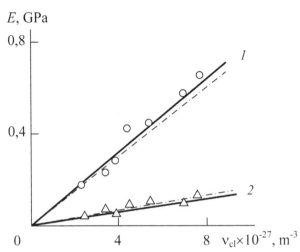

FIGURE 2.8   The dependences of elasticity modulus $E(1)$ and equilibrium modulus E_∞ (2) on macromolecular entanglements cluster network density v_{cl} for HDPE. The corresponding data for PC according to Fig. 2.5 shown by shaded lines [47].

Let us consider the reasons of structural relaxation mechanism assumed change. It has been shown earlier [55, 56], that in amorphous polymers deformation process two mechanisms are possible: the appearance and "freezing" of macromolecules nonequilibrium conformations (the mechanism I) and mutual motion of the connected by tie chains supramolecular structures (the mechanism II). As a matter of fact, relaxation processes are opposite to deformation processes and therefore, the indicated treatment can be used for the obtained results explanation. It is obvious, that at glassy loosely packed matrix structural relaxation by clusters motion is difficult (or, in any case, requires very large duration) and therefore, relaxation is realized by reversion of chains in nonequilibrium configurations to equilibrium ones (the mechanism I). This process is realized easily enough at stress application in loosely packed matrix, where values f_g are large. In case of devitrificated loosely packed matrix (PC and PAr at $T > T'_g$ and HDPE) relaxation processes in it occur very rapidly, its viscosity reduces significantly [2] and clusters motion possibility appears, that defines now structural relaxation process. In Fig. 2.8, the plots $E(v_{cl})$ and $E(v_{cl})$ for PC are traced at $T > T_g$ and, as one can see, they coincide practically with the corresponding plots for HDPE that confirms relaxation processes identity.

One more confirmation of the indicated identity can be obtained from the data of Fig. 2.9, where the dependences $E(v_{cl})$ for HDPE, received in impact tests, are adduced. Sharp notch length increase results to reduction of time up to sample fracture and, as a consequence, to E growth owing to relaxation processes incompleteness [57]. From the data of Fig. 2.9, it follows, that this also results to mechanical vitrification of loosely packed matrix part, which gives now essential contribution to elasticity modulus value (compare with the data if Fig. 2.5 and 2.6).

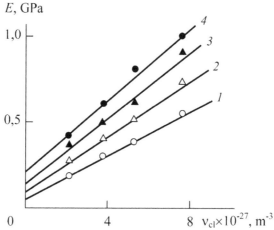

FIGURE 2.9 The dependences of elasticity modulus E on macromolecular entanglements cluster network density v_{cl} for HDPE. Received in impact test for samples with sharp notch 0.5 (1), 0.9 (2), 1.2 (3) and 1.5 mm (4) long [47].

And at last, relaxation processes proceeding depth can be expressed by the ratio E/E_∞ – the closer values E and E_∞, the more complete consummation of relaxation processes and the smaller the ratio E/E_∞. In Fig. 2.10, the dependence of E/E_∞ on temperatures difference $(T_g – T)$ (for HDPE – on difference $(T_m – T)$), that is, on approach degree of T to T_g (T_m). As it follows from the data of Fig. 2.10, relaxation processes completeness degree grows at approach to T_g (T_m). This can be due to two reasons: firstly, vitrificated loosely packed matrix viscosity reduction and, secondly, by clusters volume decrease at temperature enhancement (see Fig. 1.5) [58]. Both indicated reasons contribute to clusters motion making lighter according to the mechanism II.

In Refs. [48, 49, 59] as the index, characterizing dependence of relaxation properties on temperature, the reciprocal of relative stress decay at definite duration of its relaxation β was used:

$$\frac{1}{\beta} = \frac{\sigma_l}{\sigma_l - \sigma_r}, \tag{2.21}$$

where σ_l is stress at loading moment, σ_r is stress later definite time of relaxation proceeding.

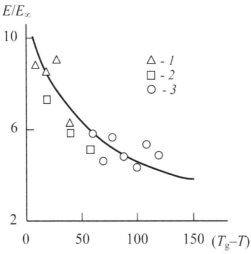

FIGURE 2.10   The dependence of ratio E/E_∞ on temperatures difference $(T_g - T)$ for PC(1), PAr (2) and HDPE (3). For HDPE the temperatures difference $(T_m - T)$ was used [47].

Within the frameworks of the considered mechanisms of structural relaxation it was to be expected, that clusters size decrease at testing temperature T growth should be facilitate process, proceeding according to the mechanism II and, consequently, the correlation between $1/\beta$ and clusters functionality F was expected (let us remind, that segments number per one cluster is equal to $F/2$ [58]). As the plot of Fig. 2.11 shows, this assumption turns out to be correct and the value $1/\beta$ extrapolates to one at T_g [49, 59], that is, at clusters complete thermofluctuation decay [19].

Hence, the cluster model of polymers amorphous state structure allows to identify structural relaxation mechanisms in them. In the case of glassy loosely packed matrix relaxation process is realized by conformational reorganizations in this structural component (mechanism I) and in the case of its devitrification – clusters mutual motions (mechanism II).

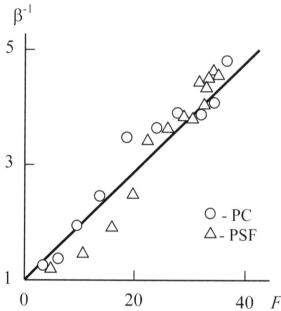

FIGURE 2.11   The relationship between $1/\beta$ and clusters functionality F for PC (1) PSF (2) [47].

KEYWORDS

- cluster functionality
- dielectric (mechanical) losses
- fractal dimension
- loosely-packed
- matrix
- molecular mobility
- polymer

REFERENCES

1. Bartenev, G. M., & Zelenev, Yu. V. (1983). Physics and Mechanics of Polymers. Moscow, Vysshaya Shkola, 391 p.

2. Sanditov, D. S., & Bartenev, G. M. (1982). Physical Properties of Disordered Structures. Novosibisrk, Nauka, 256 p.
3. Bartenev, G. M., & Frenkel, S. Ya. (1990). Physics of Polymers. Leningrad, Khimiya, 432 p.
4. Kozlov, G. V., & Novikov, V. U. (2001). The Cluster Model of Polymers Amorphous State. Uspekhi Fizicheskikh Nauk, *171(7)*, 717–764.
5. Kozlov, G. V., & Zaikov, G. E. (2004). Structure of the Polymer Amorphous State. Utrecht, Boston, Brill Academic Publishers, 465 p.
6. Novikov, V. U., Kozlov, G. V., & Yundunov, V. V. (2002). The Fractal Aspect of Molecular Mobility in Polymer Composites. Materialovedenie, *4*, 4–9.
7. Kozlov, G. V., Temiraev, K. B., Shetov, R. A., & Mikitaev, A. K. (1999). A Structural and Molecular Characteristics Influence on Molecular Mobility in Diblock-Copolymers Oligoformal *2,2*-di-*(4*-oxyphenyl*)*-propane-oligosulfone Phenolphtaleine. Materialovedenie, *2*, 34–39.
8. Vilgis, T. A. (1988). Flory Theory of Polymeric Fractals Intersection, Saturation and condensation. Physica A, *153(2)*, 341–354.
9. Alexander, S., & Orbach, R. (1982). Density of States on Fractals: *"Fractons"*. J. Phys. Lett. (Paris), *43(17)*, L625–L631.
10. Kozlov, G. V., Temiraev, K. B., Shustov, G. B., & Mashukov, N. I. (2002). Modeling of Solid State Polymer Properties at the Stage of Synthesis: Fractal Analysis. J. Appl. Polymer Sci., *85(6)*, 1137–1140.
11. Kozlov, G. V., Dolbin, I. V., & Zaikov, G. E. (2005). Fractal Physical Chemistry of Polymer Solutions. In: Chemical and Biological Kinetics. New Horizons.1. Chemical Kinetics. Ed. Burlakova, E. B., Shilov, A. E., Varfolomeev, S. D., Zaikov, G. E. Leiden, Boston, Brill Academic Publishers, 448–483.
12. Kozlov, G. V., Dolbin, I. V., & Zaikov, G. E. (2006). Fractal Physical Chemistry of Polymer Solutions. In: Focus on Natural and Synthetic Polymer Science. Vasile, C., Zaikov, G. E., Ed. New York, Nova Science Publishers Inc., 131–175.
13. Kozlov, G. V., Temiraev, K. B., & Sozaev, V. A. (1999). The Estimation of Macromolecular Coil Fractal Dimension in Diluted Solution by Viscous Characteristics. Zhurnal Fizicheskoi Khimii, *73(4)*, 766–768.
14. Kozlov, G. V., & Dolbin, I. V. (2001). Express-Method of Estimation of Biopolymers in Solution Macromolecular Coils Fractal Dimension. Biofizika, *46(2)*, 216–219.
15. Kozlov, G. V., Dolbin, I. V., Mashukov, N. I., Burmistr, M. V., & Korenyako, V. A. (2001). The Prediction of Macromolecular Coils in Diluted Solutions Fractal Dimension, based on Two-Dimensional Solubility Parameter Model. Voprosy Khimii i Khimicheskoi Tekhnologii, *6*, 71–77.
16. Kozlov, G. V., Dolbin, I. V., & Zaikov, G. E. (2002). Rapid Method of Estimating the Fractal Dimension of Macromolecular coils of Biopolymers in Solution. In: Perspectives on Chemical and Biochemical Physics. Ed. Zaikov, G. E. New York, Nova Science Publishers Inc., 217–223.
17. Kozlov, G. V., Beloshenko, V. A., Varyukhin, V. N., & Lipatov, Yu. S. (1999). Application of Cluster Model for the Description of Epoxy Polymer Structure and Properties. Polymer, *40(4)*, 1045–1051.
18. Budtov, V. P. (1992). Physical Chemistry of Polymer Solutions. Sankt-Peterburg, Khimiya, 384 p.

19. Beloshenko, V. A., Kozlov, G. V., & Lipatov, Yu. S. (1994). Glass Transition Mechanism of Gross-Linked Polymers. Fizika Tverdogo Tela, *36(10)*, 2903–2906.

20. Halvin, S., & Ben-Avraham, D. (1982). New Approach to Selfavoiding Walks as Critical Phenomenon. Phys. Rev. A, *26(3)*, 1728–1734.

21. Novikov, V. U., & Kozlov, G. V. (2000). A Macromolecules Fractal Analysis. Uspekhi Khimii, *69(4)*, 378–399.

22. Kozlov, G. V., & Novikov, V. U. (1998). Synergetics and Fractal Analysis of Gross-Linked Polymers. Moscow, Klassika, 112 p.

23. Zemlyanov, M. G., Malinovskii, V. K., Novikov, V. N., Parshin, P., & Sokolov, A. P. (1992). Fractions Study in Polymers. Zhurnal Exsperimental noi I Teoreticheskoi Fiziki, *101(1)*, 284–293.

24. Kozlov, G. V., & Zaikov, G. E. (2001). The Generalized Description of Local Order in Polymers. In: Fractals and Local Order in Polymeric Materials. Kozlov, G. V., Zaikov, G. E., Ed., New York, Nova Science Publishers Inc., 55–63.

25. Kozlov, G. V., Beloshenko, V. A., Gazaev, M. A., & Lipatov, Yu. S. (1996). A structural Changed of Gross-Linked Polymers at Heat Aging. Vysokomolek. Soed. B, *38(8)*, 1423–1426.

26. Kozlov, G. V., & Zaikov, G. E. (2002). The Concept of Quasiequilibrium State at Gross-Linked Polymers Physical Aging Description. Materialovedenie, *12*, 13–17.

27. Kozlov, G. V., Beloshenko, V. A., & Varyukhin, V. N. (1998). Similation of Cross-Linked Polymers Structure as Diffusion-Limited Aggregate. Ukrainskii Fizicheskii Zhurnal, *43(3)*, 322–323.

28. Kozlov, G. V., Shogenov, V. N., & Mikitaev, A. K. (1988). Local Order in Polymers the Description within the Frameworks of Irreversible Colloidal Aggregation Model. Inzhenerno-Fizicheskii Zhurnal, *71(6)*, 1012–1015.

29. Meakin, (1983). Diffusion-Controlled cluster Formation in *2–6*-Dimensional Space. Phys. Rev. A, *27(3)*, 1495–1507.

30. Vannimenus, J., Nadal, J. P., & Martin, H. J. (1984). On the Spreading Dimension of Percolation and Directed Percolation clusters. J. Phys. A, *17(6)*, L351–L356.

31. Meakin, P., Majid, I., Havlin, S., & Stanley, H. E. (1984). Topological Properties of Diffusion-Limited Aggregation and cluster-cluster Aggregation. J. Phys. A, *17(18)*, L975–L981.

32. Boyd, R. N. (1985). Relaxation Processes in Crystalline Polymers: Experimental Behaviour a Review. Polymer, *26(3)*, 323–347.

33. Boyd, R. N. (1985). Relaxation Processes in Crystalline Polymers: Molecular interpretation a Review. Polymer, *26(8)*, 1123–1133.

34. Kalinchev, E. L., & Sakovtseva, M. B. (1983). Properties and Processing of Thermoplastics. Leningrad, Khimiha, 288 p.

35. Berstein, V. A., & Egorov, V. M. (1990). Differential Scanning Calorimetry in Physics-Chemistry of the Polymers. Leningrad, Khimiya, 256 p.

36. Machukov, N. I., Gladyshev, G. P., & Kozlov, G. V. (1991). Structure and Properties of High Density Polyethylene Modified by High-Disperse Mixture Fe and FeO. Vysokomolek. Soed. A, *33(12)*, 2538–2546.

37. Flory, J. (1976). Spatial Configuration of Macromolecular Chains. Brit. Polymer. J., *8(1)*, 1–10.

38. Malamatov, A. Kh., Kozlov, G. V., Burya, A. I., & Mikitaev, A. K. (2006). The Fractal Analysis of Molecular Mobility in Polyethylene on Molecular and Structural Levels. Polimernyi Zhurnal, *28(1)*, 25–29.

39. Feder, F. (1989). Fractals. New York, Plenum Press, 248 p.

40. Wu, S. (1989). Chain Structure and Entanglement J. Polymer Sci.: Part B: Polymer Phys., *27(4)*, 723–741.

41. Aharoni, S. M. (1985). Correlations between Chain Parameters and Failure Characteristics of Polymers below their Glass Transition Temperature. Macromolecules, *18(12)*, 2624–2630.

42. Aharoni, S. M. (1983). On Entanglements of Flexible and Rodlike Polymers. Macromolecules, *16(9)*, 1722–1728.

43. Matveev, Yu. I., & Askadskii, A. A. (1994). The Dependence of Non-Newtonian viscosity on Polymer Molecular Weight in its Change wide Range. Vysokomolek. Soed. B, *36(10)*, 1750–1755.

44. Belousov, V. N., Kozlov, G. V., & Mashukov, N. I. (1996). The Glass Transition Temperature of Crystallizable Polymers Amorphous Regions. Doklady Adygskoi (Cherkesskoi) International AN, *2(1)*, 74–82.

45. Capaccio, J., Gibson, A. G., & Ward, I. M. (1983). Drawing and Hydrostatic Extrusion of Ultra-High-Modulus Polymers. In: Ultra-High-Modulus Polymers. Ed. Chiffery A., Ward I. Leningrad, Khimiya, 12–62.

46. Kozlov, G. V., Beloshenko, V. A., & Novikov, V. U. (1996). Mechanisms of Gross-Linked Polymers Yielding and Forced High-Elasticity. Mekhanika Kompozitnykh Materialov, *32(2)*, 270–278.

47. Kozlov, G. V., Beloshenko, V. A., & Shogenov, V. N. (1999). An Amorphous Polymers Structural Relaxation Description within the Frameworks of Cluster model. Fiziko-Khimicheskaya Mechanika Materialov, *35(5)*, 105–108.

48. Askadckii, A. (1968). A. Physics-Chemistry of Polyarylates. Moscow, Khimiya, 214 p.

49. Askadskii, A. A. (1973). Deformation of Polymers. Moscow, Khimiya, 448 p.

50. Belousov, V. N., Kotsev, B. Kh., & Mikitaev, A. K. (1983). Two-Step of Amorphous Polymers Glass Transition. Doklady AN SSSR, *270(5)*, 1145–1147.

51. Graessley, W. W., & Edwards, S. F. (1981). Entanglement Interaction in Polymers and the Chain Contour Concentration. Polymer, *22(10)*, 1329–1334.

52. Kozlov, G. V., & Sanditov, D. S. (1994). Anharmonic Effects and Physical-Mechanical Properties of Polymers. Novosibirsk, Nauka, 261 p.

53. Sanchez, I. C. (1974). Towards a Theory of Viscosity for Glass-Forming Liquids. J. Appl. Phys., *45(10)*, 4204–4215.

54. Belousov, V. N., Beloshenko, V. A., Kozlov, G. V., & Lipatov, Yu. S. (1996). Fluctuation Free Volume and Structure of Polymers. Ukrainskii Khimicheskii Zhurnal, *62(1)*, 62–65.

55. Arzhakov, S. A., & Kabanov, V. A. (1971). To Question of Stresses Relaxation in Polymers, Deformed in Forced Elasticity Regime. Vysokomolek. Soed. B, *13(5)*, 318–319.

56. Bekichev, V. I. (1974). To Question about Forsed-Elastic Deformation Mechanism. Vysokomolek. Soed. A, *16(8)*, 1745–1747.

57. Kozlov, G. V., Beloshenko, V. A., & Mikitaev, A. K. (1999). The Kinetics of Crasing in Polystyrene at Impact Testing. J. Mater. Sci. Techn., *7(1)*, 3–10.

58. Belousov, V. N., Kozlov, G. V., Mikitaev, A. K., & Lipatov, Yu. S. (1990). Entanglements in Glassy State of Linear Amorphous Polymers. Doklady AN SSSR, *313(3)*, 630–633.

59. Filyanov, E. M., Petrilenkova, E. B., & Tarakanov, O. G. (1976). About Reasons of Stress Decay at Compression Strain of Cross-Linked Polymers in Glassy State. Vysokomolek. Soed. A, *18(6)*, 1310–1315.

ELASTIC PROPERTIES

CONTENTS

The first models, describing elastic behavior of fractal structures, were used, as a rule, for simulation within the frameworks of percolation theory [1–5]. Nonhomogeneous statistical mixture of solid and liquid then only displays solid body properties (e.g., not equal to zero shear modulus G), when solid component forms percolation cluster, like at gelation in polymer solutions. If liquid component there is replaced by vacuum, then bulk modulus. B will also be equal to zero below percolation threshold [1]. Such model gives the following relationship for elastic constants [1, 3]:

$$B, G \sim (p - p_c)^h, \tag{3.1}$$

where p is volume fraction of solid component, p_c is percolation threshold, η is exponent.

The following equation was obtained for exponent η at simulation of fractal structure as "Serpinsky carpet" [1]:

$$\frac{\eta}{\nu_p} = d - 1, \tag{3.2}$$

where ν_p is correlation length index in percolation theory, d is dimension of Euclidean space, in which a fractal is considered.

The comparison of conductivity index t_p and η has shown that the condition $\eta \geq t_p$ is carried out and it contradicts to condition $\eta = t_p$, earlier proposed by de Gennes [1]. And at last, Bergman and Kantor were found [1], that the ratio of bulk modulus and shear modulus had the universal magnitude:

$$\frac{B}{G} = \frac{4}{d}. \tag{3.3}$$

The similar approach was used by Webman [3], according to which under external force F action fractal deformation occurs only on scales, exceeding some definite length L_F. In the general case elastic moduli of fractal part with specific size L_F depend on fractal skeleton structure, which in space with small d consists of both nonduplicated and repeatedly duplicated bonds. Assuming, that domain elasticity is defined by nonduplicated bonds only (i.e., considering parts, consisting of duplicated bonds, as absolutely stiff ones), can be found the lower limit of the exponent [3]:

$$\eta = d\nu_p + 1. \tag{3.4}$$

The condition $\eta > t_p$ [3] is true in this case.

Using these general notions, the authors of Ref. [6, 7] offered the fractal models for polymers elastic constants description. The quasiequilibrium state of polymers structure is characterized by the criterion $D_f = 3$ [8, 9], where D_f is dimension of excess energy localization domains. A loosely packed matrix is totality of such domains. The value D_f can be determined within the frameworks of free volume fractal theory according to the equation [9]:

$$D_f = \frac{4\pi T}{\ln\left(1/f_g\right)T_g}$$ (3.5)

From the Eq. (3.5) together with the criterion (3.4) it follows, that at arbitrary T the definite value $f_g(t_g^{qe})$ will correspond to quasiequilibrium state, that is, this parameter is a function of temperature. Then relative deviation of loosely packed matrix (in which the entire fluctuation free volume is concentrated [8, 9]) from quasiequilibrium state can be expressed as follows [6]:

$$\Delta = \frac{f_g - f_g^{qe}}{f_g^{qe}}$$ (3.6)

Since according to the indicated above reasons two order parameters are required, as a minimum, for solid-phase polymers elastic constants description, then variable percolation threshold should be introduced in the Eq. (3.1), that is, p_c should be replaced on Δ. Besides, it has been shown earlier, that for polymers structure $v_p \approx 1$ (see Table 1.1) [10] and therefore, $\eta = d_f - 1$ can be assumed in the Eq. (3.2) as the first approximation. Then the Eq. (3.1) assumes the following form [6, 7]:

$$G\left(\phi_{cl}, \Delta\right) \sim \left(\phi_{cl} - \Delta\right)^{d_f - 1},$$ (3.7)

since φ_{cl} is an order parameter of polymers structure in strict physical significance of this term [10, 11].

In Fig. 3.1, the dependence of shear modulus G on parameter $\left(\phi_{cl} - \Delta\right)^{d_f - 1}$ for amorphous glassy polyarylatesulfone (PASF) is adduced, accounting results of both quasistatic and impact tests [6]. At $\left(\phi_{cl} - \Delta\right)^{d_f - 1}$ > 0.5 a good correspondence of the indicated parameters in case of impact tests, which is worsed at smaller values $\left(\phi_{cl} - \Delta\right)^{d_f - 1}$, corresponding to temperatures range $T = 353 \div 453$ K. At the same time a good linear relationship was obtained (Fig. 3.1) for quasistatic tension tests results in the same temperatures range. This distinction is due to tests temporal scale sharp reduction at impact loading and, as consequence, to mechanical vitrification of the devitrificated loosely packed matrix part, that results to G growth [7].

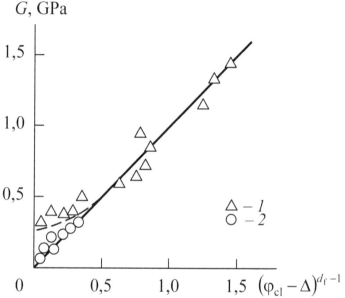

FIGURE 3.1   The relationship between shear modulus G and parameter $\left(\phi_{cl} - \Delta\right)^{d_f - 1}$ for PASF in case of impact (1) and quasistatic tension (2) tests [7].

As it is known [12], the elasticity modulus E value in polymers impact tests grows at testing temporal scale decrease (at brittle fracture – time up to failure t_f) and can be described according to the empirical relationship [12]:

$$E = E_0 t_f^{-m},\qquad(3.8)$$

where E_0 is "unit" modulus, that is, the value E at $t_f = 1$ in the chosen temporal scale, m is empirical exponent, characterizing relaxation processes completeness degree.

The description difficulty of E behavior within the frameworks of the Eq. (3.8) contains in two empirical constants (E_0 and m) availability, which can also change their values discontinuously [7]. Therefore, the authors of Refs.

[13, 14] chose an alternative method, which consists in Kohlrauch equation usage. This equation can be written as follows:

$$E = E_0' \exp\left[\left(-t_f / \tau_0\right)^{\beta}\right], 0 < \beta < 1, \qquad (3.9)$$

where E_0' is a new constant, τ_0 is the relaxation time.

Let us consider the physical significance of the included in the Eq. (3.9) parameters. The value t_f, at which parameters E_0 and m change discontinuously their values, should be accepted as τ_0. In Fig. 3.2, the dependence $E(t_f)$ was shown, corresponding to the Eq. (3.8), for samples of HDPE with sharp notch various lengths a (a increase reduces t_f) at impact testing temperature $T = 293$ K [7]. The plot of Fig. 3.2 demonstrates the graphic method of τ_0 determination. The value E_0' in context of the Eq. (3.9) determines the greatest value of elasticity modulus (E_{max}), achieved for the given polymer in case of relaxation absence. The similar value was determined in Ref. [16] at very high testing frequencies and proved to be equal to approx. $1.1 \div 2.6$ GPa for different polyethylenes. And at last, the exponent β, characterizing relaxation times spectrum width, can be chosen from the following considerations. As it is known [15], the exponent β, in the Eq. (3.9) characterizes a system spatial disorder. The same characteristics can be expressed with the aid of structure fractal dimension d_f which changes within the limits of $2 \geq d_f < 3$ [17] for nonporous solids. In such treatment $d_f = 2$ defines the structure, possessing complete local order ($\varphi_{cl} = 1.0$) and $d_f = d = 3$ – Euclidean object, characterized by full disorder (true rubber) [18]. It is obvious, that in this case the exponent β can be associated with d_f fractional part, that is, $\beta = d_f - (d - 1) = d_f - 2$ [13]. Let us note, that both β and ($d_f - 2$) are changed within the same limits. Hence, proceeding from the adduced above considerations, one can write Kohlraush equation final variant at polymers elasticity modulus determination at high-rate loading (brittle fracture) [7, 13, 14]:

$$E = E_{max0} \exp\left[-\left(t_f / \tau_0\right)^{d_f - 2}\right], 2 \leq d_f < 3. \qquad (3.10)$$

In Fig. 3.3, the comparison of experimental and calculated according to the Eq. (3.10) E values for samples HDPE at different T as a function of sharp notch length a is adduced. As one can see, a good correspondence of experiment and the offered theoretical treatment was received. The value E at $T = 293$K, obtained by superposition method, is equal to approx. 2.0 GPa, that corresponds well to corresponding values range, obtained in Ref. [16].

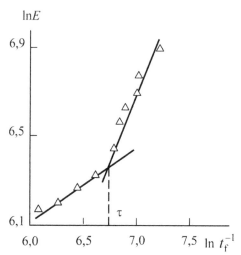

FIGURE 3.2   The relationship between elasticity modulus E and time up to fracture, corresponding to the Eq. (3.8), in double logarithmic coordinates for samples HDPE with sharp notch ($t = t_0$ is polymer relaxation time) [7].

The thermal cluster model is a complicated variant of percolation theory, assuming molecular mobility availability of elements, forming percolation cluster (in the polymers case – statistical segments). In this case the relationship is true [19]:

$$\delta = \left(\frac{T_m - T}{T_m} \right),$$ (3.11)

where δ is order parameter, T_m is critical temperature, accepted for semicrystalline polymers equal to melting temperature T_m, T is testing temperature, β_t is the exponent (index) of thermal cluster order parameter.

Assuming $\delta = \varphi_{cl}$ (where φ_{cl} is local order regions (clusters) relative fraction) and plotting the dependence, corresponding to the Eq. (3.11), in double logarithmic coordinates for HDPE (impact testing), let us receive the linear correlation, from which $\beta_t \approx 0.49$ can be determined, that is close enough to theoretical "geometrical" value $\beta_p = 0.40$ [20]. Hence, the amorphous phase HDPE cluster structure is a thermal cluster, having percolation threshold T_m. In this case φ_{cl} is an order parameter in strict physical significance of this term [21].

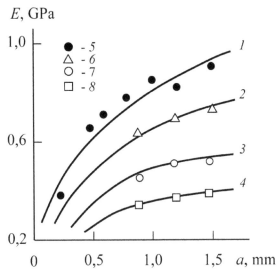

FIGURE 3.3 The experimental (1 ÷ 4) and calculated according to the Eq. (3.10) (5 8) dependences of elasticity modulus E on sharp notch length a for HDPE at testing temperatures 293 (1.5), 313 (2.6), 333 (3.7) and 353 K (4.8) [13].

Let us note, that the exponent β_t is a function of structure fractal dimension d_f [22]:

$$\beta_t = \frac{1}{d_f}. \tag{3.12}$$

Since φ_{cl} depends on strain rate (testing temporal scale) then the indicated structural parameter is dependent on the time order parameter and for it one can write [23]:

$$\phi_{cl} = \phi_{cl}^0 \exp\left(-\frac{t_f}{\tau_0}\right), \tag{3.13}$$

where ϕ_{cl}^0 is the value φ_{cl} at $t_f = 0$ (the greatest attainable theoretically one), t_f is time up to fracture, τ_0 is relaxation time.

In Fig. 3.4, the comparison of experimental and calculated according to the Eq. (3.13) values φ_{cl} and ϕ_{cl}^T, respectively, is adduced, which shows their good correspondence. This means, that at impact loading of HDPE with devitrificated amorphous phase a its definite part mechanical vitrification occurs, the fraction of which increases at testing temporal scale reduction [21].

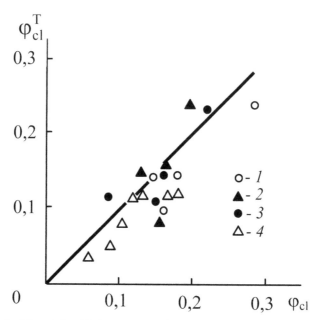

FIGURE 3.4   The relationship between experimental φ_{cl} and calculated according to the Eq. (3.13) ϕ_{cl}^T clusters relative fraction values for HDPE at testing temperatures 293 (1), 313 (2), 333 (3) and 353 K (4) [21].

As it is known [24], solid component fraction enhancement (namely such component of amorphous phase a clusters are as regards devitrificated loosely packed matrix) results to elastic constant growth. This enhancement can be described quantitatively within the frameworks of percolation theory (see the Eq. (3.7)), but in Ref. [21] the more simple variant was chosen, namely, the polymers network connectivity model [25]. Then the elasticity modulus E value is determined as follows [25]:

$$E = \frac{\eta_x a_c}{S},\qquad(3.14)$$

where η_x is network bonds formation probability ($\eta_x = \varphi_{cl}$ [24]), α_c is a value equivalent to Young's modulus for macromolecules elastic tension, S is macromolecule cross-sectional area (for HDPE $S = 14.7$ Å^2 [26]).

As the estimations have shown [26], the theoretical values E (E^T) are a little smaller than elasticity modulus experimental magnitudes. This assumes one more structural factor availability, influencing on the value E. This factor can be traditional macromolecular binary hooking network, which can be taken into account by additional probability of bonds formation $\eta_{x0} =$ constant, determined as follows [24]:

$$\eta_{x0} = n_{x_0} S, \tag{3.15}$$

where n_{x0} is density of macromolecular binary hooking network per sample cross-section, but not per volume, as the value v_1 is usually determined. Therefore, between indicated parameters the approximate relationship exists [21]:

$$n_{x_0} \approx v_e^{2/3}. \tag{3.16}$$

The comparison of the calculated according to the Eq. (3.14) at the condition $\eta_x = \varphi_{cl} + \eta_{x0} E^T$ and experimental E elasticity moduli values have shown their good correspondence (their mean discrepancy < 15%). At the greatest from the used testing temperatures $T = 353$ K the value E^T proves to be higher than on E about 40%. It is supposed [21], that this effect is due to large enhancement of chains slippage through macromolecular binary hooking at the increased temperatures, when they cannot be transferred load [27]. The calculation according to the Eq. (3.14) at $\eta_x = \varphi_{cl}$ gave an excellent correspondence between E and E^T at $T = 353$ K [21].

KEYWORDS

- **elasticity modulus**
- **excess energy localization**
- **Kolraush equation**
- **polymer**
- **relaxation**
- **thermal cluster**

REFERENCES

1. Bergman, D. L., & Kantor, Y. (1984). Critical Properties of an Elastic Fractals. Phys. Rev. Lett., *53(6)*, 511–514.
2. Sokolov, I. M. (1986). Dimensions and Other Geometrical Critical Exponents in Percolation Theory. Uspekhi Fizicheskikh Nauk, *150(2)*, 221–256.

3. Webman, I. (1986). Dynamical Properties of Random and Nonrandom Fractals. In: Fractals in Physics. Ed. Pietronero L., Tosatti E. Amsterdam, Oxford, New York, Tokyo, North-Holland, 478–497.

4. Hess, W., Vilgis, T. A., & Winter, H. H. (1988). Dynamical Critical Behavior during Chemical Gelation and Vulcanization. Macromolecules, *21(8)*, 2536–3542.

5. Adolf, D., & Martin, J. E. (1990). Time-Cure Superposition during Cross-Linking Macromolecules, *23(15)*, 3700–3704.

6. Kozlov, G. V., & Novikov, V. U. (1997). The Order Parameters of Polymers Amorphous State Structure. Prikladnaya Fizika, *1*, 85–93.

7. Novikov, V. U., & Kozlov, G. V. (2000). Fractal Analysis of Polymers Elastic Constants. Materialovedenie, *1*, 2–12.

8. Kozlov, G. V., Sanditov, D. S., & Lipatov, Yu. S. (2004). Structural Analysis of Fluctuation Free Volume in Amorphous State of Polymers. In: Achievements in Polymers Physics-Chemistry Field. Zaikov, G. E., Berlin, A. A., Zlotskii, S. S., Minsker, K. S., & Monakov, Yu. B. Ed.; Moscow, Khimija, 412–474.

9. Kozlov, G. V., Zaikov, G. E., & Lipatov, Yu. S. (2005). The Structural Treatment of Fluctuation Free Volume in Amorphous State of Polymers. In: Chemical and Biological Kinetics. New Horizons. V. 1 Chemical Kinetics. Ed. Burlakova, E. B., Shilov, A. E., Varfolomeef, S. D., Zaikov, G. E. Leiden, Boston, Brill Academic Publishers, 484–516.

10. Kozlov, G. V., Novikiv, V. U., Gazaev, M. A., & Mikitaev, A. K. (1998). Structure of Cross-Linked Polymers a Percolation System. Inzhenerno-Fizicheskii Zhurnal, *71(2)*, 241–247.

11. Kozlov, G. V., Gazaev, M. A., Novikiv, V. U., & Mikitaev, A. K. (1996). Simulation of Amorphous Polymers Structure as Percolation Cluster. Pis'ma v ZhTF, *22(16)*, 31–38.

12. Kozlov, G. V., Shetov, R. A., & Mikitaev, A. K. (1987). Poly(vinyl chloride) Mechanical Properties Change in the β-Transition Domain Vysokomolek. Soed. A, *29(1)*, 62–66.

13. Kozlov, G. V., Afaunov, V. V., Mashukov, N. I., & Zaikov, G. E. (2001). The Physical Sence of the Parameters of Kohlrauch Equation in Case of Polyethylene Brittle Fracture. In: Fractals and Local Order in Polymeric Materials Ed. Kozlov, G. V., & Zaikov, G. E. New York, Nova Science Publishers Inc., 151–156.

14. Kozlov, G. V., Sanditov, B. D., Afaunov, V. V., & Badmaev, S. S. (1999). The Physical Significance of the Kohlraush Equation Parameters at Polyethylene Brittle Failure. Mater. All-Russian Sci. Conf. "Mathematical Simulation in Synergetic Systems". 20–26 jule, Ulan-Ude, Tomsk, 293–294.

15. Shlesinger, M. F., & Klafter, J. (1986). Nature of Temporal Hierarchies, Defining Relaxation in Disordered Systems. In: Fractals in Physics. Ed. Pietronero, L., Tosatti, E. Amsterdam, Oxford, New York, Tokyo. North-Holland, 553–560.

16. Levene, A., Pullen, W. J., & Roberts, J. (1965). Sound Velocity in Polyethylene at Ultrasonic Frequencies. J. Polymer Sci., Part A-2, *3(2)*, 697–701.

17. Balankin, A. S. (1991). Synergetics of Deformable Body. Moscow, Publishers of Defence Ministry of SSSR, 404 p.

18. Kozlov, G. V., & Novikov, V. U. (1998). Synergetics and Fractal Analysis of Cross-Linked Polymers. Moscow, Klassika, 112 p.

19. Family, F. (1984). Fractal Dimension and Grand Universality of Critical Phenomena. J. Stat. Phys., *36(5/6)*, 881–896.

20. Shklowskii, B. I., & Efros, A. (1975). Theory of Percolation and Condictivity of Strongly Nonhomogeneous Mediums. Uspekhi Fizicheskikh Nauk, *117(3)*, 401–436.
21. Kozlov, G. V., Aloev, V. Z., Bazheva, R. Ch., & Zaikov, G. E. (2002). The Generalized Percolation Model of Elasticity Modulus fpr Semicrystalline Polyethylene. Proceedings of IV All-Russian Sci.-Techn. Conf. "New Chemical Technologies: Production and Application". Penza, PSU, 59–61.
22. Bobryshev, A. N., Kozomazov, V. N., Babin, L. O., & Solomatov, V. I. (1994). Synergetics of Composite Materials. Lipetsk, NPO ORIUS, 154 p.
23. Dotsenko, V. S. (1985). Fractal Dynamics of Spin Glasses. J. Phys. C: Solid State Phys., *18(15)*, 6023–6031.
24. Kozlov, G. V., Sanditov, D. S., Mil'man, L. D., & Serdyuk, V. D. (1993). The Cluster Network of Macromolecular Entanglements and Network Connectivity in Polymers. Izvestiya VUZov, Severo-Kavkazsk. Region, estestv. Nauki, *3–4*, 88–92.
25. Patrikeev, G. A. (1968). Network Connectivity and Quantitative Relationships of Macromolecular Mechanics. Doklady AN SSSR, *183(6)*, 636–639.
26. Aharoni, S. M. (1985). Correlations between Chain Parameters and Failure Characteristics of Polymers below their Glass Transition Temperature. Macromolecules, *18(12)*, 2624–2630.
27. Plummer, C. J. G., & Donald, A. M. (1990). Disentanglement and Crazing in Glassy Polymers. Macromolecules, *23(12)*, 3929–3937.

CHAPTER 4

YIELDING PROCESS

CONTENTS

A cluster model of polymers amorphous state structure allows introducing principally new treatment of structure defect (in the full sense of this term) for the indicated state [1, 2]. As it is known [3], real solids structure contains a considerable number of defects. The given concept is the basis of dislocations theory, widely applied for crystalline solids behavior description. Achieved in this field successes predetermine the attempts of authors number [4–11] to use the indicated concept in reference to amorphous polymers. Additionally used for crystalline lattices notions are often transposed to the structure of amorphous polymers. As a rule, the basis for this transposition serves formal resemblance of stress – strain ($\sigma - \varepsilon$) curves for crystalline and amorphous solids.

In relation to the structure of amorphous polymers for a long time the most ambiguous point [12–14] was the presence or the absence of the local (short-range) order in this connection points of view of various authors on this problem were significantly different. The availability of the local order can significantly affect the definition of the structure defect in amorphous polymers, if in the general case the order-disorder transition or vice versa is taken for the defect. For example, any violation (interruption) of the long-range order in crystalline solids represents a defect (dislocation, vacancy, etc.), and a monocrystal with the perfect long-range order is the ideal defect-free structure with the perfect long-range order. It is known [15], that sufficiently bulky samples of 100% crystalline polymer cannot be obtained, and all characteristics of such hypothetical polymers are determined by the extrapolation method. That is why the Flory "felt" model [16, 17] can be suggested as the ideal defect-free structure for amorphous state of polymers. This model assumes that amorphous polymers consist of interpenetrating macromolecular coils personifying the full disorder (chaos). Proceeding from this, as the defect in polymers amorphous state a violation (interruption) of full disorder must be accepted, that is, formation of the local (or long-range) order [1]. It should also be noted that the formal resemblance of the curves $\sigma - \varepsilon$ for crystalline solids and amorphous polymers appears far incomplete and the behavior of these classes of materials displays principal differences, which will be discussed in detail below.

Turning back to the suggested concept of amorphous polymer structural defect, let us note that a segment including in the cluster can be considered as the linear defect – the analog of dislocation in crystalline solids. Since in the cluster model the length of such segment is accepted equal to the length of statistical segment, l_{st}, and their amount per volume unit is equal to the density of entanglements cluster network, v_{cl}, then the density of linear defects, ρ_d, per volume unit of the polymer can be expressed as follows [1]:

$$\rho_d = v_{cl} \times l_{st}. \tag{4.1}$$

The offered treatment allows application of well developed mathematical apparatus of the dislocation theory for the description of amorphous polymers properties. Its confirmation by the X-raying methods was stated in Ref. [18].

Further on, the rightfulness of application of the structural defect concept to polymers yielding process description will be considered. As a rule, previously assumed concepts of defects in polymers were primarily used for the description of this process or even exclusively for this purpose [4–11]. Theoretical shear strength of crystals was first calculated by Frenkel, basing on a simple model of two atoms series, displaced in relation to one another by the shear stress (Fig. 4.1a) [3]. According to this model, critical shear stress τ_0 is expressed as follows [3]:

$$\tau_0 = \frac{G}{2\pi}, \tag{4.2}$$

where G is the shear modulus.

Slightly changed, this model was used in the case of polymers yielding [6], wherefrom the following equation was obtained:

$$\tau_{0Y} = \frac{G}{\pi\sqrt{3}}, \tag{4.3}$$

where τ_{0Y} is a theoretical value of the shear stress at yielding.

Special attention should be paid to the fact that characterizes principally different behavior of crystalline metals compared with polymers. As it is known [3, 19], τ_{0Y}/τ_Y ratio (where τ_Y is experimentally determined shear stress at yielding) is much higher for metals than for polymers. For five metals possessing the face-centered cubic or hexagonal lattices the following ratios were obtained: $\tau_{0Y}/\tau_Y = 37400 \div 22720$ (according to the data of Ref. [3]), whereas for five polymers this ratio makes $2.9 \div 6.3$ [6]. In essence, sufficient closeness of τ_{0Y}/τ_Y ratio values to one may already be the proof for the possibility of realizing of Frenkel mechanism in polymers (in contrast with metals), but it will be shown below that for polymers a small modification of the law of shear stress τ periodic change used commonly gives τ_{0Y}/τ_Y values very close to one [20].

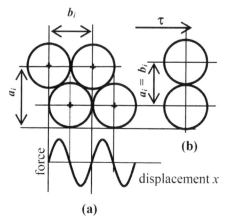

(a)

FIGURE 4.1   Schematic representation of deformation of two sequences of atoms according to the Frenkel model. Positions before (a) and after (b) deformation [3].

As it has been shown in Ref. [21], dislocation analogies are also true for amorphous metals. In essence, the authors of Ref. [21] consider the atoms construction distortion (which induces appearance of elastic stress fields) as a linear defect (dislocation) being practically immovable. It is clear that such approach correlates completely with the offered above structural defect concept. Within the frameworks of this concept, Fig. 4.1a may be considered as a cross-section of a cluster (crystallite) and, hence, the shear of segments in the latter according to the Frenkel mechanism – as a mechanism limiting yielding process in polymers. This is proved by the experimental data [32], which shown that glassy polymers yielding process is realized namely in densely packed regions. Other data [23] indicate that these densely packed regions are clusters. In other words, one can state that yielding process is associated with clusters (crystallites) stability loss in the shear stresses field [24].

In Ref. [25], the asymmetrical periodic function is adduced, showing the dependence of shear stress τ on shear strain γ_{sh} (Fig. 4.2). As it has been shown before [19], asymmetry of this function and corresponding decrease of the energetic barrier height overcome by macromolecules segments in the elementary yielding act are due to the formation of fluctuation free volume voids during deformation (that is the specific feature of polymers [26]). The data in Fig. 4.2 indicate that in the initial part of periodic curve from zero up to the maximum dependence of τ on displacement x can be simulated by a sine-shaped function with a period shorter than in Fig. 4.1. In this case, the function $\tau(x)$ can be presented as follows:

$$\tau(x) = k \sin\left(\frac{6\pi x}{b_i}\right),\qquad(4.4)$$

that is fully corresponds to Frenkel conclusion, except for arbitrarily chosen numerical coefficient in brackets (6 instead of 2).

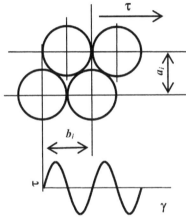

FIGURE 4.2   Schematic picture of shear deformation and corresponding stress – strain (τ – γ_{sh}) function [25].

Further calculation of τ_0 by method, described in Ref. [3], and its comparison with the experimental values τ_x indicated their close correspondence for nine amorphous and semicrystalline polymers (Fig. 4.3), which proves the possibility of realization of the above-offered yielding mechanism at the segmental level [18].

Inconsistency of τ_{oY} and τ_Y values for metals results to a search for another mechanism of yielding realization. At present, it is commonly accepted that this mechanism is the motion of dislocations by sliding planes of the crystal [3]. This implies that interatomic interaction forces, directed transversely to the crystal sliding plane, can be overcome in case of the presence of local displacements number, determined by stresses periodic field in the lattice. This is strictly different from macroscopic shear process, during which all bonds are broken simultaneously (the Frenkel model). It seems obvious that with the help of dislocations total shear strain will be realized at the

applying much lower external stress than for the process including simultaneous breakage of all atomic bonds by the sliding plane [3].

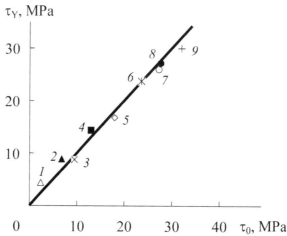

FIGURE 4.3   The relationship between theoretical τ_0 and experimental τ_Y shear stresses
at yielding for polytetrafluoroethylene (1), HDPE (2), polypropelene (3), polyamide-6 (4),
poly(vinil chloride) (5), polyhydroxiester (6), PC (7), polysulfone (8) and PAr (9) [20].

Payerls and Nabarro [3] were the first who calculated the shear stress
necessary for the dislocations motion, τ_{dm}. They used a sinusoidal approximation and deduced the expression for τ_{dm} as follows:

$$\tau_{dm} = \frac{2G}{1-v}e^{-2\pi a_i/b_i(1-v)},\qquad\qquad(4.5)$$

where v is the Poisson's ratio and parameters a_i and b_i are of the same meaning as in Fig. 4.2.

By substituting reasonable v value, for example, 0.35 [27], and assuming
$a_i = b_i$, the following value for τ_{dm} is obtained: $\tau_{dm} = 2 \times 10^{-4}\, G$. Though for
metals this value is higher than the observed τ_Y, it is much closer to them
than the stress calculated using simple shear model (the Frenkel model, Fig.
4.1).

However, for polymers the situation is opposite: analogous calculation
indicates that their τ_{dm} does not exceed 0.2 MPa, which is by two orders of
magnitude, approximately, lower than the observed τ_Y values.

Let us consider further the free path length of dislocations, λ_d. As it is known for metals [3], in which the main role in plastic deformation belongs to the mobile dislocations, λ_d assesses as ~ 10^4 Å. For polymers, this parameter can be estimated as follows [28]:

$$\lambda_Y = \frac{\varepsilon_Y}{b\rho_d}, \tag{4.6}$$

where ε_Y is the yield strain, b is Burgers vector, ρ_d is the density if linear defects, determined according to the Eq. (4.1).

The value ε_Y assesses as ~0.10 [29] and the value of Burgers vector b can be estimated according to the equation [30]:

$$b = \left(\frac{60,7}{C_\infty} \right)^{1/2}, \text{Å.} \tag{4.7}$$

The values for different polymers, λ_d assessed by the Eq. (4.6) is about 2.5 Å. The same distance, which a segment passes at shearing, when it occupies the position, shown in Fig. 4.1 b, that can be simply calculated from purely geometrical considerations. Hence, this assessment also indicates no reasons for assuming any sufficient free path length of dislocations in polymers rather than transition of a segment (or several segments) of macromolecule from one quasiequilibrium state to another [31].

It is commonly known [3, 25] that for crystalline materials Baily Hirsh relationship between shear stress, τ_Y, and dislocation density, ρ_d, is fulfilled:

$$\tau_Y = \tau_{in} + \alpha Gb\rho_d^{1/2}, \tag{4.8}$$

where τ_{in} is the initial internal stress, α is the efficiency constant.

The Eq. (4.8) is also true for amorphous metals [21]. In Ref. [20] it was used for describing mechanical behavior of polymers on the example of these materials main classes representatives. For this purpose, the data for amorphous glassy PAr [32], semicrystalline HDPE [33] and cross-linked epoxy polymers of amine and anhydride curing types (EP) were used [34]. Different loading schemes were used: uniaxial tension of film samples [32], high-speed bending [33] and uniaxial compression [34]. In Fig 4.4, the relations between calculated and experimental values τ_Y for the indicated polymers are adduced, which correspond to the Eq. (4.8). As one can see, they are linear and pass through the coordinates origin (i.e., $\tau_{in} = 0$), but α values for linear and cross-linked polymers are different. Thus in the frameworks of the offered defect concept the Baily Hirsh relationship is also true for polymers. This means that dislocation analogs are true for any linear defect, distorting the material ideal structure and creating the elastic stresses field [20]. From this point of view high defectness degree of polymers will be noted: $\rho_d \approx 10^9$ 10^{14} cm^{-2} for amorphous metals [21], and $\rho_d \approx 10^{14}$ cm^{-2} for polymers [28].

Hence, the stated above results indicate that in contrast with metals, for polymers realization of the Frenkel mechanism during yielding is much more probable rather than the defects motion (Fig. 4.1). This is due to the above-discussed (even diametrically opposed) differences in the structure of crystalline metals and polymers [1].

As it has been shown above using position spectroscopy methods [22], the yielding in polymers is realized in densely packed regions of their structure. Theoretical analysis within the frameworks of the plasticity fractal concept [35] demonstrates that the Poisson's ratio value in the yielding point, v_Y, can be estimated as follows:

$$v_Y = v\chi + 0.5 (1 - \chi), \qquad (4.9)$$

where v is Poisson's ratio value in elastic strains field, χ is the relative fraction of elastically deformed polymer.

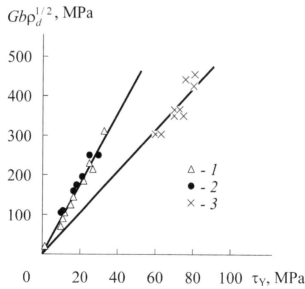

FIGURE 4.4   The relation between calculated and shear stress at yielding τ_Y, corresponding to the Eq. (4.8), for PAr (1), HDPE (2) and EP (3) [20].

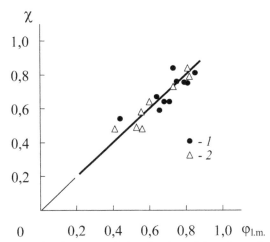

FIGURE 4.5   The relation between relative fraction of loosely packed matrix $\varphi_{l.m.}$ and probability of elastic state realization χ for PC (1) and PAr (2) [23].

In Fig. 4.5, the comparison of values χ and $\varphi_{l.m.}$ for PC and PAr is adduced, which has shown their good correspondence. The data of this figure assume that the loosely packed matrix can be identified as the elastic deformation region and clusters are identified as the region of inelastic (plastic) deformation [23]. These results prove the conclusion made in the Ref. [22] about proceeding of inelastic deformation processes in dense packing regions of amorphous glassy polymers and indicate correctness of the plasticity fractal theory at their description.

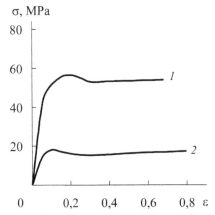

FIGURE 4.6   The stress-strain ($\sigma - \varepsilon$) diagrams for PAr at testing temperatures 293 (1) and 453 K (2) [24].

The yielding process of amorphous glassy polymers is often considered as their mechanical devitrification [36]. However, if typical stress – strain ($\sigma - \varepsilon$) plot for such polymers is considered (Fig. 4.6), then one can see, that behind the yield stress σ_Y the forced elasticity (cold flow) plateau begins and its stress σ_p is practically equal to σ_Y, that is, σ_p has the value of order of several tens MPa, whereas for devitrificated polymer this value is, at least, on the order of magnitude lower. Furthermore, σ_p is a function of the temperature of tests T, whereas for devitrificated polymer such dependence must be much weaker what is important, possess the opposite tendency (σ_p enhancement at T increase). This disparity is solved easily within the frameworks of the cluster model, where cold flow of polymers is associated with devitrivicated loosely packed matrix deformation, in which clusters "are floating." However, thermal devitrification of the loosely packed matrix occurs at the temperature T_g' – which is approximately 50 K lower than T_g. That is why it should be expected that amorphous polymer in the temperature range $T_g' \div T_g$ will be subjected to yielding under the application of even extremely low stress (of about 1 MPa). Nevertheless, as the plots in Fig. 6 show, this does not occur and $\sigma - \varepsilon$ curve for PAr in $T_g' \div T_g$ range is qualitatively similar to the plot $\sigma - \varepsilon$ at $T < T_g'$ (curve 1). Thus is should be assumed that devitrification of the loosely packed matrix is the consequence of the yielding process realization, but not its criterion. Taking into account realization of inelastic deformation process in the clusters (Fig. 4.5) one can suggest that the sufficient condition of yield in the polymer is the loss of stability by the local order regions in the external mechanical stress field, after which the deformation process proceeds without increasing the stress σ (at least, nominal one), contrary to deformation below the yield stress, where a monotonous increase of σ is observed (Fig. 4.6).

Now using the model suggested by the authors of Ref. [37] one can demonstrate that the clusters lose their stability, when stress in the polymer reaches the macroscopic yield stress, σ_Y. Since the clusters are postulated as the set of densely packed collinear segments, and arbitrary orientation of cluster axes in relation to the applied tensile stress σ should be expected, then they can be simulated as "inclined plates" (IP) [37], for which the following expression is true [37]:

$$\tau_Y < \tau_{IP} = 24 G_{cl} \varepsilon_0 \left(1 + v_2\right) / \left(2 - v_2\right), \qquad (4.10)$$

where τ_Y is the shear stress in the yielding point, τ_{IP} is the shear stress in IP (cluster), G_{cl} is the shear modulus, which is due to the clusters availability and determined from the plots similar to the ones from Figs. 2.5 and 2.6, that is, $G(v_{cl})$, ε_o is the IP proper strain, v_2 is Poisson's ratio for clusters.

Since the Eq. (4.10) characterizes inelastic deformation of clusters, the following can be accepted: $v_2 = 0.5$. Further on, under the assumption that $\tau_Y = \tau_{IP}$, the expression for the minimal (with regard to inequality in the left part of the Eq. (4.10)) proper strain ε_0^{min} is obtained [24]:

$$\varepsilon_0^{min} = \frac{\tau_Y}{\sqrt{24G_{cl}}} \tag{4.11}$$

The condition for IP (clusters) stability looks as follows [37]:

$$q = \sqrt{\frac{3}{2} \cdot \frac{\varepsilon_0}{\tau_Y}} \left\{ \left| 1 + \frac{\tilde{\varepsilon}_0}{\varepsilon_0} \right| - \sqrt{\frac{3}{8} \frac{\tau_Y}{G_{cl}\varepsilon_0 (1 + v_2)}} \right\}, \tag{4.12}$$

where q is the parameter, characterizing plastic deformation, $\tilde{\varepsilon}_0$ is the proper strain of the loosely packed matrix.

The cluster stability violation condition is fulfillment of the following inequality [37]:

$$q \leq 0. \tag{4.13}$$

Comparison of the Eqs. (4.12) and (4.13) gives the following criterion of stability loss for IP (clusters) [24]:

$$\left| 1 + \frac{\tilde{\varepsilon}_0}{\varepsilon_0} \right| = \sqrt{\frac{3}{8} \frac{\tau_Y^T}{G_{cl}\varepsilon_0 (1 + v_2)}}, \tag{4.14}$$

from which theoretical stress τ_Y (τ_Y^T) can be determined, after reaching of which the criterion (4.13) is fulfilled.

To perform quantitative estimations, one should make two simplifying assumptions [24]. Firstly, for IP the following condition is fulfilled [37]:

$$0 \leq \sin^2 \theta_{IP} (\tilde{\varepsilon}_0 / \varepsilon_0) \leq 1, \tag{4.15}$$

where θ_{IP} is the angle between the normal to IP and the main axis of proper strain.

Since for arbitrarily oriented IP (clusters) $\sin^2\theta_{IP} = 0.5$, then for fulfillment of the condition (4.15) the assumption is enough. Secondly, the Eq. (4.11) gives the minimal value of ε_o, and for the sake of convenience of calculations parameters τ_Y and G_{cl} were replaced by σ_Y and E, respectively. E value is greater than the elasticity modulus E_{cl} due to the availability of clusters, as it follows from the plots of Fig. 2.5 and 2.6. That is why to compensate two mentioned effects the strain ε_0, estimated according to the Eq. (4.11), was twice increased. The final equation looks as follows [24]:

$$\varepsilon_0 \approx 0,64\frac{\sigma_Y}{E} = 0,64\varepsilon_{cl},\qquad(4.16)$$

where ε_{cl} is the elastic component of macroscopic yield strain [38], which corresponds to strains ε_0 and $\widetilde{\varepsilon}_0$ by the physical significance [37].

Combination of the Eqs. (4.14) and (4.16) together with the plots similar to the ones shown in Figs. 2.5 and 2.6, wherefrom $E_{cl}(G_{cl})$ can be determined, allows to estimate theoretical yield stress σ_Y^T and compare it with experimental values σ_Y. Such comparison is adduced in Fig. 4.7, which demonstrates satisfactory conformity between σ_Y^T and σ_Y that proves the suggestion made in the Ref. [24] and justifies the above-made assumptions.

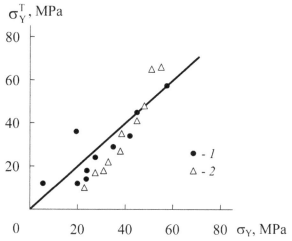

FIGURE 4.7 The relation between experimental σ_Y and calculated according to the Eq. (4.14) yield stress values for PAr (1) and PC (2) [24].

Hence, realization of the yielding process in amorphous glassy polymers requires clusters stability loss in the mechanical stress field, after which mechanical devitrification of the loosely packed matrix proceeds. Similar criterion was obtained for semicrystalline polymers [24].

As results obtained in Refs. [34, 39] have shown, the behavior of cross-linked polymers is just slightly different from the above-described one for linear PC and PAr. However, further progress in this field is quite difficult due to, at least, two reasons: excessive overestimation of the chemical cross-links role and the quantitative structural model absence. In the Ref. [39] the yielding mechanism of cross-linked polymer has been offered, based on the application of the cluster model and the latest developments in the deformable solid body synergetics field [40] on the example of two already above-mentioned epoxy polymers of amine (EP-1) and anhydrazide (EP-2) curing type.

Figure 4.8 shows the plots $\sigma - \varepsilon$ for EP-2 under uniaxial compression of the sample up to failure (curve 1) and at successive loading up to strain ε exceeding the yield strain ε_Y (curves 2–4). Comparison of these plots indicates consecutive lowering of the "yield tooth" under constant cold flow stress, σ_p. High values of σ_p assume corresponding values of stable clusters network density v_{cl}^{st}, which is much higher than the chemical cross-links network density v_c [34]. Thus though the behavior of a cross-linked polymer on the cold flow plateau is described within the frameworks of the rubber high-elasticity theory, the stable clusters network in this part of $\sigma - \varepsilon$ plots is preserved. The only process proceeding is the decay of instable clusters, determining the loosely packed matrix devitrification. This process begins at the stress equal to proportionality limit that correlates with the data from [41], where the action of this stress and temperature $T_2 = T_g'$ is assumed analogous. The analogy between cold flow and glass transition processes is partial only: the only one component, the loosely packed matrix, is devitrivicated. Besides, complete decay of instable clusters occurs not in the point of yielding reaching at σ_Y, but at the beginning of cold flow plateau at σ_p. This can be observed from $\sigma - \varepsilon$ diagrams shown in Fig. 4.8. As a consequence, the yielding is regulated not by the loosely packed matrix devitrification, but by other mechanism. As it is shown above, as such mechanism the stability loss by clusters in the mechanical stress field can be assumed, which also follows from the well-known fact of derivative $d\sigma/d\varepsilon$ turning to zero in the yield point [42]. According to the Ref. [40] critical shear strain γ_* leading to the loss of shear stability by a solid is equal to:

$$\gamma_* = \frac{1}{mn}, \tag{4.17}$$

where *m* and *n* are exponents in the Mie equation [27] setting the interconnection between the interaction energy and distance between particles. The value of parameter 1/mn can be expressed via the Poisson's ratio, v [27]:

$$\frac{1}{mn} = \frac{1-2v}{6(1+v)}.$$ (4.18)

From the Eqs. (2.20) and (4.18) it follows [18]:

$$\frac{1}{mn} = \gamma_* = \frac{\sigma_Y}{E}.$$ (4.19)

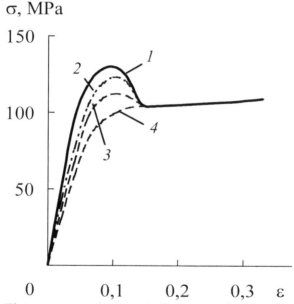

FIGURE 4.8 The stress – strain (σ – ε) diagrams at loading up to failure (1) and at cyclic load exertion (2–4) for EP-2: 2 – the first loading cycle; 3 – the second loading cycle; 4 – the third loading cycle [39].

The Eq. (4.19) gives the strain value with no regard to viscoelastic effect, that is, diagram σ – ε deviation from linearity behind the proportionality limit. Taking into account that tensile strain is twice greater, approximately, than the corresponding shear strain [42], theoretical yield strain ε_Y^T, corresponding to the stability loss by a solid, can be calculated. In Fig. 4.9 the comparison of experimental ε_Y and ε_Y^T yield strain magnitudes is fulfilled.

Approximate equality of these parameters is observed that assumes association of the yielding with the stability loss by polymers. More precisely, we
are dealing with the stability loss by clusters, because parameter v depends
upon the cluster network density v_{cl} (the Eq. (1.10)) and ε_Y value is proportional to v_{cl} [32].

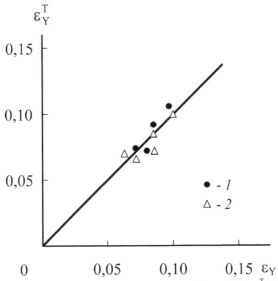

FIGURE 4.9   The relation between experimental ε_Y and theoretical ε_Y^T yield strain values
for epoxy polymers EP-1 (1) and EP-2 (2) [39].

The authors of Ref. [43] consider the possibility of intercommunication
of polymers yield strain and these materials suprasegmental structure evolution, which is constituent part of hierarchical systems behavior [44, 45].
It is supposed that within the frameworks of this general concept polymer
suprasegmental structures occupy their temporal and energetic "niches" in
general hierarchy of real world structures [46]. As a structure quantitative
model the authors of Ref. [43] use a cluster model of polymers amorphous
state structure [18, 24]. Ten groups of polymers, belonging to different deformation schemes in wide range of strain rates and temperatures, were used
for obtaining possible greater results community. The yield strain ε_Y was
chosen as the parameter, characterizing suprasegmental structures stability
in the mechanical stresses field.

The Gibbs function of suprasegmental (cluster) structure self-assembly at temperature $T = T_g - \Delta T$ was calculated as follows [45]:

$$\Delta \tilde{G}^{im} = \Delta S \Delta T,$$ (4.20)

where ΔS is entropy change in this process course, which can be estimated as follows [47]:

$$\Delta S = (3 \div 5) \times k \times f_g \times \ln f_g$$ (4.21)

In the Eq. (4.21), the coefficient $(3 \div 5)$ takes into account conformational molecular changes contribution to ΔS, k is Boltzmann constant, f_g is a polymer relative fluctuation free volume.

Let us consider now the results of the concept [44, 45] application to polymers yielding process description. The yielding can be considered as polymer structure loss of its stability in the mechanical stresses field and the yield strain ε_Y is measure of this process resistance. In Ref. [44], it is indicated that specific lifetime of suprasegmental structures t^{im} is connected with $\Delta \tilde{G}^{im}$ as follows:

$$t^{im} \sim \exp\left(-\Delta \tilde{G}^{im}/RT\right)$$ (4.22)

where R is the universal gas constant.

Assuming, that $t^{im} \sim t_Y$ (t_Y is time, necessary for yield stress σ_Y achievement) and taking into account, that $\varepsilon_Y \sim t_Y$, it can be written [43]:

$$\varepsilon_Y \sim \exp\left(-\Delta \tilde{G}^{im}/RT\right)$$ (4.23)

Let us note, that the Eq. (4.23) correctness in reference to polymers means, that yielding process in them is controlled by supra segmental structures thermodynamical stability.

In Fig. 4.10, the dependence of ε_Y on $\exp\left(-\Delta \tilde{G}^{im}/RT\right)$, corresponding to the Eq. (4.23), for all groups of the considered in Ref. [43] polymers. Despite definite (and expected) data scattering it is obvious, that all data break down into two branches, the one of which is approximated by a straight line. Such division reasons are quite obvious: the data with negative value of $\Delta \tilde{G}^{im}$ cover one (right) branch and with positive ones – the other (left) one. The last group consists of semicrystalline polymers with devitrificated at testing temperature loosely packed matrix (polytetrafluoroethylene, polyethylenes, polypropylene). The cluster model [18, 24] postulates thermofluctuation character of clusters formation and their decay at $T \geq T_g$. Therefore, such clusters availability in the indicated semicrystalline polymers devitrificated amorphous phase has quite another origin, namely, it is due to amorphous chains tightness in crystallization process [48]. In practice this effect results to the condition $\Delta \tilde{G}^{im} > 0$ realization, which in the low-molecular substances case belongs to hypothetical non existent in reality transitions "an overheated liquid $\to$ solid body" [45].

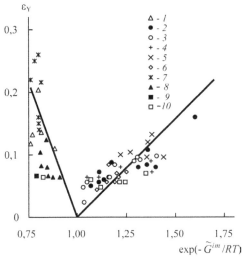

FIGURE 4.10 The correlation of yield strain e_Y and parameter $\exp\left(-\Delta \tilde{G}^{im}/RT\right)$ for polymers with devitrificated (1, 7, 8, 9) and glassy (2, 3, 4, 5, 6, 10) loosely packed matrix [43].

Let us note one more feature, confirming existence reality of Fig. 4.10 plot left branch. At present ε_Y increase at T growth for polyethylenenes and ε_Y decrease at the same conditions – for amorphous glassy polymers are well-known [18]. One can see easily, that this experimental fact is explained completely by two branches availability in the plot of Fig. 4.10, which confirms again existence reality of suprasegmental structures, which are in quasiequilibrium with "free" segments. This quasiequilibrium is characterized by $\Delta \tilde{G}^{im}$ so, that in this case $\Delta \tilde{G}^{im} > 0$.

As it was to be expected the value $\varepsilon_Y = 0$ at $\exp\left(-\Delta \tilde{G}^{im}/RT\right)$ or $\Delta \tilde{G}^{im} = 0$. The last condition is achieved at $T = T_g$, where the yield strain is always equal to zero. Let us also note, that the data for semicrystalline polymers with vitrificated amorphous phase (polyamide-6, poly(ethylene terephthalate)) cover the right branch of the Fig. 4.10 plot. This means, that suprasegmental structures existence ($\Delta \tilde{G}^{im} > 0$) is due just to devitrificated amorphous phase availability, but no crystallinity. At $T < T_g$ amorphous phase viscosity, increases sharply and amorphous chains tightness cannot be exercised its action, displacing macromolecules parts, owing to that local order formation has thermofluctuation character. The attention is paid to the obtained plot community, if to remember, that the values t_Y for impact and quasistatic tests are differed by five orders. This circumstance is explained simply enough, since the value ε_Y can be written as follows [43]:

$$\varepsilon_Y = t_Y \dot{\varepsilon},$$ (4.24)

where $\dot{\varepsilon}$ is strain rate and substitution of the Eq. (4.24) in the Eq. (4.23) shows, that in the right part of the latter the factor $\dot{\varepsilon}^{-1}$ appears, which for the considered loading schemes changes by about five orders.

The Gibbs specific function notion for nonequilibrium phase transition "overcooled liquid $\rightarrow$ solid body" is connected closely to local order notion (and, hence, fractality notion, see chapter one), since within the frameworks of the cluster model the indicated transition is equivalent to cluster formation start. In Fig. 1.1, the dependence of clusters relative fraction φ_{cl} on the value $\left| \Delta \tilde{G}^{im} \right|$ for PC and PAr is adduced. As one can see, this dependence is linear, φ_{cl} growth at $\left| \Delta \tilde{G}^{im} \right|$ increasing is observed and at $\left| \Delta \tilde{G}^{im} \right| = 0$ (i.e., for the selected standard temperature $T = T_g$) the cluster structure complete decay ($\varphi_{cl} = 0$) occurs.

The adduced above results can give one more, at least partial, explanation of "cell's effect." As it has been shown in Ref. [18], the following approximate relationship exists between ε_Y and Grüneisen parameter γ_L:

$$\varepsilon_Y = \frac{1}{2\gamma_L}$$ (4.25)

Using this relationship and the plots of Fig. 4.10, it is easy to show, that the decrease and, hence, φ_{cl} reduction results to γ_L growth, characterizing intermolecular bonds anharmonicity level. This parameter shows, how fast intermolecular interaction weakens at external (e.g., mechanical one [49]) force on polymer and the higher γ_L the faster intermolecular interaction weakening occurs at other equal conditions. In other words, the greater $\left| \Delta \tilde{G}^{im} \right|$ and φ_{cl}, the smaller γ_L and the higher polymer resistance to external influence. In Fig. 4.11, the dependence of γ_L on mean number of statistical segments per one cluster n_{cl}, which demonstrates clearly the said above.

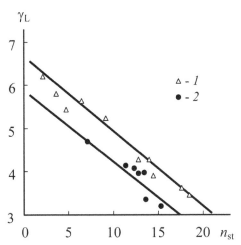

FIGURE 4.11   The dependence of Grüneisen parameter γ_L on mean number of statistical segments per one cluster n_{cl} for PC (1) and PAr (2) [43].

Hence, the stated above results shown that polymer yielding process can be described within the frameworks of the macrothermodynamical model. This is confirmed by the made in Ref. [44] conclusion about thermodynamical factor significance in those cases, when quasiequilibrium achievement is reached by mechanical stresses action. The existence possibility of structures with $\left|\Delta\tilde{G}^{im}\right| > 0$ (connected with transition "overheated liquid → solid body" [45] was shown and, at last, one more possible treatment of "cell's effect" was given within the frameworks of intermolecular bonds anharmonicity theory for polymers [49].

If to consider the yielding process as polymer mechanical devitrification [36], then the same increment of fluctuation free volume f_g is required for the strain ε_Y achievement. This increment Δf_g can be connected with ε_Y as follows [50]:

$$\Delta f_g = \varepsilon_Y(1-2v). \tag{4.26}$$

Therefore, f_g decrease results to Δf_g growth and respectively, ε_Y enhancement.

Let us consider, which processes result to necessary for yielding realization fluctuation free volume increasing. Theoretically (within the frameworks of polymers plasticity fractal concept [35] and experimentally (by positrons annihilation method [22]) it has been shown, that the yielding process is realized in densely packed regions of polymer. It is obvious, that the clusters will be such regions in amorphous glassy polymer and in semi-crystalline one – the clusters and crystallites [18]. The Eq. (4.9) allows to estimate relative fraction χ of polymer, which remains in the elastic state.

In Fig. 4.12, the temperature dependence of χ for HDPE is shown, from which one can see its decrease at T growth. The absolute values χ change within the limits of $0.516 \div 0.364$ and the determined by polymer density crystallinity degree K for the considered HDPE is equal to 0.687 [51]. In other words, the value χ in all cases exceeds amorphous regions fraction and this means the necessity of some part crystallites melting for yielding process realization. Thus, the conclusion should be made, that a semicrystal-line HDPE yielding process includes its crystallites partial mechanical melt-ing (disordering). For the first time Kargin and Sogolova [52] made such conclusion and it remains up to now prevalent in polymers mechanics [53]. The concept [35] allows to obtain quantitative estimation of crystallites frac-tion χ_{cr}, subjecting to partial melting – recrystallization process, subtracting amorphous phase fraction in HDPE from χ. The temperature dependence of χ_{cr} was also shown in Fig. 4.12, from which one can see, that χ_{cr} value is changed within the limits of $0.203 \div 0.051$, decreasing at T growth.

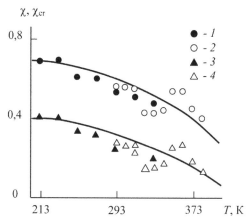

FIGURE 4.12 The dependences of elastically deformed regions fraction χ (1, 2) and crystallites fraction, subjected to partial melting, χ_{cr} (3, 4) on testing temperature T in impact (1, 3) and quasistatic (2, 4) tests for HDPE [51].

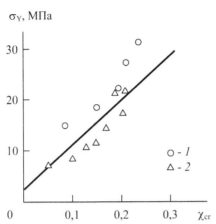

FIGURE 4.13 The dependence of yield stress σ_Y on crystallites fraction, subjecting to partial melting χ_{cr}, in impact (1) and quasistatic (2) tests for HDPE [51].

It is natural to assume, that the yield stress is connected with parameter χ_{cr} as follows: the greater χ_{cr}, the larger the energy, consumed for melting and the higher σ_Y. The data of Fig. 4.13 confirm this assumption and the dependence $\sigma_Y(\chi_{cr})$ is extrapolated to finite σ_Y value, since not only crystallites, but also clusters participate in yielding process. As it was to be expected [54], the crystalline regions role in yielding process realization is much larger, than the amorphous ones.

Within the frameworks of the cluster model [55] it has been assumed, that the segment joined to cluster means fluctuation free volume microvoid "shrinkage" and vice versa. In this case the microvoids number ΔN_h, forming in polymer dilation process, should be approximately equal to segments number ΔN_f, subjected to partial melting process. These parameters can be estimated by following methods. The value ΔN_h is equal to [51]:

$$\Delta N_h = \frac{\Delta f_g}{V_h}, \qquad (4.27)$$

where V_h is free volume microvoid volume, which can be estimated according to the kinetic theory of free volume [27].

Macromolecules total length L per polymer volume unit is estimated as follows [56]:

$$L = \frac{1}{S}, \qquad (4.28)$$

where S is macromolecule cross-sectional area.

The length of macromolecules L_{cr}, subjected to partial melting process, per polymer volume unit is equal to [51]:

$$L_{cr} = L\chi_{cr}$$ (4.29)

Further the parameter ΔN_χ can be calculated [51]:

$$\Delta N_\chi = \frac{L_{cr}}{l_{st}} = \frac{L\chi_{cr}}{l_0 C_\infty}$$ (4.30)

The comparison of ΔN_h and ΔN_Y values is adduced in Fig. 4.14, from which their satisfactory correspondence follows. This is confirmed by the conclusion, that crystallites partial mechanical melting (disordering) is necessary for fluctuation free volume f_g growth up to the value, required for polymers mechanical devitrification realization [51].

Whenever work is done on a solid, there is also a flow of heat necessitated by the deformation. The first law of thermodynamics:

$$dU = dQ + dW$$ (4.31)

states that the internal energy change dU of a sample is equal to the sum of the work dW performed on the sample and the heat flow dQ into the sample. This relation is valid for any deformation, whether reversible or irreversible. There are two thermodynamically irreversible cases for which dQ and dW are equal by absolute value and opposite by sign: uniaxial deformation of a Newtonian fluid and ideal elastic-plastic deformation. For amorphous glassy polymers deformation has essentially different character: the ratio $dQ/dW \neq 1$ and changes within the limits of $0.36 \div 0.75$ depending on testing condition [57]. In other words, the thermodynamically ideal plasticity is not realized for these materials. The reason of this effect is thermodynamical nonequilibrium of polymers structure. Within the frameworks of the fractal analysis it has been shown that it results to polymers yielding process realization not only in entire sample volume, but also in its part (see the Eq. (4.9)) [35]. Besides, it has been demonstrated experimentally and theoretically that amorphous glassy polymer structural component, in which the yielding is realized, is densely packed local order regions (clusters) [22, 23].

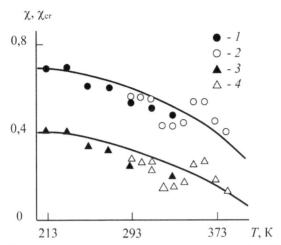

FIGURE 4.14 The relation between increment of segments number ΔN_Y in crystallites, subjecting to partial melting, and increment of free volume microvoids number ΔN_h, which is necessary for yielding process realization, for HDPE [51].

Lately the mathematical apparatus of fractional integration and differentiation [58, 59] was used for fractal objects description, which is amorphous glassy polymers structure. It has been shown [60] that Kantor's set fractal dimension coincides with an integral fractional exponent, which indicates system states fraction, remaining during its entire evolution (in our case deformation). As it is known [61], Kantor's set ("dust") is considered in one-dimensional Euclidean space ($d = 1$) and therefore, its fractal dimension obey the condition $d_f \leq 1$. This means, that for fractals, which are considered in Euclidean spaces with $d > 2$ ($d = 2, 3, \ldots$) the fractional part of fractal dimension should be taken as fractional exponent ν_{fr} [62, 63]:

$$\nu_{fr} = d_f - (d - 1). \tag{4.32}$$

The value ν_{fr} characterizes that states (structure) part of system (polymer), which remains during its entire evolution (deformation). In Fig. 4.15, the dependence of latent energy fraction dU at PC and poly(methyl methacrylate) (PMMA) deformation on $\nu_{fr} = d_f - (d - 1)$ [64] is shown. The value dU was estimated as $(W - Q)/W$. In Fig. 4.15, the theoretical dependence $dU(\nu_{fr})$ is adducted, plotted according to the conditions $dU = 0$ at $\nu_{fr} = 0$ or $d_f = 2$ and $dU = 1$ at $\nu_{fr} = 1$ or $d_f = 3$ ($d = 3$), that is, at ν_{fr} or d_f limiting values [64]. The experimental data correspond well to this theoretical dependence, from which it follows [64]:

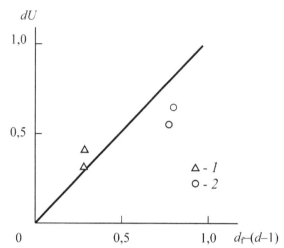

FIGURE 4.15 The dependence of relative fraction of latent energy dU on fractional exponent $v_{fr} = d_f - (d-1)$ value for PC (1) and PMMA (2) [64].

$$dU = v_{fr} = d_f - (d-1). \tag{4.33}$$

Let us consider two limiting cases of the adduced in Fig. 4.15 dependence at $v_{fr} = 0$ and 1.0, both at d = 3. In the first case ($d_f = 2$) the value $dU = 0$ or, as it follows from dU definition (the Eq. (4.31)), $dW = dQ$ and polymer possesses an ideal elastic-plastic deformation. Within the frameworks of the fractal analysis $d_f = 2$ means, that $\varphi_{cl} = 1.0$, that is, amorphous glassy polymer structure represents itself one gigantic cluster. However, as it has been shown above, the condition $d_f = 2$ achievement for polymers is impossible in virtue of entropic tightness of chains, joining clusters, and therefore, $d_f > 2$ for real amorphous glassy polymers. This explains the experimental observation for the indicated polymers: $dU \neq 0$ or $|dW| \neq |dQ|$ [57]. At $v_{fr} = 1.0$ or $d_f = d = 3$ polymers structure loses its fractal properties and becomes Euclidean object (true rubber). In this case from the plot of Fig. 4.15 $dU = 1.0$ follows. However, it has been shown experimentally, that for true rubbers, which are deformed by thermodynamically reversible deformation $dU = 0$. This apparent discrepancy is explained as follows [64]. Figure 4.15 was plotted for the conditions of inelastic deformation, whereas at $d_f = d$ only elastic deformation is possible. Hence, at $v_{fr} = 1.0$ or $d_f = d = 3$ deformation type discrete (jump – like) change occurs from $dU \to 1$ up to $dU = 0$. This point becomes an initial one for fractal object deformation in Euclidean space with the next according to the custom dimension $d = 4$, where $3 \leq d_f \leq$

4 and all said above can be repeated in reference to this space: at $d = 4$ and $d_f = 3$ the value $v_{fr} = 0$ and $dU = 0$. Let us note in conclusion that exactly the exponent v_{fr} controls the value of deformation (fracture) energy of fractal objects as a function of process length scale. Let us note that the equality $dU = \varphi_{l.m.}$ was shown, from which structural sense of fractional exponent in polymers inelastic deformation process follows: $v_{fr} = \varphi_{l.m.}$ [64].

Mittag-Lefelvre function [59] usage is one more method of a diagrams $\sigma - \varepsilon$ description within the frameworks of the fractional derivatives mathematical calculus. A nonlinear dependences, similar to a diagrams $\sigma - \varepsilon$ for polymers, are described with the aid of the following equation [65]:

$$\sigma(\varepsilon) = \sigma_0 \left[1 - E_{v_{fr},1}\left(-\varepsilon^{v_{fr}}\right)\right], \tag{4.34}$$

where σ_0 is the greatest stress for polymer in case of linear dependence $\sigma(\varepsilon)$ (of ideal plasticity), is the Mittag-Lefelvre function [65]:

$$E_{v_{fr},1}\left(-\varepsilon^{v_{fr}}\right) = \sum_{k=0}^{\infty} \frac{\varepsilon^{v_{fr}k}}{\tilde{A}\left(v_{fr}k + \beta\right)}, \quad v_{fr} > 0, \beta > 0, \tag{4.35}$$

where Γ is Eiler gamma-function.

As it follows from the Eq. (4.34), in the considered case $\beta = 1$ and gamma-function is calculated as follows [40]:

$$\tilde{A}\left(v_{fr}k + 1\right) = \sqrt{\frac{\pi}{2}}\left(v_{fr}k - \varepsilon^{v_{fr}k}\right)^{v_{fr}k - \varepsilon^{-v_{fr}}} e^{-\left(v_{fr}k - \varepsilon^{v_{fr}}\right)}. \tag{4.36}$$

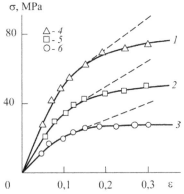

FIGURE 4.16 The experimental ($1 \div 3$) and calculated according to the Eqs. (4.34) ÷ (4.36) ($4 \div 6$) diagrams $\sigma - \varepsilon$ for PAr at $T = 293$ (1, 4), 353 (2, 5) and 433 K (3, 6). The shaded lines indicate calculated diagrams $\sigma - \varepsilon$ for forced high-elasticity part without v_{fr} change [66].

In Fig. 4.16, the comparison of experimental and calculated according to the Eqs. (4.34) ÷ (4.36) diagrams $\sigma - \varepsilon$ for PAr at three testing temperatures is adduced. The values σ_0 were determined as the product $E\varepsilon_Y$ [66]. As it follows from the data of Fig. 4.16, the diagrams $\sigma - \varepsilon$ on the part from proportionality limit up to yield stress are well described well within the frameworks of the Mittag-Lefelvre function. Let us note that two necessary for these parameters (σ_0 and v_{fr}) are the function of polymers structural state, but not filled parameters. This is a principal question, since the usage in this case of empirical fitted constants, as, for example, in Ref. [67], reduces significantly using method value [60, 65].

In the initial linear part (elastic deformation) calculation according to the Eq. (4.34) was not fulfilled, since in it deformation is submitted to Hooke law and, hence, is not nonlinear. At stresses greater than yield stress (high-elasticity part) calculation according to the Eq. (4.34) gives stronger stress growth (stronger strain hardening), than experimentally observed (that it has been shown by shaded lines in Fig. 4.16). The experimental and theoretical dependences $\sigma - \varepsilon$ matching on cold flow part within the frameworks of the Eq. (4.34) can be obtained at supposition $v_{fr} = 0.88$ for $T = 293$ K and $v_{fr} = 1.0$ for the two remaining testing temperatures. This effect explanation was given within the frameworks of the cluster model of polymers amorphous state structure [39], where it has been shown that in a yielding point small (instable) clusters, restraining loosely packed matrix in glassy state, break down. As a result of such mechanical devitrification glassy polymers behavior on the forced high-elasticity (cold flow) plateau is submitted to rubber high-elasticity laws and, hence, $d_f \rightarrow d = 3$ [68]. The stress decay behind yield stress, so-called "yield tooth," can be described similarly [66]. An instable clusters decay in yielding point results to clusters relative fraction φ_{cl} reduction, corresponding to d_f growth (the Eq. (1.12)) and v_{fr} enhancement (the Eq. (4.33)) and, as it follows from the Eq. (4.34), to stress reduction. Let us note in conclusion that the offered in Ref. [69] techniques allow to predict parameters, which are necessary for diagrams $\sigma - \varepsilon$ description within the frameworks of the considered method, that is, E, ε_Y and d_f.

In Ref. [70], it has been shown that rigid-chain polymers can be have a several substates within the limits of glassy state. For polypiromellithimide three such substates are observed on the dependences of modulus $d\sigma/d\varepsilon$, determined according to the slope of tangents to diagram $\sigma - \varepsilon$, on strain ε [70]. However, such dependences of $d\sigma/d\varepsilon$ on ε have much more general character: in Fig. 4.17 three similar dependences for PC are adduced, which were plotted according to the data of Ref. [49]. If in Ref. [70] the transition from part I to part II corresponds to polypiromellithimide structure phase change with axial crystalline structure formation [71], then for the adduced in Fig. 4.17 similar dependences for PC this transition corresponds to deformation type change from elastic to inelastic one (proportionality limit [72]). The dependences $(d\sigma/d\varepsilon)(\varepsilon)$ community deserves more intent consideration that was fulfilled by the authors of Ref. [73] within the frameworks of fractal analysis.

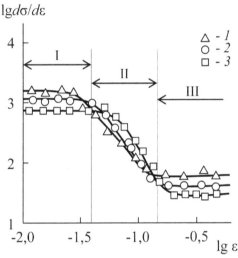

FIGURE 4.17 The dependences of modulus $d\sigma/d\varepsilon$, determined according to the slope of tangents to diagram $\sigma - \varepsilon$, on strain ε in double logarithmic coordinates for PC at $T = 293$ (1), 343 (2) and 373 K (3) [73].

In Fig. 1.2, the dependence of physical fractal density ρ on measurement scale L in double logarithmic coordinates was shown. For $L < L_{min}$ and $L > L_{max}$ Euclidean behavior is observed and within the range of $L = L_{min} \div L_{max}$ – fractal one [40]. Let us pay attention to the complete analogy of the plots of Figs. 1.2 and 4.17.

There is one more theoretical model allowing nonstandard treatment of the shown in Fig. 4.17 dependences $d\sigma/d\varepsilon(\varepsilon)$. Kopelman was offered the fractal descriptions of chemical reactions kinetics, using the following simple relationship [74]:

$$k \sim t^{-h}, \tag{4.37}$$

where k is reaction rate, t is its duration, h is reactionary medium nonhomo-geneity (heterogeneity) exponent ($0 < h < 1$), which turns into zero only for Euclidean (homogeneous) mediums and the Eq. (4.37) becomes classical one: $k = $ const.

From the Eq. (4.37) it follows, that in case $h \neq 0$, that is, for heteroge-neous (fractal) mediums the reaction rate k reduces at reaction proceeding. One should attention to qualitative analogy of curves $\sigma - \varepsilon$ and the depen-dences of conversion degree on reaction duration $Q(t)$ for a large number of polymers synthesis reactions [75]. Still greater interest for subsequent theoretical developments presents complete qualitative analogy of diagrams $\sigma - \varepsilon$ and strange attractor trajectories, which can be have "yield tooth," strain hardening and so on [76].

If to consider deformation process as polymer structure reaction with supplied, from outside mechanical energy to consider, then the modulus $d\sigma/d\varepsilon$ will be k analog (Fig. 4.17). The said above allows to assume, that de-formation on parts I and III (elasticity and cold flow) proceeds in Euclidean space and on part II (yielding) – in fractal one. The comparison with the schematic plot of Fig. 1.2 assumes also, that transitions from one part to another were due to measurement scale change, induced by deformation.

The fractal analysis main rules usage for polymers structure and proper-ties description [68, 77] allows to make quantitative estimation of measure-ment scale L change at polymer deformation. There are a several methods of such estimation and the authors of Ref. [73] use the simplest from them as ensuring the greatest clearness. As it was noted in chapter one, the self-similarity (fractality) range of amorphous glassy polymers structure coin-cides with cluster structure existence range: the lower scale of self-similarity corresponds to statistical segment length l_{st} and the upper one – to distance between clusters R_{cl}. The simplest method of measurement scale L estima-tion is the usage of well-known Richardson equation – Eq. (2.12). For PC at testing temperature $T = 293$ K included in the Eq. (2.12) parameters are equal to: $L_{ch} = L_{cl} = 76.5$ Å, $R_{ch} = R_{cl} = 30.1$ Å [78] and the value D_{ch} can be calculated according to the Eq. (2.5).

In the case of affine deformation the value R_{cl} will be changed propor-tionally to drawing ratio λ [70]. This change value can be estimated from the equation [79]:

$$R_{cl}\lambda = l_{st}\frac{2(1-\nu)}{(1-2\nu)}. \tag{4.38}$$

From the Eq. (4.38) ν increase follows and from the Eq. (1.9) – the d_f increase at drawing ratio λ growth. In its turn, the value λ is connected with the strain ε by a simple relationship (in the case of affine deformation) [80]:

$$\lambda = 1 + \varepsilon. \qquad (4.39)$$

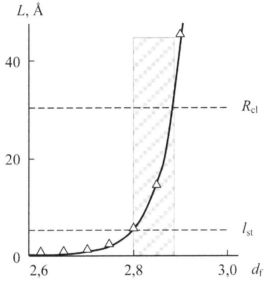

FIGURE 4.18   The dependence of measurement scale L on structure fractal dimension d_f for PC at $T = 293$ K. Horizontal shaded lines indicate nondeformed PC structure self-similarity boundaries (l_{st} and R_{cl}) and the shaded region – deformation fractal behavior range [73].

In Fig. 4.18, the dependence of L on d_f is adduced, which is calculated according to the Eqs. (2.5), (2.12), (2.15), (4.38) and (4.39) combination and at the condition, that the parameters $C_\infty d_f$ are connected with each other as follows [18]:

$$C_\infty = \frac{2d_f}{d(d-1)(d-d_f)} + \frac{4}{3}. \qquad (4.40)$$

As it follows from the data of Fig. 4.18, L growth at d_f increase is observed and within the range of $d_f \approx 2.80 \div 2.89$ polymer deformation proceeds in fractal space. At $d_f > 2.90$ the deformation space transition from fractal to Euclidean one is observed (PC yielding is achieved at $d_f = 2.85$ [23]) and structure PC approaching to true rubber state ($d_f = d = 3$) induces very fast L growth. Let us consider the conditions of transition from part II to part III (Fig. 4.17). Replacing in the Eq. (2.12) the value R_{cl} by $R_{cl}\lambda$ according to the indicated above reasons it assumption $D_{ch} = 1.0$, that is, part of chain, stretched completely between clusters, let us obtain [73]:

$$\frac{L_{cl}}{l_{st}} = \frac{R_{cl}\lambda}{l_{st}},$$

(4.41)

That is, drawing ratio critical value λ_{cr}, corresponding to the transition from part II to part III (from fractal behavior to Euclidean one) or the transition from yielding to cold flow, is equal to [73]:

$$\lambda_{cr} = \frac{L_{cl}}{R_{cl}l_{st}},$$

(4.42)

that corresponds to the greatest attainable molecular draw [81].

For PC at $T = 293$ K and the indicated above values L_{cl} and R_{cl} λ_{cr} will be equal to 2.54. Taking into account, that drawing ratio at uniaxial deformation $\lambda''_{cr} = \lambda_{cr}^{1/3}$, let us obtain $\lambda''_{cr} = 1.364$ or critical value of strain of transition. Let us note that within the frameworks of the cluster model of polymers amorphous state structure [18] chains deformation in loosely packed matrix only is assumed and since the Eq. (4.42) gives molecular drawing ratio, which is determined in the experiment according to the relationship [81]:

$$\varepsilon_{cr} = \varepsilon''_{cr}\left(1 - \phi_{cl}\right).$$

(4.43)

Let us obtain $\varepsilon_{cr} = 0.117$ according to the Eq. (4.43). The experimental value of yield strain ε_{Y} for PC at $T = 293$ K is equal to 0.106. This means, that the transition to PC cold flow begins immediately beyond yield stress that is observed experimentally [23].

Therefore, the stated above results show, that the assumed earlier substates within the limits of glassy state are due to transitions from deformation in Euclidean space to deformation in fractal space and vice versa. These transitions are controlled by deformation scale change, induced by external load (mechanical energy) application. From the physical point of view this postulate has very simple explanation: if size of structural element, deforming deformation proceeding, hits in the range of sizes $L_{min} - L_{max}$ (Fig. 1.2), then deformation proceeds in fractal space, if it does not hit – in Euclidean one. In part I intermolecular bonds are deformed elastically on scales of 3 ÷ 4Å ($L < L_{min}$), in part II – cluster structure elements with sizes of order of 6

÷ 30Å [18, 24] ($L_{min} < L < L_{max}$) and in part III chains fragments with length of L_{cl} or of order of several tens of Ånströms ($L > L_{max}$). In Euclidean space the dependence $\sigma - \varepsilon$ will be linear ($d\sigma/d\varepsilon$ = const) and in fractal one – curvilinear, since fractal space requires deformation deceleration with time. The yielding process realization is possible only in fractal space. The stated model of deformation mechanisms is correct only in the case of polymers structure presentation as physical fractal.

In the general case polymers structure is multifractal, for behavior description of which in deformation process in principle its three dimensions knowledge is enough: fractal (Hausdorff) dimension d_f, informational one d_I and correlation one d_c [82]. Each from the indicated dimension describes multifractal definite properties change and these dimensions combined application allows to obtain more or less complete picture of yielding process [73].

As it is known [83], a glassy polymers behavior on cold flow plateau (part III in Fig. 4.17) is well described within the frameworks of the rubber high-elasticity theory. In Ref. [39] it has been shown that this is due to mechanical devitrification of an amorphous polymers loosely packed matrix. Besides, it has been shown [82, 84] that behavior of polymers in rubber-like state is described correctly under assumption, that their structure is a regular fractal, for which the identity is valid:

$$d_I = d_c = d_f. \tag{4.44}$$

A glassy polymers structure in the general case is multifractal [85], for which the inequality is true [82, 84]:

$$d_c < d_I < d_f. \tag{4.45}$$

Proceeding from the said above and also with appreciation of the known fact, that rubbers do not have to some extent clearly expressed yielding point the authors of Ref. [73] proposed hypothesis, that glassy polymer structural state changed from multifractal up to regular fractal, that is, criterion (4.44) fulfillment, was the condition of its yielding state achievement. In other words, yielding in polymers is realized only in the case, if their structure is multifractal, that is, if it submits to the inequality Eq. (4.45).

Let us consider now this hypothesis experimental confirmations and dimensions d_I and d_c estimation methods in reference to amorphous glassy polymers multifractal structure. As it is known [82], the informational dimension d_I characterizes behavior Shennone informational entropy $I(\varepsilon)$:

$$I(\varepsilon) = \sum_{i}^{M} P_i \ln P_i , \qquad (4.46)$$

where M is the minimum number of d-dimensional cubes with side ε, necessary for all elements of structure coverage, P_i is the event probability that structure point belongs to i-th element of coverage with volume ε^d.

In its turn, polymer structure entropy change ΔS, which is due to fluctuation free volume f_g, can be determined according to the Eq. (4.21). Comparison of the Eqs. (4.21) and (4.46) shows that entropy change in the first from them is due to f_g change probability and to an approximation of constant the values $I(\varepsilon)$ and ΔS correspond to each other. Further polymer behavior at deformation can be described by the following relationship [84]:

$$\Delta S = -c \left(\sum_{j=1}^{d} \lambda_i^{d_1} - 1 \right), \qquad (4.47)$$

where c is constant, λ_1 is drawing ratio.

Hence, the comparison of the Eqs. (4.21), (4.46) and (4.47) shows that polymer behavior at deformation is defined by change f_g exactly, if this parameter is considered as probabilistic measure. Let us remind, that f_c such definition exists actually within the frameworks of lattice models, where this parameter is connected with the ratio of free volume microvoids number N_h and lattice nodes number N (N_h/N) [49]. The similar definition is given and for P_i in the Eq. (4.46) [86].

The value d_1 can be determined according to the following equation [82]:

$$d_1 = \lim_{\varepsilon \to 0} \left[\frac{\sum_{i=1}^{M} P_i(\varepsilon) \ln P_i(\varepsilon)}{\ln \varepsilon} \right]. \qquad (4.48)$$

Since polymer structure is a physical fractal (multifractal), then for it fractal behavior is observed only in a some finite range of scales (Fig. 1.2) and the statistical segment length l_{st} is accepted as lower scale. Hence, assuming $P_i(\varepsilon) = f_g$ and $\varepsilon \to l_{st}$ the Eq. (4.48) can be transformed into the following one [73]:

$$d_I = c_1 \frac{R f_g \ln f_g}{\ln l_{st}}, \tag{4.49}$$

where c_1 is constant.

As it is known [23], in yielding point for PC $d_f = 2.82$ (or $\nu = 0.41$), that allows to determine the value f_g according to the Eq. (2.19) and constant c_1 – from the condition (4.44). Then, using the known equality $\lambda_I = 1 + \varepsilon_I$, similar to the Eq. (4.39), the yield strain ε_Y theoretical value can be calculated, determining constant c by matching method and the value Δf_g for ΔS calculation – as values f_g difference for nondeformed polymer and polymer in yielding point. Since in Ref. [73] thin films are subjected to deformation, then as the first approximation $d = 1$ is assumed in the Eq. (4.47) sum sign. In Fig. 4.19, comparison of the temperature dependences of experimental and calculated by the indicated method values ε_Y for PC is adduced. As one can see, the good correspondence of theory and experiment is obtained by both the dependences $\varepsilon_Y(T)$ course and ε_Y absolute values and discrepancy of experimental and theoretical values ε_Y at large T is due to the fact, that for calculation simplicity the authors of Ref. [73] did not take into consideration transverse strains influence on value ε_Y in the Eq. (4.47).

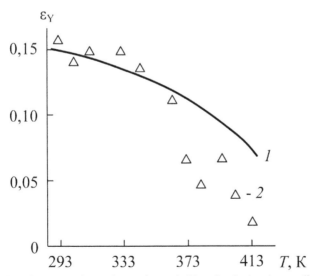

FIGURE 4.19   The comparison of experimental (1) and calculated according to the Eq. (4.47) temperature dependences of yield strain ε_Y for PC [73].

Williford [87] proposed, that the value d_1 of multifractal corresponded to its surface dimension (the first subfractal) – either sample surface or fracture surface. For this supposition checking the authors of Ref. [73] calculate a fracture surface fractal dimension for brittle (d_{fr}^{br}) and ductile (d_{fr}^{duc}) failure types according to the equations [40]:

$$d_{fr}^{br} = \frac{10(1+v)}{7-3v} \qquad (4.50)$$

and

$$d_{fr}^{duc} = \frac{2(1+4v)}{1+2v}. \qquad (4.51)$$

In Fig. 4.20, comparison of the temperature dependences of d_1, d_{fr}^{duc} and d_{fr}^{br} for PC is adduced. As one can see, at low T ($T < 373$ K) the value d_1 corresponds to d_{fr}^{br} well enough and at higher temperatures ($T > 383$ K) it is close to d_{fr}^{duc}. In Fig. 4.20, the shaded region shows the temperature range corresponding to the brittle-ductile transition for PC [73]. It is significant that this interval beginning ($T = 373$ K) coincides with loosely packed matrix devitrification temperature T_g, which is approximately 50 K lower than polymer glass transition temperature T_g [18], for PC equal to ~423 K [88].

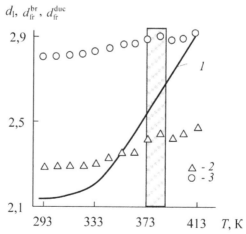

FIGURE 4.20   Comparison of the temperature dependences of informational dimension d_1 (1), fractal dimensions of fracture surface at brittle d_{fr}^{br} (2) and ductile d_{fr}^{duc} (3) failure for PC. The temperature range of brittle-ductile transition is shown by shaded region [73].

The correlation dimension d_c is connected with multifractal structure internal energy U [61] and it can be estimated according to the equation [82]:

$$\Delta U = -c_2 \left(\lambda_F^{d-d_c} \right),$$

(4.52)

where ΔU is internal energy change in deformation process, c_2 is constant, λ_F is macroscopic drawing ratio, which in the case of uniaxial deformation is equal to λ_1. The value λ_F is determined as follows [82]:

$$\lambda_F = \lambda_1 \lambda_2 \lambda_3,$$

(4.53)

where λ_2 and λ_3 are transverse drawing rations, connected with λ_1 by the simple relationships [82]:

$$\lambda_2 = 1 + \varepsilon_2,$$

(4.54)

$$\lambda_3 = 1 + \varepsilon_3,$$

(4.55)

$$\varepsilon_2 = \varepsilon_3 = \nu \varepsilon_1.$$

(4.56)

The temperature dependence of d_c can be calculated, as and earlier, assuming from the condition (4.44) that at yielding $d_c = d_f = 2.82$, estimating ΔU as one half of product $\sigma_Y \varepsilon_Y$ (with appreciation of practically triangular form of curve $\sigma - \varepsilon$ up to yield stress) and determining the constant c_2 by the indicated above mode. Comparison of the temperature dependences of multifractal three characteristic dimensions d_c, d_1 and d_f, calculated according to the Eqs. (4.52), (4.47) and (1.9), respectively, is adduced in Fig. 4.21. As it follows from the plots of this figure, for PC the inequality Eq. (4.45) is confirmed, which as a matter of fact is multifractal definition. The dependences $d_c(T)$ and $d_1(T)$ are similar and their absolute values are close, that is explained by the indicated above intercommunication of f_g and U change [89]. Let us note, that dimension d_1 controls only yield strain ε_Y and dimension d_c – both ε_Y and yield stress σ_Y. At approaching to glass transition temperature, that is, at $T \circledR T_g$, the values d_c, d_1 and d_f become approximately equal, that is, rubber is a regular fractal. Thus, with multifractal formalism positions the glass transition can be considered as the transition of structure from multifractal to the regular fractal. Additionally it is easy to show the fulfillment of the structure thermodynamical stability condition [90]:

$$d_f(d - d_1) = d_c(d - d_c).$$

(4.57)

Hence, the authors of Ref. [73] considered the multifractal concept of amorphous glassy polymers yielding process. It is based on the hypothesis, that the yielding process presents itself structural transition from multifractal to regular fractal. In a amorphous glassy polymers Shennone informational entropy to an approximation of constant coincides with entropy, which is due to polymer fluctuation free volume change. The approximate quantitative estimations confirm the offered hypothesis correctness. The postulate about yielding process realization possibility only for polymers fractal structure, presented as a physical fractal within the definite range of linear scales, is the main key conclusion from the stated above results. What is more, the fulfilled estimations strengthen this definition – yielding is possible for polymer multifractal structure only and represents itself the structural transition multifractal – regular fractal [73].

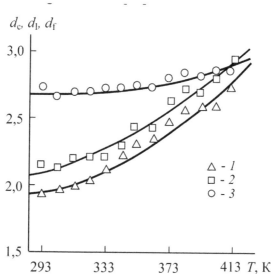

FIGURE 4.21   The temperature dependences of correlation d_c (1), informational d_l (2) and Hausdorff d_f (3) dimensions of PC multifractal structure [73].

Let us consider in the present chapter conclusion the treatment of dependences of yield stress on strain rate and crystalline phase structure for semicrystalline polymers [77]. As it known [91], the clusters relative fraction φ_{cl} is an order parameter of polymers structure in strict physical significance of this term and since the local order was postulated as having thermofluctuation origin, then φ_{cl} should be a function of testing temporal scale in virtue of super-position temperature-time. The dependent on time t the order parameter $\varphi_{cl}(t)$ is determined according to the Eq. (3.13).

Having determined value t as duration of linear part of diagram load – time $(P – t)$ in impact tests and accepting is equal to clusters relative fraction in quasistatic tensile tests [92], the values $\varphi_{cl}(t)$ can be estimated. In Fig. 4.22, the dependences $\varphi_{cl}(t)$ on strain rate for HDPE and polypropylene (PP) are shown, which demonstrate φ_{cl} increase at strain rate growth, that is, tests temporal scale decrease.

The elasticity modulus E value decreases at $\dot{\varepsilon}$ growth (Fig. 4.23). This effect cause (intermolecular bonds strong anharmonicity) is described in detail in Refs. [49, 94, 95]. In Fig. 4.24, the dependences of elasticity modulus E on structure fractal dimension d_f are adduced for HDPE and PP, which turned out to be linear and passing through coordinates origin. As it is known [84], the relation between E and d_f is given by the equation:

$$E = Gd_f,\qquad(4.58)$$

where G is a shear modulus.

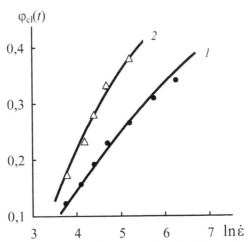

FIGURE 4.22 The dependences of clusters relative fraction $\varphi_{cl}(t)$ on strain rate $\dot{\varepsilon}$ in logarithmic coordinates for HDPE (1) and PP (2) [93].

From the Eq. (4.58) and Fig. 4.24 it follows, that for both HDPE and PP G is const. This fact is very important for further interpretation [77].

Let us estimate now determination methods of crystalline σ_Y^{cr} and non-crystalline σ_Y^{nc} regions contribution to yield stress σ_Y of semicrystalline polymers. The value σ_Y^{cr} can be determined as follows [96]:

$$\sigma_Y^{cr} = \frac{Gb}{2\pi}\left(\frac{K}{S}\right)^{1/2},$$ (4.59)

where b is Burgers vector, determined according to the Eq. (4.7), K is crystallinity degree, S is cross-sectional area, which is equal to 14.35 and 26.90 Å² for HDPE and PP, accordingly [97].

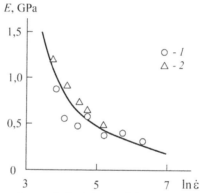

FIGURE 4.23 The dependence of elasticity modulus E on strain rate $\dot{\varepsilon}$ in logarithmic coordinates for HDPE (1) and PP (2) [93].

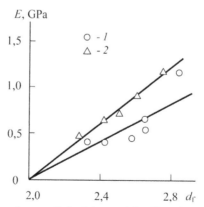

FIGURE 4.24 The dependences of elasticity modulus E on fractal dimension of structure d_f for HDPE (1) and PP (2) [93].

As one can see, all included in the Eq. (4.59) parameters for each polymer are constant, from which it follows, that σ_Y^{cr} = const. The value σ_Y^{cr} = 16.3 MPa for HDPE and 17.6 MPa – for PP. Thus, the change, namely, increase σ_Y at $\dot{\varepsilon}$ growth is defined by polymers noncrystalline regions contribution σ_Y^{nc} [77].

The Grist dislocation model [98] assumes the formation in polymers crystalline regions of screw dislocation (or such dislocation pair) with Burgers vector b and the yield process is realized at the formation of critical nucleus domain with size u:

$$u^* = \frac{Bb}{2\pi\tau_Y},$$

(4.60)

where B is an elastic constant, τ_Y shear yield stress.

In its turn, the domain with size U^* is formed at energetic barrier ΔG^* overcoming [98]:

$$\Delta G^* = \frac{Bl^2 l_d}{2\pi}\left[\ln\left(\frac{u^*}{r_0}\right)-1\right],$$

(4.61)

where l_d is dislocation length, which is equal to crystalline thickness, r_0 is dislocation core radius.

In the model [98] it has been assumed, that nucleus domain with size u^* is formed in defect-free part of semicrystalline polymer, that is, in crystallite. Within the frameworks of model [1] and in respect to these polymers amorphous phase structure such region is loosely packed matrix, surrounding a local order region (cluster), whose structure is close enough to defect-free polymer structure, postulated by the Flory "felt" model [16, 17]. In such treatment the value u^* can be determined as follows [43]:

$$u^* = R_{cl} - r_{cl},$$

(4.62)

where R_{cl} is one half of distance between neighboring clusters centers, r_{cl} is actually cluster radius.

The value R_{cl} is determined according to the equation [18]:

$$R_{cl} = 18\left(\frac{2v_e}{F}\right)^{-1/3}, \text{ Å},$$

(4.63)

where v_e is macromolecular binary hooking network density, values of which for HDPE and PP are adduced in Ref. [99], $F = 4$.

The value r_{cl} can be determined as follows [77]:

$$r_{cl} = \left(\frac{n_{st} S}{\pi \eta_{pac}} \right),$$ (4.64)

where n_{st} is statistical segments number per one cluster ($n_{st} = 12$ for HDPE and 15 – for PP [77]), η_{pac} is packing coefficient, for the case of dense packing equal to 0.868 [100].

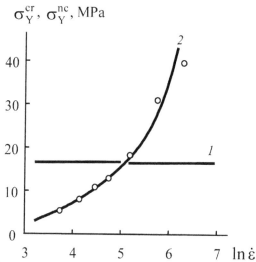

FIGURE 4.25   The dependences of crystalline σ_Y^{cr} (1) and noncrystalline σ_Y^{nc} (2) regions contributions in yield stress value σ_Y on strain rate in logarithmic coordinates for HDPE [93].

The shear modulus G, which, has been noted above, is independent on $\dot{\varepsilon}$, is accepted as B. Now the value σ_Y^{nc} can be determined according to the Eq. (4.60), assuming, that n_{st} changes proportionally to φ_{cl} (Fig. 4.22). In Fig. 4.25, the dependences of σ_Y^{cr} and σ_Y^{nc} on strain rate $\dot{\varepsilon}$ are adduced for HDPE (the similar picture was obtained for PP). As it follows from the data of this figure, at the condition $\sigma_Y^{c} = \text{const}$ the value σ_Y^{nc} grows at $\dot{\varepsilon}$ increase and at $\dot{\varepsilon} \approx 150$ s-¹ $\sigma_Y^{nc} > \sigma_Y^{cr}$, that is, a noncrystalline regions contribution in σ_Y begins to prevail [93].

In Fig. 4.26, the comparison of the experimental σ_Y and calculated theoretically σ_Y^T as sum ($\sigma_Y^{cr} + \sigma_Y^{nc}$) dependences of yield stress on $\dot{\varepsilon}$ for HDPE and PP are adduced. As one can see, a good correspondence of theory and experiment is obtained, confirming the offered above model of yielding process for semicrystalline polymers.

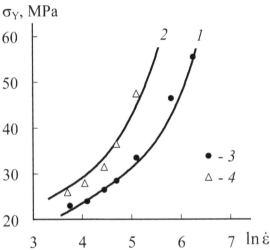

FIGURE 4.26 Comparison of the experimental (1, 2) and theoretical (3, 4) dependences of yield stress σ_Y on strain rate $\dot{\varepsilon}$ in logarithmic coordinates for HDPE (1, 3) and PP (2, 4) [93].

Let us note in conclusion the following. As it follows from the comparison of the data of Figs. 4.23, 4.24 and 4.26, σ_Y increase occurs at E reduction and G constancy, that contradicts to the assumed earlier σ_Y and E proportionality [101]. Such proportionality absence is assumed even by the Eq. (2.20), since the value v is a function of several structural factors (see, e.g., the Eq. (1.10)). Therefore, the postulated in Ref. [101] σ_Y and E proportionality is only an individual case, which is valid either at invariable structural state or at the indicated state, changing by definite monotonous mode [77].

Hence, the stated above results demonstrate nonzero contribution of noncrystalline regions in yield stress even for such semicrystalline polymers, which have devitrificated amorphous phase in testing conditions. At definite conditions noncrystalline regions contribution can be prevailed. Polymers yield stress and elastic constants proportionality is not a general rule and is fulfilled only at definite conditions.

The authors of Ref. [102] use the considered above model [93] for branched polyethylenes (BPE) yielding process description. As it is known [103], the crystallinity degree K, determined by samples density, can be expressed as follows:

$$K = \alpha_c + \alpha_{if},$$ (4.65)

where α_c and α_{if} are chains units fractions in perfect crystallites and anisotropic interfacial regions, accordingly.

The Eq. (4.59) with replacement of K by α_c was used for the value σ_Y^{cr} estimation. Such estimations have shown that the value σ_Y^{cr} is always smaller than macroscopic yield stress σ_Y. In Fig. 4.27, the dependence of crystalline phase relative contribution in yield stress σ_Y^{cr}/σ_Y on α_c is adduced. This dependence is linear and at $\alpha_c = 0$ the trivial result $\sigma_Y^{cr} = 0$ is obtained. Let us note that this extrapolation assumes $\sigma_Y \neq 0$ at $\alpha_c = 0$. At large α_c the crystalline phase contribution is prevailed and at $\alpha_c = 0.75$ $\sigma_Y^{cr}/\sigma_Y = 1$ (Fig. 4.27).

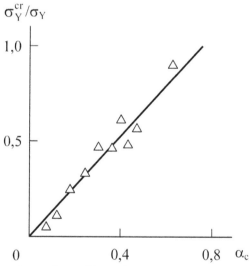

FIG. 4.27 The dependence of crystalline regions relative contribution σ_Y^{cr}/s_Y in yield stress on perfect crystallites fraction a_c for series of BPE [102].

The dependence $\alpha_{if}(\alpha_c)$ showed α_{if} linear reduction at α_c growth. Such α_{if} change and simultaneous σ_Y^{cr}/σ_Y increasing (Fig. 4.27) at α_c growth assume local order degree reduction, determining noncrystalline regions contribution in σ_Y, at crystallinity degree enhancement. Besides, the correlation

$\alpha_{if}(\alpha_c)$ shows that local order regions of BPE noncrystalline regions are concentrated mainly in anisotropic interfacial regions, as it has been assumed earlier [18]. At $\alpha_c = 0.82$ $\alpha_{if} = 0$, that corresponds to the data of Fig. 4.27.

The common fraction of the ordered regions (clusters and crystallites) φ_{ord} can be determined according to the percolation relationship [91]:

$$\varphi_{ord} = 0.03(T_m - T)^{0.55}, \qquad (4.66)$$

where T_m and T are melting and testing temperatures, respectively.

Then clusters relative fraction φ_{cl} is estimated from the obvious relationship [102]:

$$\varphi_{cl} = \varphi_{ord} - \alpha_c. \qquad (4.67)$$

The comparison of the experimental and estimated as a sum of crystalline and noncrystalline regions contributions in yield stress theoretical values α_Y shows their good correspondence for considered BPE.

It has been noted earlier [104, 105], that stress decay beyond stress ("yield tooth") for polyethylenes is expressed the stronger the greater value α_c is. For amorphous polymers it has been shown that the indicated "yield tooth" is due to instable clusters decay in yielding process and this decay is expressed the clearer the higher instable clusters relative fraction [39]. By analogy with the indicated mechanism the authors of Ref. [102] assume that "yield tooth" will be the stronger the larger crystallites fraction is subjected to mechanical disordering (partial melting) in yielding process. The indicated fraction of crystallites χ_{cr} is determined as difference [102]:

$$\chi_{cr} = \chi - \alpha_{am}, \qquad (4.68)$$

where χ is polymer fraction, subjecting to elastic deformation, α_{am} is fraction of amorphous phase.

The value χ can be determined within the frameworks of polymers plasticity fractal concept [35] according to the Eq. (4.9). The dependence of χ_{cr} on α_c shows χ_{cr} increasing at α_c growth [102]. At small α_c values all crystallites are subjected to disordering owing to that "yield tooth" in BPE curves stress-strain is absent and these curves are acquired the form, which is typical for rubbers. Hence, stress decay beyond yield stress intensification is due to χ_{cr} growth [103].

KEYWORDS

- anharmonicity
- latent energy
- Mittag-Lefelvre function
- multifractal formalism
- polymer
- yield stress

REFERENCES

1. Kozlov, G. V., Belousov, V. N., Serdyuk, V. D., & Kuznetsov, E. N. (1995). Defects of Polymers Amorphous State Structure. Fizika i Technika Yysokikh Davlenii, *5(3)*, 59–64.
2. Kozlov, G. V., Beloshenko, V. A., & Varyukhin, V. N. (1996). Evolution of Dissipative Structures in Yielding Process of Cross-Linked Polymers. Prikladnaya Mekhanika i Tekhnicheskaya Fizika, *37(3)*, 115–119.
3. Honeycombe, R. W. K. (1968). The Plastic Deformation of Metals. London, Edward Arnold Publishers, Ltd, 398 p.
4. Argon, A. S. (1988). A theory for the Low-Temperature Plastic Deformation of Glassy Polymers. Phil. Mag., 1974, *29(1)*, 149–167.
5. Argon, A. S. (1973). Physical Basis of Distortional and Dilational Plastic Flow in Glassy Polymers. J. Macromol. Sci.-Phys., *88(3–4)*, 573–596.
6. Bowden, B., & Raha, S. A. (1974). Molecular Model for Yield and Flow in Amorphous Glassy Polymers Making use of a Dislocation Analoque. Phil. Mag., *29(1)*, 149–165.
7. Escaig, B. (1978). The Physics of Plastic Behaviour of Crystalline and Amorphous Solids. Ann. Phys., *3(2)*, 207–220.
8. Pechhold, W. R., & Stoll, B. (1982). Motion of Segment Dislocations as a Model for Glass Relaxation. Polymer Bull., *7(4)*, 413–416.
9. Sinani, A. B., & Stepanov, V. A. (1981). Prediction of Glassy Polymer Deformation Properties with the Aid of Dislocation Analogues. Mekhanika Kompozitnykh Materialov, *17(1)*, 109–115.
10. Oleinik, E. F., Rudnev, S. M., Salamatina, O. B., Nazarenko, S. I., & Grigoryan, G. A. (1986). Two Modes of Glassy Polymers Plastic Deformation. Doklady AN SSSR, *286(1)*, 135–138.
11. Melot, D., Escaig, B., Lefebvre, J. M., Eustache, R. R., & Laupretre, F. (1994). Mechanical Properties of Unsaturated Polyester Resins in Relation to their Chemical Strcucture: 2. Plastic Deformation Behavior. J. Polymer Sci.: Part B: Polymer Phys., *32(11)*, 1805–1811.
12. Boyer, R. F. (1976). General Reflections on the Symposium on Physical Structure of the Amorphous State. J. Macromol. Sci., Phys., B*12 (12)(2)*, 253–301.
13. Fischer, E. W., Dettenmaier, M. (1978). Structure of Polymer Glasses and Melts. J. Non-Cryst. Solids, *31(1–2)*, 181–205.

14. Wendorff, J. H. (1982). The Structure of Amorphous Polymers. Polymer, *23(4)*, 543–557.
15. Nikol skii, V. G., Plate, I. V., Fazlyev, F. A., Fedorova, E. A., Filippov, V. V., & Yudaeva, L. V. (1983). Structure of Polyolefins Thin Films, Obtained by Quenching of Melt up to 77K. Vysokomolek. Soed. A, *25(11)*, 2366–2371.
16. Flory, J. (1984). Conformations of Macromolecules in Condenced Phases. Pure Appl. Chem., *56(3)*, 305–312.
17. Flory, J. (1976). Spatial Configuration of Macromolecular Chains. Brit. Polymer J., *8(1–10)*.
18. Kozlov, G. V., & Zaikov, G. E. (2004). Structure of the Polymer Amorphous State. Utrecht, Boston, Brill Academic Publishers, 465 p.
19. Kozlov, G. V., Shogenov, V. N., & Mikitaev, A. K. (1988). A Free Volume Role in Amorphous Polymers Forced Elasticity Process. Doklady AN SSSR, *298(1)*, 142–144.
20. Kozlov, G. V., Afaunova, Z. I., & Zaikov, G. E. (2002). The Theoretical Estimation of Polymers Yield Stress. Electronic Zhurnal "Issledovano v Rossii", *98*, 1071–1080. http://zhurnal.ape relarn ru/articles /2002/ 098 pdf.
21. Liu, R. S., & Li, J. Y. (1989). On the Structural Defects and Microscopic Mechanism of the High Strength of Amorphous Alloys. Mater. Sci. Engng., A *114(1)*, 127–132.
22. Alexanyan, G. G., Berlin, A. A., Gol danskii, A. V., Grineva, N. S., Onitshuk, V. A., Shantarovich V. P., & Safonov, G. P. (1986). Study by the Positrons Annihilation Method of Annealing and Plastic Deformation Influence on Polyarylate Microstructure. Khimicheskaya Fizika, *5(9)*, 1225–1234.
23. Balankin, A. S., Bugrimov, A. L., Kozlov, G. V., Mikitaev, A. K., Sanditov, D. S. (1992). The Fractal Structure and Physical-Mechanical Properties of Amorphous Glassy Polymers. Doklady, A. N., *326(3)*, 463–466.
24. Kozlov, G. V., & Novikov, V. U. (2001). The Cluster Model of Polymers Amorphous State. Uspekhi Fizicheskikh Nauk, *171(7)*, 717–764.
25. McClintock, F. A., & Argon, A. S. (1966). Mechanical Behavior of Materials. Massachusetts, Addison-Wesley Publishing Company Inc., 432 p.
26. Wu, S. (1992). Secondary Relaxation, Brittle-Ductile Transition Temperature, and Chain Structure. J. Appl. Polymer Sci. *46(4)*, 619–624.
27. Sanditov, D. S., & Bartenev, G. M. (1982). Physical Properties of Disordered Structures. Novosibirsk, Nauka, 256 p.
28. Mil man, L. D., & Kozlov, G. V. (1986). Polycarbonate Nonelastic Deformation Simulation in Impact Loading Conditions with the Aid of Dislocations Analogues. In: Polycondencation Processes and Polymers, Nal chik, KBSU, 130–141.
29. Shogenov, V. N., Kozlov, G. V., & Mikitaev, A. K. (1989). Prediction of Forsed Elasticity of Rigid-Chain Polymers. Vysokomolek. Soed. A, *31(8)*, 1766–1770.
30. Sanditov, D. S., & Kozlov, G. V. (1993). About Correlation Nature between Elastic Moduli and Glass Transition Temperature of Amorphous Polymers. Fizika i Khimiya Stekla, *19(4)*, 593–601.
31. Peschanskaya, N. N., Bershtein, V. A., & Stepanov, V. A. (1978). The Connection of Glassy Polymers Creep Activation Energy with Cohesion Energy. Fizika Tverdogo Tela, *20(11)*, 3371–3374.

32. Shogenov, V. N., Belousov, V. N., Potapov, V. V., Kozlov, G. V., & Prut, E. V. (1991). The Glassy Polyarylatesulfone Curves stress-strain Description within the Frameworks of High-Elasticity Concepts. Vysokomolek. Soed. A, *33(1)*, 155–160.

33. Mashukov, N. I., Gladyshev, G. P., & Kozlov, G. V. (1991). Structure and Properties of High Density Polyethylene Modified by High-Disperse Mixture Fe and FeO. Vysokomolek. Soed. A, *33(12)*, 2538–2546.

34. Beloshenko, V. A., & Kozlov, G. V. (1994). The Cluster Model Application for Epoxy Polymers Yielding Process. Mechanika Kompozitnykh Materialov, *30(4)*, 451–454.

35. Balankin, A. S., & Bugrimov, A. L. (1992). The Fractal Theory of Polymers Plasticity. Vysokomolek. Soed. A, *34(10)*, 135–139.

36. Andrianova, G. P., & Kargin, V. A. (1970). To Necking Theory at Polymers Tension. Vysokomolek. Soed. A, *12(1)*, 3–8.

37. Kachanova, I. M., & Roitburd, A. L. (1989). Plastic Deformation Effect on New Phase Inclusion Equilibrium Form and Thermodynamical Hysteresis. Fizika Tverdogo Tela, *31(4)*, 1–9.

38. Hartmann, B., Lee, G. F., & Cole, R. F. (1986). Tensile Yield in Polyethylene. Polymer Engng. Sci., *26(8)*, 554–559.

39. Kozlov, G. V., Beloshenko, V. A., Gazaev, M. A., & Novikov, V. U. (1996). Mechanisms of Yielding and Forced High-Elasticity of Cross-Linked Polymers. Mekhanika Kompozitnykh Materialov, *32(2)*, 270–278.

40. Balankin, A. S. (1991). Synergetics of Deformable Body. Moscow, Publishers of Ministry of Defence SSSR, 404 p.

41. Filyanov, E. M. (1987). The Connection of Activation Parameters of Cross-Linked Polymers Deformation in Transient Region with Glassy State Properties Vysokomolek. Soed. A, *29(5)*, 975–981.

42. Kozlov, G. V., & Sanditov, D. S. (1992). The Activation Parameters of Glassy Polymers Deformation in Impact Loading Conditions. Vysokomolek. Soed. B, *34(11)*, 67–72.

43. Kozlov, G. V., Shustov, G. B., Zaikov, G. E., Burmistr, M. V., & Korenyako, V. A. (2003). Polymers Yielding Description within the Franeworks of Thermodynamical Hierarchical Model. Voprosy Khimii I Khimicheskoi Tekhnologii, *1*, 68–72.

44. Gladyshev, G. P. (1988). Thermodynamics and Macrokinetics of Natural Hierarchical Processes. Moscow, Nauka, 290 p.

45. Gladyshev, G. P., & Gladyshev, D. P. (1994). The Approximate Thermodynamical Equation for Nonequilibrium Phase Transitions. Zhurnal Fizicheskoi Khimii, *68(5)*, 790–792.

46. Kozlov, G. V., & Zaikov, G. E. (2003). Thermodynamics of Polymer Structure Formation in an Amorphous State. In: Fractal Analysis of Polymers: From Synthesis to Composites. Kozlov, G. V., Zaikov, G. E., Novikov, V. U. Ed.; New York, Nova Science Publishers Inc., 89–97.

47. Matsuoka, S., & Bair, H. E. (1977). The Temperature Drop in Glassy Polymers during Deformation. J. Appl. Phys., *48(10)*, 4058–4062.

48. Kozlov, G. V., & Zaikov, G. E. (2003). Formation Mechanisms of Local Order in Polymers Amorphous State Structure. Izvestiya KBNC RAN, *1(9)*, 54–57.

49. Kozlov, G. V., & Sanditov, D. S. (1994). Anharmonic Effects and Physical-Mechanical Properties of Polymers. Novosibirsk, Nauka, 261 p.

50. Matsuoka, S., Aloisio, C. Y., & Bair, H. E. (1973). Interpretation of Shift of Relaxation Time with Deformation in Glassy Polymers in Terms of Excess Enthalpy. J. Appl. Phys., *44(10)*, 4265–4268.

51. Kosa, N., Serdyuk, V. D., Kozlov, G. V., & Sanditov, D. S. (1995). The Comparative Analysis of Forced Elasticity Process for Amorphous and Semicrystalline Polymers. Fizika I Tekhnika Vysokikh Davlenii, *5(4)*, 70–81.

52. Kargin, V. A., & Sogolova, I. I. (1953). The Studu of Crystalline Polymers Mechanical Properties. I. Polyamides Zhurnal Fizicheskoi Khimii, *27(7)*, 1039–1049.

53. Gent, A. N., & Madan, S. (1989). Plastic Yielding of Partially Crystalline Polymers. J. Polymer Sci." Part B: Polymer Phys., *27(7)*, 1529–1542.

54. Mashukov, N. I., Belousov, V. N., Kozlov, G. V., Ovcharenko, E. N., & Gladychev, G. P. (1990). The Connection of Forced Elasticity Stress and Structure for Semicrystalline Polymers. Izvestiya AN SSSR, seriya khimicheskaya, *9*, 2143–2146.

55. Sanditov, D. S., Kozlov, G. V., Belousov, V. N., & Lipatov, Yu. S. (1994). The Cluster Model and Fluctuation Free Volume Model of Polymeric Glasses. Fizika i Khimiya Stekla, *20(1)*, 3–13.

56. Graessley, W. W., & Edwards, S. F. (1981). Entanglement Interactions in Polymers and the Chain Contour Concentration. Polymer, *22(10)*, 1329–1334.

57. Adams, G. W., & Farris, R. Y. (1989). Latent Energy of Deformation of Amorphous Polymers. 1. Deformation Calorimetry. Polymer, *30(9)*, 1824–1828.

58. Oldham, K., & Spanier, J. (1973). Fractional Calculus. London, New York, Academic Press, 412 p.

59. Samko, S. G., Kilbas, A. A., & Marishev, O. I. (1987). Integrals and Derivatives of Fractional Order and their some Applications. Minsk, Nauka i Tekhnika, 688 p.

60. Nigmatullin, R. R. (1992). Fractional Integral and its Physical Interpretation. Teoreticheskaya i Matematicheskaya Fizika, *90(3)*, 354–367.

61. Feder, F. (1989). Fractals. New York, Plenum Press, 248 p.

62. Kozlov, G. V., Shustov, G. B., & Zaikov, G. E. (2002). Polymer Melt Structure Role in Heterochain Polyeters Thermooxidative Degradation Process. Zhurnal Prikladnoi Khimii, *75(3)*, 485–487.

63. Kozlov, G. V., Batyrova, H. M., & Zaikov, G. E. (2003). The Structural Treatment of a number of Effective Centres of Polymeric Chains in the Process of the Thermooxidative Degradation. J. Appl. Polymer Sci., *89(7)*, 1764–1767.

64. Kozlov, G. V., Sanditov, D. S., & Ovcharenko, E. N. (2001). Plastic Deformation Energy and Structure of Amorphous Glassy Polymers. Proceeding of Internat. Interdisciplinary Seminar "Fractals and Applied Synergetics FiAS-01", 26–30 November, Moscow, 81–83.

65. Meilanov, R. P., Sveshnikova, D. A., & Shabanov, O. M. (2001). Sorption Kinetics in Systems with Fractal Structure. Izvestiya VUZov, Severo-Kavkazsk. Region, estestv nauki, *1*, 63–66.

66. Kozlov, G. V., & Mikitaev, A. K. (2007). The Fractal Analysis of Yielding and Forced High-Elasticity Processes of Amorphous Glassy Polymers. Mater. I-th All-Russian Sci-Techn. Conf. "Nanostructures in Polymers and Polymer Nanocomposites". Nal chik, KBSU, 81–86.

67. Kekharsaeva, E. R., Mikitaev, A. K., & Aleroev, T. S. (2001). Model of Stress-Strain Characteristics of Chlor-Containing Polyesters on the Basis of Derivatives of Fractional Order. Plast. Massy, *3*, 35.

68. Novikov, V. U., & Kozlov, G. V. (2000). Structure and Properties of the Polymers within the Frameworks of Fractal Approach. Uspekhi Khimii, *69(6)*, 572–599.

69. Shogenov, V. N., Kozlov, G. V., & Mikitaev, A. K. (2007). Prediction of Mechanical Behavior, Structure and Properties of Film polymer Samples at Quasistatic Tension. In: Polycondensation Reactions and Polymers. Selected Works. Nal chik, KBSU, 252–270.

70. Lur e, E. G., & Kovriga, V. V. (1977). To Question about Unity of Deformation Mechanism and Spontaneous Lengthening of Rigid-Chain Polymers. Mekhanika Polimerov, *4*, 587–593.

71. Lur e, E. G., Kazaryan, L. G., Kovriga, V. V., Uchastkina, E. L., Lebedinskaya, M. L., Dobrokhotova, M. L., & Emel yanova, L. M. (1970). The Features of Crystallization and Deformation of Polyimide Film PM. Plast. Massy. *8*, 59–63.

72. Shogenov, V. N., Kozlov, G. V., & Mikitaev, A. K. (1989). Prediction of Rigid-Chain Polymers Mechanical Properties in Elasticity Region. Vysokomolek. Soed. B, *31(7)*, 553–557.

73. Kozlov, G. V., Yanovskii, Yu. G., & Karnet, Yu. N. (2008). The Generalized Fractal Model of Amorphous Glassy Polymers Yielding Process. Mekhanika Kompozitsionnykh Materialov i Konstruktsii, *14(2)*, 174–187.

74. Kopelman, R. (1986). Excitons Dynamics Mentioned Fractal one: Geometrical and Energetic Disorder. In: Fractals in Physics. Ed. Pietronero L., Tosatti E. Amsterdam, Oxford, New York, Tokyo, North-Holland, 524–527.

75. Korshak, V. V., & Vinogradova, S. V. (1972). Nonequilibrium Polycondensation. Moscow, Nauka, 696 p.

76. Vatrushin, V. E., Dubinov, A. E., Selemir, V. D., & Stepanov, N. V. (1995). The Analysis of SHF Apparatus Complexity with Virtual Cathode as Dynamical Objects. In: Fractals in Applied Physics. Ed. Dubinov A.E. Arzamas-16, VNIIEF, 47–58.

77. Mikitaev, A. K., & Kozlov, G. V. (2008). Fractal Mechanics of Polymer Materials. Nal chik, Publishers KBSU, 312 p.

78. Kozlov, G. V., Belousov, V. N., & Mikitaev, A. K. (1998). Description of Solid Polymers as Quasitwophase Bodies. Fizika i Tekhika Vysokikh Davlenii, *8(1)*, 101–107.

79. Aloev, V. Z., Kozlov, G. V., & Beloshenko, V. A. (2001). Description of Extruded Componors Structure and Properties within the Frameworks of Fractal Analysis. Izvestiya VUZov, Severo-Kavkazsk. Region, estesv. nauki, 1, 53–56.

80. Argon, A. S., & Bessonov, M. I. (1977). Plastic Deformation in Polyimides, with New Implications on the Theory of Plastic Deformation of Glassy Polymers. Phil. Mag. *35(4)*, 917–933.

81. Beloshenko, V. A., Kozlov, G. V., Slobodina, V. G., Prut, E. U., & Grinev, V. G. (1995). Thermal Shrinkage of Extrudates of Ultra-High-Molecular Polyethylene and Polymerization-Filled Compositions on its Basis. Vysokomolek. Soed. B, *37(6)*, 1089–1092.

82. Balankin, A. S., Izotov, A. D., & Lazarev, V. B. (1993). Synergetics and Fractal Thermomechanics of Inorganic Materials. I. Thermomechanics of Multifractals. Neorganicheskie Materialy, *29(4)*, 451–457.

83. Haward, R. N. (1993) Strain Hardening of Thermoplastics. Macromolecules, *26(22)*, 5860–5869.

84. Balankin, A. S. (1992). Elastic Properties of Fractals and Dynamics of Solids Brittle Fracture. Fizika Tverdogo Tela, *34(36)*, 1245–1258.
85. Kozlov, G. V., Afaunova, Z. I., & Zaikov, G. E. (2005). Experimental Estimation of Multifractal Characteristics of Free Volume for Poly (vinyl acetate). Oxidation Commun. *28(4)*, 856–862.
86. Hentschel, H. G. E., & Procaccia, I. (1983). The Infinite Number of Generalized Dimensions of Fractals and Strange Attractors Phys. D., *8(3)*, 435–445.
87. Williford, R. E. (1988). Multifractal Fracture. Scripta Metallurgica, *22(11)*, 1749–1754.
88. Kalinchev, E. L., & Sakovtseva, M. B. (1983). Properties and Processing of Thermoplastics. Leningrad, Khimiya, 288 p.
89. Sanditov, D. S., & Sangadiev, S. Sh. (1988). About Internal Pressure and Microhardness of Inorganic Glasses. Fizika i Khimiya Stekla, *24(6)*, 741–751.
90. Balankin, A. S. (1991). The Theoty of Elasticity and Entropic High-Elasticity of Fractals. Pis ma v ZhTF, *17(17)*, 68–72.
91. Kozlov, G. V., Gazaev, M. A., Novikov, V. U., & Mikitaev, A. K. (1996). Simulation of Amorphous Polymers Structure as Percolation cluster. Ris ma v ZhTF, *22(16)*, 31–38.
92. Serdyuk, V. D., Kosa, N., & Kozlov, G. V. (1995). Simulation of Polymers Forced Elasticity Process with the Aid of Dislocation Analoques. Fizika i Tekhnika Vysokikh Davlenii, *5(3)*, 37–42.
93. Kozlov, G. V. (2002). The Dependence of Yield Stress on Strain Rate for Semicrystalline Polymers. Manuscript disposed to V I N I I I RAN, Moscow, November 1, *(1884-B2002)*.
94. Kozlov, G. V., Shetov, R. A., & Mikitaev, A. K. (1987). Methods of Elasticity Modulus Measurement in Polymers Impact Tests. Vysokomolek. Soed. A, *29(5)*, 1109–1110.
95. Sanditov, D. S., & Kozlov, G. V. (1995). Anharmonicity of Interatomic and Intermolecular Bonds and Physical-Mechanical Properties of Polymers. Fizika i Khimiya Stekla, *21(6)*, 547–576.
96. Belousov, V. N., Kozlov, G. V., Mashukov, N. I., & Lipatov, Yu. S. (1993). The Application of Dislocation Analogues for Yieldin Process Description in Crystallizable Polymers. Doklady AN, *328(6)*, 706–708.
97. Aharoni, S. M. (1985). Correlations between Chain Parameters and Failure Characteristics of Polymers below their Glass Transition Temperature. Macromolecules, *18(12)*, 2624–2630.
98. Crist, B. (1989). Yielding of Semicrystalline Polyethylene: a Quantitive Dislocation Model. Polymer Commun. *30(3)*, 69–71.
99. Wu, S. (1989). Chain Structure and Entanglement. J. Polymer Sci.: Part B: Polymer Phys., *27(4)*, 723–741.
100. Kozlov, G. V., Beloshenko, V. A., & Varyukhin, V. N. (1988). Simulation of Gross-Linked Polymers Structure as Diffusion-Limited Aggregate. Ukrainskii Fizicheskii Zhurnal, *43(3)*, 322–323.
101. Brown, N. (1971). The relationship between Yield Point and Modulus for Glassy Polymers. Mater. Sci. Engng. *8(1)*, 69–73.
102. Kozlov, G. V., Shustov, G. B., Aloev, V. Z., & Ovcharenko, E. N. (2002). Yielding Process of Branched Polyethylenes. Proceedings of II All-Russian Sci-Pract. Conf. "Innovations in Mechanical Engineering". Penza, PSU, 21–23.

103. Mandelkern, L. (1985). The Relation between Structure and Properties of Crystalline Polymers. Polymer J. *17(1)*, 337–350.

104. Peacock, A. J., & Mandelkern, L. (1990). The Mechanical Properties of Random Copolymers of Ethylene: Force-Elongation Relations. J. Polymer Sci.: Part B: Polymer Phys. *28(11)*, 1917–1941.

105. Kennedy, M. A., Reacock, A. J., & Mandelkern, L. (1994). Tensile Properties of Crystalline Polymers: Linear Polyethylene. Macromolecules, *27(19),* 5297–5310.

CHAPTER 5

LOCAL PLASTICITY

CONTENTS

As it is known [1], the main characteristics of local plasticity zone ("shear lips" or crazes) are its length η_p and critical opening δ_c. Both indicated parameters are connected with polymer structure fractal dimension d_f. The authors of Ref. [2] obtained the following relationship between δ_c and d_f:

$$\delta_c = \frac{2\pi G_{\mathrm{Ic}} d_f}{Eb\left(\rho_d\right)^{1/2}},\qquad (5.1)$$

where G_{Ic} is deformation energy release critical rate, b is Burgers vector, determined according to the Eq. (4.7), ρ_d is density of structure linear defects.

The value ρ_d is determined as linear defects length per polymer volume unit (see the Eq. (4.1)). Since in the cluster model a segment included in densely packed regions (crystallites or clusters) is assumed as linear defect (dislocation analog) then value ρ_d is determined as follows [3]:

$$\rho_d = \frac{K + \phi_{\mathrm{cl}}}{S},\qquad (5.2)$$

where K is crystallinity degree, φ_{cl} is clusters relative fraction, S is macromolecule cross-sectional area.

The authors of Ref. [1, 2] obtained a good correspondence of experimental and calculated according to the Eq. (5.1) δ_c values for HDPE at different testing temperatures.

The local plasticity zone is the most important element of the deformed polymer at quasibrittle (quasiductile) fracture. A local plasticity zone type defines fracture type: if the craze is formed at critical defect tip, then polymer is destroyed by quasibrittle mode, and if local shearing yielding zone ("shear lips"), then by quasiductile mode [4]. An inelastic deformation mechanism change is considered usually as brittle-ductile transition [5]. Length increase of both craze [6, 7] and shear yielding zone [8] results to polymer fracture toughness increase. As a matter of fact the sole theoretical method of local plasticity zone length r_p estimation is the Dugdale-Barenblatt equation [4]:

$$r_p = \frac{\pi}{8}\frac{K_{\mathrm{Ic}}^2}{\sigma_Y^2},\qquad (5.3)$$

where K_{Ic} is stress intensity critical factor, σ_Y is yield stress.

Although the Eq. (5.3) gives for polymers good correspondence to experiment in the case of both crazes [6] and local shear zones [9], but it does not connect the parameter η_p with any structural characteristics of polymers.

It is obvious, that this circumstance creates obstacles to the value η_p prediction and, respectively, to prediction of polymers quasibrittle (quasiductile) failure parameters. Therefore, the authors of Refs. [10–12] established intercommunication between η_p and polymers structural characteristics within the frameworks of percolation theory and fractal analysis on the example of two polymers – HDPE and polystyrene (PS). These polymers were chosen because they belong to different classes (semicristalline and amorphous glassy ones, respectively) and are deformed by different mechanisms (shear yielding and crazing, respectively). Therefore, receiving of identical results for them allows to assume high enough degree of the proposed below treatment community.

The percolation theory [13, 14] supposes formation at percolation threshold infinite cluster, joining the system. Polymers structure itself can be considered as percolation system at high enough molecular weights, since exactly entanglements network formation on the entire length of sample gives to it strength and deformability [15]. Such percolation network formation at temperature reduction up to T_g was shown in Refs. [16, 17] with the cluster model using. In such treatment probability of particle (statistical segment) belonging to infinite percolation network is equal to φ_{cl} [17]. At assumption, that the value r_p is correlation length of such percolation system it can be written [14]:

$$r_p \sim \left(\phi_{cl} - \phi_0\right)^{-\nu_p},\qquad(5.4)$$

where φ_o is percolation threshold, ν_p is percolation index.

As it has been shown [16, 17], the glass transition (melting) temperature is percolation threshold for polymers structure on temperature scale. At approaching to it from below the φ_{cl} value decreases and can become any amount of small, that allows to accept in the first approximation $\varphi_o = 0$. In Fig. 5.1, the dependence $r_p(\varphi_{cl})$ in double logarithmic coordinates for HDPE and PC is shown, which corresponds to the Eq. (5.4).

As one can see, the data for both considered polymer lie on one straight line, that allows to determine the value ν_p, which is equal to 0.76. This magnitude corresponds well to classical value of correlation length percolation index ν_p, which is equal to ~0.80 [14].

Let us consider the critical index ν_p physical significance. In Ref. [18] it has been shown that percolation cluster is a fractal object with dimension d_f, for which the following relationship is valid:

$$v_p = \frac{2}{d_f} \tag{5.5}$$

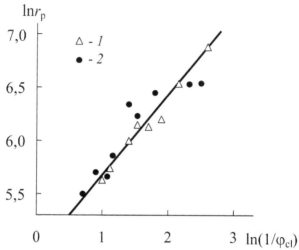

FIGURE 5.1   The relation between local plasticity zone length h_p and clusters relative fraction j_{cl} in double logarithmic coordinates, corresponding to the Eq. (5.4) for HDPE (1) and PS (2) [10].

Thus, the value v_p receives the clear physical significance. The estimation of d_f according to the Eq. (5.5) gives magnitude ~2.63, that corresponds well to magnitudes d_f, obtained by other methods [1].

Let us note, that index v_p characterizes that part of percolation system, which surrounds percolation network [18]. In respect to polymer this means, that index v_p characterizes loosely packed matrix, in which entire free volume is concentrated [19]. Hence, as it was to be expected, local plasticity zone formation is connected with polymer structure loosely packed component and its free volume. Let us note, that combination of the Eqs. (5.4) and (5.5) demonstrates the typical example of usage necessity for polymers structure and properties description, as a minimum, of two order parameters. In the considered case φ_{cl} and d_f are these order parameters.

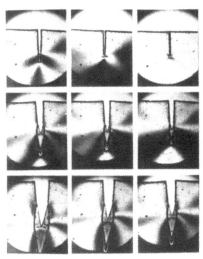

FIGURE 5.2 The optical microphotographs, showing ZD evolution in PASF film, received from solution in chloroform, at sample strain growth Enlargement 18′ [23].

Lately the large attention is given to macromolecular entanglements fluctuation network influence on crazes formation process in polymers [20]. Much less is known about entanglements network influence on shear deformation zones (ZD) formation processes, which by their essence are crazes, without microvoids. The authors of Ref. [21] fulfilled quantitative estimation of such influence on deformation processes in ZD on the example of polyarylatesulfone (PASF) samples in the form of films with a sharp notch. In this analysis two models of macromolecular entanglements network were used: of binary hooking's [20] and cluster ones [22]. The PASF films were prepared with the aid of nine various solvents that allows their structure wide enough variation [21].

In Fig. 5.2, ZD evolution at sample strain increase for PASF film, prepared from solution in chloroform [23], is shown and in Fig. 5.3 electron microphotographs of ZD surfaces and elastically deformed part of sample (Figs. 5.3a and 5.3b, respectively) are adduced. As it follows from the comparison of Figs. 5.3a and 5.3b, in ZD considerable drawing of polymer is observed – globules as if "floating" in the deformed material. The draw ratio in ZD λ_{ZD} can be determined as follows [24]:

$$\lambda_{ZD} = 1 + \ln\left(\frac{r_p + \delta}{r_p - \delta}\right), \tag{5.6}$$

where r_p and δ are length and opening of ZD respectively. Both indicated parameters can be measured easily by sample photographs of ZD (see Fig. 5.2).

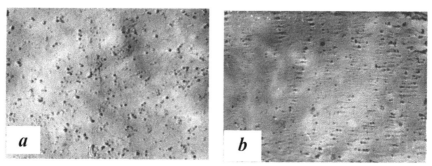

FIGURE 5.3 The electron microphotographs of ZD surfaces (a) and sample elastically deformed part (b) for PASF film, prepared from chloroform. Enlargement 5625× [21].

It is well-known [25], that limiting strain value for polymers grows at molecular weight of polymer chain part between entanglements nodes increase. In Fig. 5.4, the correlation between λ_{ZD} and M_{cl} is adduced: having the indicated character. The shown in this figure dependence of λ_{ZD} on molecular weight M_e between macromolecular binary hooking's reveals the opposite tendency. This allows to assume, that in PASF glassy state ZD parameters are controlled by macromolecular entanglements cluster network. The comparison of λ_{ZD} values and determined experimentally draw ratio at failure λ_e for these PASF samples without notch, and tensile-deformed [26] shows (Table 5.1), that both absolute values and change tendencies of these parameters are very close. It is known [20], that both crazes and ZD are formed by locally oriented polymer and differed by availability and absence of microvoids. The last availability in crazes defines entanglements "geometrical loss" necessity [20], in virtue of that draw ratio in craze fibrils is higher than λ_e. The equality of λ_e and λ_{ZD} testifies, that in ZD entanglements loss does not occur. Thus, polymer drawing in ZD is practically never not differed from the same sample without notch drawing with the exception of its local concentration at notch tip, which is stress concentrator.

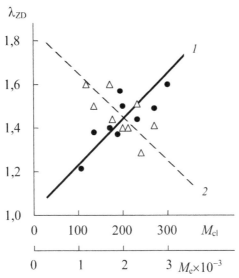

FIGURE 5.4   The dependences of limiting draw ratio in ZD λ_{ZD} on molecular weight of polymer chain par between nodes of cluster network M_{cl} (1) and macromolecular binary hooking's network M_e (2) for PASF films, prepared from various solvents [21].

There are several methods of limiting draw ratio theoretical estimation and the authors of Ref. [21] use the two of them. Within the frameworks of rubber high-elasticity concept the limiting draw ratio λ_1 is determined as follows [25]:

$$\lambda_1 = n_{st}^{1/2} , \tag{5.7}$$

where n_{st} is macromolecule independently mobile kinetic units number between entanglements nodes.

For polymers the statistical segments is accepted as such kinetic unit [35]. Then the value n_{st} can be calculated according to the Eq. (2.17).

Donald and Kramer [20] assumed, that the greatest draw ratio λ_2 could be determined according to the relationship:

$$\lambda_2 = \frac{L_e}{d_e} , \tag{5.8}$$

where L_e is polymer chain part length between entanglements nodes, d_e is distance between entanglements nodes in indeformed polymer. These parameters for the cluster model (L_{cl} and $2R_{cl}$) are calculated according to the Eqs. (2.13) and (4.63), accordingly.

In Table 5.1, the values λ_1 and λ_2, calculated according to the Eqs. (5.7) and (5.8), are adduced. For all samples the good correspondence of experimental and theoretical data is obtained. Attention is attracted to the fact, that the values λ_2 exceed λ_{ZD} in about one and half times. This is connected with local plasticity zone type: for crazes the ratio λ_2/λ_{ZD} is equal to about one, for ZD it makes up ~1.40 [20].

TABLE 5.1   The Values of Experimental and Calculated Draw Ratio in ZD For PASF Films, Prepared From Various Solvents [21]

Solvent	λ_{ZD}	λ_1	λ_1	λ_2
Chlorobenzene	1.62	—	1.98	3.07
Tetrahydrofuran	1.58	1.55	1.60	2.07
N, N-dimethylformamide	1.58	1.62	1.36	1.75
Dichloroethane	1.55	—	1.38	1.79
1,4-dioxane	1.50	—	1.90	2.47
N, N-dimethylacetamide	1.44	1.46	1.79	2.33
Methylene chloride	1.43	1.57	1.49	1.94
Chloroform	1.29	1.49	1.20	1.56
Tetrachloroethane	1.70	1.73	—	—

It is known [5], that at elastoplastic behavior a system crack-local deformation zone deviates from thermodynamical equilibrium and for its analysis a principles, correct for close to equilibrium systems, for example, Griffith theory, are inapplicable. Besides, prefailure zone structure is differed from elastically deformed material structure (Fig. 5.3) that complicates additionally process analysis. As it was noted above, for polymers this effect is displayed as the formation of local deformation zones near crack tip, containing microvoids and oriented material (crazes) or oriented material only (ZD) [20]. Therefore, for fracture analysis in such cases fracture fractal theory is applied, using fractal analysis and general principles of synergetics [28].

The stable crack in PASF film samples with a sharp notch has triangle form (Fig. 5.2) and, as calculations have shown, the ratio of crack length l_c

to its opening δ_c remains approximately constant during its growth up to failure. This allows to consider the crack as self-similar stochastic fractal with dimension D_{cr}, determined according to the equation [29]:

$$\delta_{\tilde{n}} \sim l_c^{D_{cr}/2}.$$ (5.9)

The stress concentration coefficient K_c of sharp triangle crack is expressed as follows [30]:

$$K_c = 1 + \left(\frac{l_c}{\delta_c}\right)^{1/2}.$$ (5.10)

Comparing the Eqs. (5.9) and (5.10) one can see, that the value D_{cr} reflects stress concentration by the crack, moreover, the higher D_{cr} the weaker stress concentration (the smaller K_c).

The draw ratio λ_{ZD} in deformation zone can be determined according to the Eq. (5.6). In Fig. 5.5, the dependence of λ_{ZD} on fractal dimension of chain part between clusters D_{ch}, which characterizes molecular mobility level, is adduced. As one can see, deformability of chain increase or D_{ch} enhancement results to λ_{ZD} growth and corresponds to mechanical energy dissipation in ZD intensification. A self-similar fractal is not only crack, but also ZD itself. If to estimate its fractal dimension according to the Eq. (5.9), then it is obvious, that λ_{ZD} characterization, that is, polymer deformability degree in ZD, will be its physical significance [28].

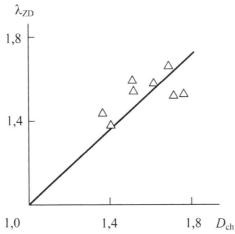

FIGURE 5.5   The dependence of draw ratio in deformation zone λ_{ZD} on fractal dimension of chain part between clusters D_{ch} for PASF [28].

The greatest draw ratio λ_{max} of polymer can be estimated according to the Eq. (5.8). In Fig. 5.6, the comparison of λ_{max} and λ_{ZD} is adduced, from which it follows, that λ_{max} exceeds systematically λ_{ZD} (see also Table 5.1). The reasons of this discrepancy are obvious – strictly speaking, the Eq. (5.8) is valid for rubbers only ($D_{ch} = 2.0$) and does not take into consideration steric hindrances, induced D_{ch} reduction in glassy state, that is, availability of "frozen" local order (clusters). These hindrances can be taken into account by a simple semiempirical mode: in glassy state corrected value λ_{max} (λ_{fr}) is equal to [28]:

$$\lambda_{fr} = \lambda_{max}^{D_{ch}/2} \tag{5.11}$$

Since for rubbers $D_{ch} = 2.0$, then for them the Eq. (5.11) gives trivial identity. The comparison of calculated by the indicated mode values λ_{fr} and λ_{ZD}, adduced in Fig. 5.6, shows their good correspondence.

Thus, the stated above results demonstrated the fractal analysis possibilities at polymers local deformation description. In each from the described cases fractal dimension of either element has simple and clear physical significance that allows to obtain both empirical and analytic correlations between different structural levels in polymers and also describe their evolution in polymers deformation and failure processes.

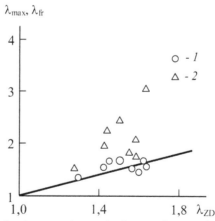

FIGURE 5.6 The relation between draw ratios λ_{ZD} and λ_{max} (1), λ_{ZD} and λ_{fr} (2) for PASF [28].

At local plasticity zones formation near defect, in particular, a notch, polymer volume, which is in yielding state, is surrounded by elastically deformed material. This results to local yield stress-enhancement and the given effects called plastic constraint [31]. The authors of Ref. [32] have under-taken an attempt of quantitative estimation of plastic constraint influence on polymers mechanical properties at impact loading on the example of HDPE.

The plastic constraint factor k_{cons} is defined as the ratio of local σ_Y^{loc} and macroscopic σ_Y yield stresses [32]:

$$k_{cons} = \frac{\sigma_Y^{loc}}{\sigma_Y},$$ (5.12)

where the value σ_Y^{loc} is determined according to technique [33].

The HDPE plasticity can be estimated with the aid of deformation energy release critical rate G_{Ic}, which is determined as follows [4]:

$$G_{Ic} = \frac{\pi a \sigma_f^2}{E},$$ (5.13)

where a is sharp notch length, σ_f is fracture stress.

The dependence k_{cons} on G_{Ic} is presented in Fig. 5.7 and shows that for polymer, as and for metals [31], G_{Ic} enhancement increases plastic con-straint. At $G_{Ic} = 0$ plastic constraint is absent, that is quite obvious, and at the greatest value of sharp notch the value k_{cons} reaches 2.74, that is close enough to Hill's theoretical prediction – 2.571 [31]. The discrepancy of the adduced values k_{cons} does not exceed 6%, in addition polymer is impossible to assume absolutely strictly as elastoplastic body [34]. Hence, the adduced in Fig. 5.7 correlation $k_{cons}^{1/2}(G_{Ic})$ has an expected character.

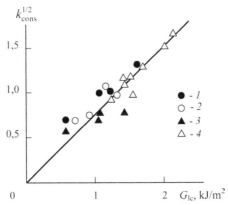

FIGURE 5.7   The dependence of plastic constraint factor k_{cons} on deformation energy release critical rate G_{Ic} at testing temperature: 293 (1), 313 (2), 333 (3) and 353 K (4) for HDPE [32].

Within the frameworks of plasticity fractal concept [35] it has been shown that yielding process is realized not in the entire polymer sample volume, but only in its part, the fraction of which makes up $(1 - \chi)$. The Poisson's ratio value for the deformed up to yielding state polymer gives the Eq. (4.9). The value χ characterizes polymer fraction, which does not participate in yielding process, but subjects to elastic deformation. For semi-crystalline polymers in the used testing temperatures range $T - 293 \div 353$ K this fraction includes devitrificated amorphous phase and crystalline phase part, subjected to partial mechanical melting (disordering). In other words, the parameter χ characterizes structural state of the deformed polymer. It is clear, that the indicated above structure components, possessing very small stiffness (their elasticity modulus makes up several megapackals), cannot influence on plastic constraint and consequently it can be supposed, that the value k_{cons} will be the smaller the higher fraction of elastically deforming component χ for HDPE. The relation of the values k_{cons} and χ, adduced in Fig. 5.8, confirms completely the made above suggestion.

The size of shear deformation local zone r_p to a great extent defines impact toughness A_p of HDPE samples with a sharp notch [36]. The values k_{cons} enhancement at notch length a increase, as a matter of fact meaning local yield stress σ_Y^{loc} growth, should result to value r_p reduction, if to issue from general postulates of Dugdale model (see the Eq. (5.3)) [4]. The dependence $r_p (k_{cons})$, adduced in Fig. 5.9, shows the expected r_p reduction at k_{cons} growth and demonstrates correctness (at any rate, qualitatively) of Dugdale model application for the description of local plasticity in HDPE.

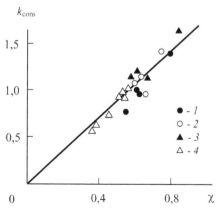

FIGURE 5.8 The relation of plastic constraint factor k_{cons} and elastically deforming regions fraction χ at testing temperature 293 (1), 313 (2), 333 (30 and 353 K (4) for HDPE [32].

It is well-known [31], that stress concentration at notch tip intensification results to HDPE samples strength decrease in impact tests. The stress concentration coefficient K_s for the notch of length a can be estimated according to the Noiber formula [4], which is similar to the Eq. (5.10):

$$K_s = 1 + 2\left(\frac{a}{R}\right)^{1/2},$$ (5.14)

where R is notch tip radius.

Despite the fact, that HDPE testing samples have nominally a sharp notch, it is well-known [37], that local plasticity zone availability results to a sharp notch blunting and as the first approximation the value R can be accepted equal to notch critical opening δ_c [37], determined according to Dugdale model [4]:

$$\delta_c = \frac{\pi a \sigma_f^2}{\sigma_Y^{loc} E}$$ (5.15)

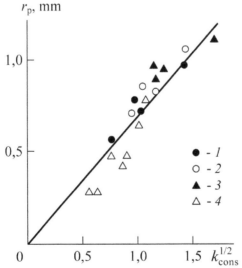

FIGURE 5.9 The relation of local shear deformation zone length r_p and plastic constraint factor k_{cons} at testing temperature: 293 (1), 313 (2), 333 (3) and 353 K (4) for HDPE [32].

In Fig. 5.10, the dependence of stress concentration coefficient K_s on value $k_{cons}^{1/2}$ is shown, which is linear and passes through coordinates origin. Thus, plastic constraint intensification results to values K_s growth and, hence, to HDPE samples strength reduction.

Hence, the stated above results have shown that plastic constraint factor k_{cons} is dependent on structural characteristics of polymer and their change in deformation process and influences essentially on its mechanical properties in impact tests. Growth k_{cons} results to reduction of both plasticity and strength of polymer samples [32].

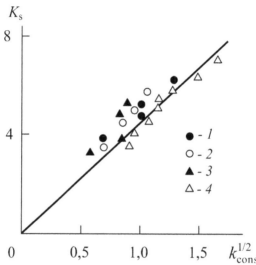

FIGURE 5.10   The relation between stress concentration coefficient K_s and plastic constraint factor k_{cons} at testing temperature 293 (1), 313 (2), 333 (3) and 353 K (4) for HDPE [32].

Lately a works number has appeared, unequivocally connecting surfaces of fracture fractal dimension with materials limiting characteristics, mainly with stress intensity critical factor K_{Ic} [29, 38–40]. These works have both experimental [38, 39] and theoretical character [29, 40]. However, there is another point of view, obtaining more and more confirmations [41–44], which supposes, as a minimum, similar correlations community absence. Ivanova and Vstovskii [45] were pointed, that with the aid of only one fractal dimension of fracture surface it was impossible to describe it by an exhaustive mode. In this connection they offered to apply multifractal formalism for fracture processes simulation. The multifractality notion implies that their geometry can be described in general case by infinite number of fractal dimensions, parameterized properly. This suggestion was realized in Ref.

[46]. The authors of Ref. [47] considered this problem on concrete example of HDPE samples with sharp notch fracture process.

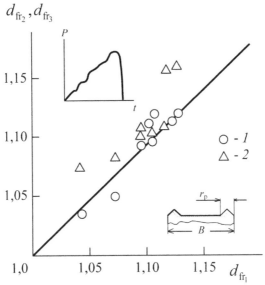

FIGURE 5.11 The relation between fractal dimensions of fracture surface d_{fr_1}, d_{fr_2}(1) and d_{fr_3} (2), calculated according to the Eqs. (5.16), (4.50) and (5.17), respectively, for HDPE. In insertion overhead-diagram $P - t$ for HDPE sample with sharp notch. In insertion below – schematic cross-section of sample with "shear lips" [47].

In the insertion of Fig. 5.11, the typical diagram load – time (P – t) for HDPE with a sharp notch is adduced, illustrating quasibrittle fracture process. Such diagram $P - t$ allows to give the following treatment of the indicated samples deformation and fracture processes [4]. At sample bending elastic energy accumulation occurs, local deformation zone is formed at notch tip (stationary crack), which retards main crack propagation. At the greatest load P_{max} achievement accumulated elastic energy becomes large enough for sample fracture and this process is realized by instable crack propagation, occurring almost momentarily. This postulate is confirmed by vertical line of load decay immediately beyond P_{max}. Thus, all mechanical properties of sample are defined on its deformation stage, but not fracture. So, impact toughness A_p is a linear function of shear local plastic deformation zones ("shear lips") size r_p, visually observed on samples surface [33]. The same can be said about other parameters of fracture process [47]. The fracture energy is defined by the area under diagram $P - t$ and load vertical decay beyond P_{max} means practical absence of energetic consumptions on actually fracture process, if such one is considered as new surfaces formation. Therefore, strictly speaking, any direct connection between fracture surface characteristics (including fractal ones) and fracture process parameters cannot exists. Nevertheless, indirect connection can be observed at fracture of samples, similar to polyethylene, if to suppose, that the main contribution in fractality (roughly speaking, roughness) of fracture surface gives exactly those "shear lips," which, as it has been noted above, form macroscopic limiting properties of samples. To such treatment of fracture surface fractality of HDPE samples quite definite physical significance can be added. If schematically "shear lips" can be presented like this, as it is made in lower insertion of Fig. 5.11, then they can be simulated by Koch figure elements [48], presented in top insertion of Fig. 5.12. In this case the fractal dimension of Koch figures is determined as $\ln 4/\ln 3 = 1.263$, that is, as the ratio of natural logarithms after conversion and before it. From schematic cross-section of fracture surface it is obvious, that since "shear lips" size r_p can be arbitrary, then elements number owing to this can be fractional one. Then fractal dimension of fracture surface d_{fr_1} by analogy with Koch figures is written as follows [47]:

$$d_{fr_1} = \frac{\ln\left(B + 2r_{p_1}\right)}{\ln B},$$

(5.16)

where B is sample width and coefficient 2 at value r_p reflects two "shear lips" availability on fracture surface.

The value $d_{fr.1}$ estimation correctness can be verified by two methods, which allow calculation of fracture surface fractality degree by the known values of Poisson's ratio ν. The first from the indicated methods assumes the Eq. (4.50) usage and the second one – the formula [49]:

$$d_{fr_3} = 1 + \nu^m,$$

(5.17)

where m is the exponent, characterizing solid body fracture mechanism.

In subsequent calculations with appreciation of diagram $P - t$ form the value $m = 2$ is used (Griffith fractal crack, quasibrittle fracture) [49]. Since the Eq. (4.50) gives the value for three-dimensional space, but in Ref. [47] fractal dimension for two-dimensional spaces was calculated, then for the comparison the value $d_{fr_2}/2$ was used [49]. The value v was determined with the aid of the Eq. (2.20).

In Fig. 5.11, the comparison of calculated by the indicated methods dimension d_{fr_1}, d_{fr_2} and d_{fr_3} is adduced. As it follows from the data of this figure, between them the good correspondence is obtained, that confirms the offered treatment correctness.

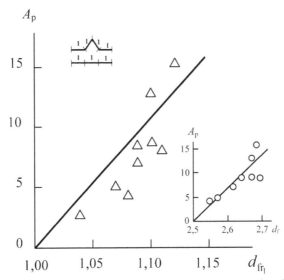

FIGURE 5.12   The dependence of impact toughness A_p on fracture surface fractal dimension d_{fr_1} for HDPE samples with sharp notch. In insertion overhead – Koch figures elements, illustrating fractal dimension determination for "shear lips" (Fig. 5.11, lower insertion). In insertion below – the dependence of impact toughness A_p on structure fractal dimension d_f for HDPE samples with sharp notch [47].

It is obvious, that putting in fracture surface fractality, namely, in dimension d_{fr_1} physical significance assumes this parameter and A_p correlation, moreover, d_{fr_1} increase should be resulted to A_p growth. Actually, the dependence $A_p(d_{fr_1})$ shown in Fig. 5.12, confirms this conclusion. Nevertheless, one should notice, that empirical correlation, similar to adduced one in Fig. 5.12, not are common, since density of the dissipated in "shear lips" energy varies from polymer to polymer.

If to take into consideration, that polymer structure fractal dimension d_f is defined also by Poisson's ratio value according to the Eq. (1.9), then the correlation $A_p(d_f)$ was expected from the adduced above results, which is shown in Fig. 5.12 lower insertion. Although the correlations A_p and $A_p(d_f)$ have similar character, but the latter bears in itself considerably greater physical significance, since it is clear, that polymer local plasticity level at loading fixed conditions is defined by its structure.

Thus, the direct usage of fractal dimension of failure surface, formed by instable crack, for polymers limiting properties characterization is deprived of sense. The multifractal treatment application [46] in the given case does not change the situation. Nevertheless, if fractal dimension reflects local plastic deformation level, then similar correlations has, at any rate, applied significance [47].

KEYWORDS

- crack
- crazing
- local placticity
- plastic constraint
- polymer
- shear

REFERENCES

1. Novikov, V. U., & Kozlov, G. V. (2001). Polymers Fracture Analysis within the Frameworks of Fractals Concept Moscow Publishers MSOU, *136* p.
2. Kozlov, G. V., Beloshenko, V. A., Serdyuk, V. D., & Sanditov, D. S. (1995). The Relation between Local Plasticity and Fractality for High Density Polyethylene. Problemy Prochnosti, *8*, 38–41.
3. Belousov, V. N., Kozlov, G. V., Mashukov, N. I., & Lipatov, Yu. S. (1993). The application of Dislocation Analigues for Yielding Process Description in Crystallizable Polymers Doklady AN, *328(6)*, 706–708.
4. Bucknall, C. B. (1977). Toughened Plastics. London, Applied Science, 318 p.
5. Narisawa, Y. (1987). The Strength of Polymeric Materials Moscow, Khimiya: 400 p.
6. Kozlov, G. V., Shetov, R. A., & Mikitaev, A. K. (1988). Mechanismen der Plastischen Deformation in Reinem und in Kautschuk Modifiziertem Polystyren by Schbagbean spruching Plaste und Kautschuk, Bd. *35(7)*, 261–263.

7. Kozlov, G. V., & Mikitzev, A. K. (1987). The Dependence of Crazing Stress on Testing Temperature at Polystyrene Impact Fracture Izvestiya VUZov Severo-Kavkazsk Region estestv nauk, *3*, 66–69.
8. Kozlov, G. V., Beloshenko, V. A., & Gazaev, M. A. (1996). The Fractal Treament of Polyethylene Fracture at Impact, Loading Problemy Prochnosti, *5*, 109–113.
9. Kozlov, G. V., Shetov, R. A., & Mikitaev, A. K. (1986). The Brittle-Ductile Transition in Linear Homopolymers Vysokomolek Soed A., *28(9)*, 1848–1852.
10. Kozlov, G. V., Afaunov, V. V., & Lipatov, Yu. S. (2000). Size of Local Plasticity Zone as Correlation Length of Polymer Structure at Inelastic Deformation Inzhenerno-Fizicheskii Zhurnal, *73(2)*, 439–442.
11. Kozlov, G. V., Afaunov, V. V., & Novikov, V. U. (2000). The Analysis of Polymers Local Plasticity within the Frameworks of Percolation and Fractal Models, Materialovedenie. *9*, 19–21.
12. Kozlov, G. V., Afaunov, V. V., & Novikov, V. U. (1999). The Analysis of Polymers Local Plasticity within the Frameworks of Percolation and Fractal Models, Proceedings of First Interdisciplinary Seminar "Fractals and Applied Synergetics, FaPS-99", Moscow, 18–21 October; 129–130.
13. Sokolov, I. M. (1986). Dimensions and Other Geometrical Critical Exponents in Percolation Theory Uspekhi Fizicheskikh Nauk, *150(2)*, 221–256.
14. Shklovskii, B. I., & Efros, A. L. (1974). The Percolation Theory and Strongly Heteroheneous Mediums Conductivity Uspekhi Fizicheskikh Nauk, *177(3)*, 401–436.
15. Kozlov, G. V., & Aloev, V. Z. (2005). The Percolation Theory in Physics-Chemictry of Polymers Nal chik Polygraphservice and T, 148 p.
16. Kozlov, G. V., Novikov, V. U., Gazaev, M. A., & Mikitaev, A. K. (1998). Cross-Linked Polymers Structure as Percolation System Inzhenerno-Fizicheskii Zhurnal, *71(2)*, 241–247.
17. Kozlov, G. V., Gazaev, M. A., Novikov, V. U., & Mikitaev, A. K. (1996) Simulaion of Amorphous Polymers Structure as Percolation Cluster Pis'ma v ZhTF, *22(16)*, 31–38.
18. Bobryshev, A. N., Kozomazov, V. N., Babin, L. O., & Solomatov, V. I. (1994). Synergetics of Composite Materials Lipetsk NPO PRIUS, 154 p.
19. Belousov, V. N., Beloshenko, V. A., Kozlov, G. V., & Lipatov, Yu. S. (1996). Fluctuation Free Volume and Polymers Structure. Ukrainskii Khimicheskii Zhurnal, *62(1)*, 62–65.
20. Donald, A. M., & Kramer, E. J. (1981). Macromechanics and Kinetics of Deformation Zones at Crack Tips in Polycarbonate. J. Mater. Sci., *16(12)*, 2977–2987.
21. Kozlov, G. V., Beloshenko, V. A., Shogenov, V. N., & Lipatov, Yu. S. (1995). Local Deformation of Polyarylatesulfone Films Doklady NAN Ukraine, *5*, 100–102.
22. Kozlov, G. V., Ovcharenko, E. N., & Mikitaev, A. K. (2009). Structure of Polymers Amorphous State, Moscow, Publishers D.I. Mendeleev RKhTU; 392 p.
23. Shogenov, V. N., Burya, A. I., Shusov, G. B., & Kozlov, G. V. (2000). Stable Cracks Fractality in Deformaion Zones of Glassy Polyerylatesulfone Fiziko-Khimicheskaya Mekhanika Materialov, *36(1)*, 51–55.
24. Baskes, M. I. (1974). The Prediction of K_{Ic} from Tensile Data Enghg Fracture Mech., *6(1)*, 11–18.
25. Haward, R. N., & Thackaray, G. (1968). The Use of a Mathematical Model to Describe Isothermal Stress-Strain Curves in Glassy Thermoplastics Proc. Roy Soc. London, A *302(1471)*, 453–472.

26. Shogenov, V. N., Belousov, V. N., Potapov, V. V., Kozlov, G. V., & Prut, E. V. (1991). The Glassy Polyarylatesulfone Corves Stress Strain Description within the Frameworks of High-Elasticity Concepts Vysokomolek Soed A, *31(1),* 155–160.

27. Bershtein, V. A., & Egorov, V. M. (1990). Differential Scanning Colorimetry in Physic Chemistry of Polymer, Leningrad Khimiya, 256 p.

28. Shogenov, V. N., Novikov, V. U., & Kozlov, G. V. (1999). Local Deformation of Polyarylatesulfone Film Samples: Fractal Analysis. Proceedings of First Interdisciplinary Seminar "Fractals and Applied Synergetics, FaPS-*99*" Moscow, 18–21 October, 144–146.

29. Mosolov, A. B. (1991). Fractal Griffith Crack Zhurnal Tekhnicheskoii Fiziki, *67(7),* 57–60.

30. Kozlov, G. V., & Novikov, V. U. (1997). The Physical Significance of Dissipation Processes in Impact Tests of Semicrystalline Polymers Prikladnaya Fizika, *1,* 77–84.

31. Weiss, V. (1971). Fracture Analysis in Conditions of Stresses Concentration. In: Fracture. An Advanced Treatise Ed. Liebowitz 3, New York, London, Academic Press, 263–302.

32. Kozlov, G. V., Serdyuk, V. D., & Beloshenko, V. A. (1994). Plastic Constraint Factor and Mechanical Properties of High Density Polyethylene at Impact Loading Mekhanika Kompozitnykh Materialov *30(5),* 691–695.

33. Kozlov, G. V., Serdyuk, V. D., & Mil'man, L. D. (1993). Determination of Polyethylene Forced Elasticity Local Stress in Impact Loading Conditions. Vysokomolek Soed. B. *35(12),* 2049–2050.

34. Narisawa, I., Ishikawa, M., & Ogawa, H. (1977). Yielding of a Notched Poly(vinyl chloride) sheet. J. Polymer Sci.: Polymer Phys. *Ed., 15(12),* 2227–2237.

35. Balankin, A. S., & Bugrimov, A. L. (1992). The Fractal Theory of Polymers Plasticity Vysokomolek Soed A, 34(10), 135–139.

36. Mashukov, N. I., Serduyk, V. D., Kozlov, G. V., Ovcharenko, E. N., Gladyshev, G. P., & Vodakhov, A. B. (1990). Stabilization and Modification of Polyethylene by Oxygen Acceptors (Preprint) Moscow IKhF AN SSSR, 64 p.

37. Kinloch, A. J., & Williams, J. G. (1980). Crack Blunting Mechanisms in Polymers J Mater Sci *15(5),* 987–996.

38. Mu, Z. Q., & Lung, C. W. (1988). Studies on the Fractal Dimension and Fractal Toughness of Steel J Ohys D: Appl. Phys., *21(5),* 848–850.

39. Mosolov, A. B., & Dinariev, O. Yu. (1988). Fracture Automodelity and Fractal Geometry Problemy Prochnosti, 1, 3–7.

40. Goldstein, R. V., & Mosolov, A. B. (1991). A Crack with Fractal Surface Doklady AN SSSR, 3*19(4),* 840–844.

41. Richards L.E., Dempsey B.D. Fractal Characterization of Fractured Surfaces. Scripta Metal., 1988, *22(5),* 687–689.

42. Zhenyi, Ma., Langford, S. C., Dickinson, J. I., Endelhard, M. N., & Baer, D. R. (1991). Fractal Character of Crack Propagation in Epoxy and Epoxy Composites as Revealed by Photon Emission during Fracture. J. Mater. Res. *6(1),* 183–195.

43. Baran, G. R., Claude, R. C., Wehbi, D., & Degrange, M. (1992). Fractal Characteristics of Fracture Surfaces J Amer. Ceram Soc., *75(10),* 2687–2691.

44. Davidson, D. L. (1989). Fracture Surface Roughness: As a Gauge of Fracture Toughness: Aluminium-Particulate SiC composites. J. Mater. Sci., *24(3),* 681–687.

45. Ivanova, V. S., & Vstovskii, G. V. (1990). Mechanical Properties of Metals and Alloys with Synergetics Positions In: The General Results of Science and Technology. Metal Science and Thermal Processing 24. Moscow, VINITI, 43–98.
46. Vstovskii, G. V., Kolmakov, L. G., & Terentev, V .F. (1993). Multifractal Analysis of Fracture Features of Molybdenum near Surface Layers Metally, *4,* 164–178.
47. Kozlov, G. V., Zaikov, G. E., Mikitaev, A. K. (1999). Physical Sence of Fractality of Polyethylene Fracture Surfaces. Intern. J. Polymeric Mater., 43(1–2), 13–18.
48. Smirnov, B. M. (1991). The Physics of Fractal Clusters Moscow Nauka, 136 p.
49. Balankin, A. S. (1992). Fractal Mechanics of Deformable Mediums and Solid Bodies Fracture Topology Doklady AN, *322(5)*, 869–874.

CHAPTER 6

COLD FLOW (FORCED HIGH-ELASTICITY)

CONTENTS

As it has been shown in Refs. [1, 2], although from the geometrical point of view the multiple sliding is the simplest way of plastic deformation realization in metals, it is not advantageous energetically and it gives the smallest efficiency of elastic energy dissipation. The deformation along sliding systems limited number, accompanied by rotational turns of different structural levels, is more effective. This is connected with the general postulate about larger efficiency of energy dissipation at motion whirl character. Therefore, in case of large enough plastic strains of solid bodies the transition to plastic flow turbulent regime is always realized [2].

At high enough stresses, exceeding the critical one, the energy, received by medium, is accumulated in localized strongly nonequilibrium regions [3], the distribution of which has fractal character, that ensures energy effective pumping. This process was considered in details for metals, but the question arises about the given rule transference over amorphous polymers plastic flow. Here principal difficulties are contained not in the indicated materials structure (polycrystalline for metals and amorphous ones for polymers) distinction, but in their more profound their distinction more precise, in their ideal structures diametrical opposition [4].

As it has been noted in chapter four, for metals the ideal structure is the defect-free monocrystal [5], possessing perfect long-range order, and for polymers – interpenetrating macromolecular coils set, personifying complete disorder (Flory "felt" model [6]). In conformity with these definitions structure defect treatment will vary – for metals the same will be long-range order violation (dislocations, vacancies and so on) and for polymers – disorder (chaos) violation, that is, local (short-range) or long-range order formation [4]. The indicated opposition results to distinctions in plastic flow mechanism definition – for metals such is dissipative structure (dislocation substructure) formation and for polymers – already existed dissipative structures (local order regions). The authors of Ref. [7] showed that in spite of the indicated fundamental structural distinctions for polymers, the described above statement about energy larger effectiveness of energy dissipation at motion whirl character was fulfilled or, in other words, in them plastic flow turbulent regime was realized.

It is known [2], that the turbulent structures fractality is one of the key properties of turbulent motions. It is accepted to assume, that energy dissipation in three-dimensional turbulent currents is concentrated on multitude with fractional (fractal) dimension. However, the experimental data about rates fluctuation moments testify, that small-scale properties of turbulent current cannot be described with the aid of self-similar fractal [2]. Therefore, a "heterogeneous fractals" were used for turbulent dissipative structures description. Such fractals formation rules on each scales hierarchy step are chosen randomly in correspondence with some probability distribution. Besides energy transfer is described with the aid of random fragmentation model in supposition, that between process different stages a correlation is absent. In this case fractal volume has no global invariance properties in respect to similarity transformation. Nevertheless, there exists the fractal relationship between active whirls number on nth fragmentation steps $<N_n>$ and turbulent structures characteristic scale L_n [2]:

$$\langle N_n \rangle \sim L_n^{d_f} ,$$

(6.1)

where d_f is fractal dimension.

If to consider the cluster network density v_{cl} as $<N_n>$ and to choose statistical segment length l_{st} as the scale, then the Eq. (6.1) is changed as follows [7]:

$$v_{cl} \sim l_{st}^{d_f} .$$

(6.2)

The value d_f of amorphous polymers cluster structure was determined according to the mechanical tests results (the Eqs. (1.9) and (2.20)).

In Fig. 6.1, the relation ship between v_{cl} and $l_{st}^{d_f}$ is adduced for PC and PAr, which turns out to be linear and, hence, corresponds to the Eqs. (6.1) and (6.2). Thus, the value v_{cl} (or φ_{cl}) can be considered as an analog of active whirls number on segmental level.

Goldstein and Mosolov [8] offered the general relationship for estimation of fractal liquid viscosity $\eta(l)$:

$$\eta(l) \sim \eta_0 l^{2-d_f} ,$$

(6.3)

where l is a flow characteristic scale, η_0 is constant.

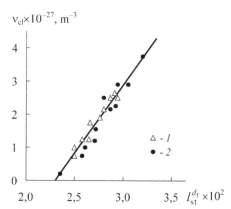

FIGURE 6.1 The relationship between macromolecular entanglements cluster network density ν_{cl} and parameter $l_{st}^{d_f}$ for PC(1) and PAr (2) [7].

Let us suppose, that the forced high-elasticity (cold flow) plateau stress σ_p of amorphous polymers will be the higher the larger polymer viscosity $\eta(l_{st})$ will be. In this case the following relationship should be fulfilled [7]:

$$\sigma_p \sim l_{st}^{2-d_f} . \tag{6.4}$$

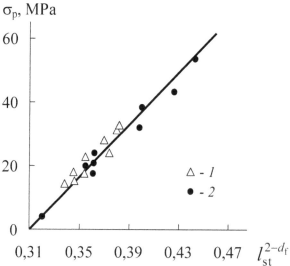

FIGURE 6.2 The dependence of forced high-elasticity plateau stress s_p on parameter $l_{st}^{2-d_f}$ value for PC (1) and PAr (2) [7].

In Fig. 6.2, the dependence $\sigma_p(l_{st}^{2-d_f})$ is shown for PC and PAr, which is linear and, hence, the Eq. (6.4) is valid at any rate for the considered polymers. Since the Eq. (6.3), which was served as the basis for the Eq. (6.4) receiving, was derived for fractal turbulent liquid, then the dependence $\sigma_p(l_{st}^{2-d_f})$ linearity is the confirmation of turbulent character of amorphous polymers cold flow.

As it is known [9], polymer viscosity at viscous flow depends on relative fluctuation free volume f_g value, which can be estimated according to the Eq. (2.19). In Fig. 6.3 the dependence of parameter $l_{st}^{2-d_f}$ characterizing polymer viscosity in cold flow region (the Eq. (6.3)) on f_g reciprocal value is adduced. This dependence is linear and has the expected character $-f_g$ increase results to $l_{st}^{2-d_f}$ decrease and, hence, to viscosity η reduction.

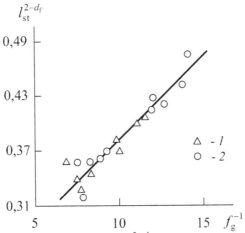

FIGURE 6.3   The dependence of parameter $l_{st}^{2-d_f}$, characterizing fractal liquid viscosity, on relative fluctuation free volume f_g for PC (1) and PAr (2) [7].

In Ref. [8] the dependence of viscous stress τ, which acts on solid body, moving in fractal liquid with rate υ, was given:

$$\tau \sim L^{1/d_f-1}\upsilon^{2-1/d_f}, \tag{6.5}$$

where L is this solid body characteristic linear scale.

It is obvious, that stress τ can be associated with σ_p and L, and as earlier, with l_{st} and υ – with strain rate $\dot{\varepsilon}$. Hence, for the considered case the Eq. (6.5) can be written as follows [7]:

$$\sigma_p \sim l_{st}^{d_f-1} \dot{\varepsilon}^{2-1/d_f} . \tag{6.6}$$

In Fig. 6.4 the dependence, corresponding to the Eq. (6.6) for PC and PAr is adduced. This dependence linearity is one more confirmation of amorphous polymers cold flow turbulence. Thus, glassy polymers high-elasticity process can be considered as motion of objects with length characteristic scale l_{st}, which occurs in turbulent regime.

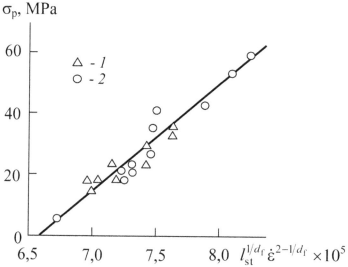

FIGURE 6.4   The dependence of forced high-elasticity plateau stress σ_p on parameter $l_{st}^{d_f-1} \dot{\varepsilon}^{2-1/d_f}$ value for PC (1) and PAr (2) [7].

The processes, occurring at glassy amorphous polymers yielding and local flow, can be identified within the frameworks of cluster model of polymers amorphous state structure [4, 9–12]. According to this model polymers amorphous state contains two types of clusters – relatively stable ones with a large number of segments per on cluster and instable, with a small number of segments [13]. The latter keep the loosely packed matrix in glassy state. At application to polymer of stress above proportionality stress on diagram stress – strain ($\sigma - \varepsilon$) instable clusters decay begins, as a result leading to loosely packed matrix mechanical devitrification. This process is completed by the forced high-elasticity (cold flow) plateau beginning and at the given point of diagram $\sigma - \varepsilon$ polymer presents itself a network of stable clusters, connected by the chains and introduced into devitrificated loosely packed matrix. This treatment explains the reason, why on the indicated part of diagram $\sigma - \varepsilon$ glassy polymer behavior is described excellently within the frameworks of rubber high-elasticity concept [14]. Thus, polymers deformation processes on high-elasticity plateau within the frameworks of cluster model are considered as the motion of connected with each other by the chains stable clusters (with characteristic scale l_{st}) in devitrificated loosely packed matrix (whose viscosity depends on relative fraction of fluctuation free volume). The similar qualitative model of amorphous glassy polymers deformation was proposed in Refs. [15, 16]. Let us also note, that according to the model [17] external load cannot induce failure of crystallites with stretched chains (CSC), having molecules axis parallel to tensile direction, but can be induced randomly oriented instable clusters decay in loosely packed matrix. Since the clusters are CSC analog [18], then this defines their motion on loosely packed matrix possibility, not subjecting to decay. It is easy to see, that the offered treatment gives physical picture, explaining all dependences, adduced in Figs. 6.1–6.4.

The considered model of amorphous glassy polymers cold flow allows to make two main conclusions [7].

Firstly, the cause of transition to turbulent regime is necessary to search in loosely packed matrix high viscosity, owing to this its fraction, rejected by cluster at its motion, gets into influence field of subsequent moving cluster is, rejected by it and so on, that results to flow turbulent regime. It is obvious, that the higher cluster fraction is (or v_{cl}, see the Eq. (1.11)) the stronger process turbulence is expressed, that the plot of Fig. 6.1 reflects.

Secondly, preservation of stable clusters network with high density v_{cl} on this part of diagram $\sigma - \varepsilon$ explains high values σ_p, comparable with yield stress value.

The authors of Ref. [19] used the stated above treatment of polymers cold flow with application of Witten-Sander model of diffusion-limited aggregation [20] on the example of PC. As it has been shown in Refs. [21, 22], PC structure can be simulated as totality of Witten-Sander clusters (WS clusters) large number. These clusters have compact central part, which in the model [18, 23] is associated with notion "cluster." Further to prevent misunderstandings the term "cluster" will be understood exactly as a compact local order region. At translational motion of such compact region in viscous medium molecular friction coefficient ξ_0 of each cluster, a particle, having radius a, is determined as follows [24]:

$$\xi_0 = 6\pi\eta_0 a, \tag{6.7}$$

where η_o is viscosity of medium, in which particle is moved.

The friction stress σ_{fric}^0 for cluster from n_{cl} particles (in the considered case-statistical segments) can be expressed as follows [24]:

$$\sigma_{fric}^0 = \xi_0 c n_{cl}^{1/d},$$ (6.8)

where d is cluster dimension, c is coefficient, determined according to the equation [24]:

$$c = \frac{1}{a\rho^{1/d}},$$ (6.9)

where ρ is the polymer density.

Substitution of the Eqs. (6.7) and (6.9) into the Eq. (6.8) gives the following relationship [19]:

$$\sigma_{fric}^0 = 6\pi\eta_0 \left(\frac{n_{cl}}{\rho}\right)^{1/d},$$ (6.10)

That is, molecular friction value is independent on the size of particles, forming a cluster (in the considered case – macromolecule cross-sectional area).

For determination of cold flow (forced high-elasticity) macroscopic stress, it is necessary to multiply σ_{fric}^0 by cluster number, performing motion in viscous medium. The clusters number N_{cl} per polymer volume unit can be determined according to the relationship [23]:

$$N_{cl} = \frac{\nu_{cl}}{n_{cl}},$$ (6.11)

where ν_{cl} is macromolecular entanglements cluster network density, as the first approximation equal to a number of segments in clusters per polymer volume unit.

As estimations according to the Eq. (6.11) with the data of Ref. [23] by the temperature dependences of ν_{cl} and n_{cl} drawing have shown, the value N_{cl} is practically independent on T. Since σ_p is calculated per cross-sectional area of sample and N_{cl} is accepted per volume unit, then it is obvious, that for σ_p calculation the value $N_{cl}^{2/3}$ should be used. Besides, since in the given model the motion of incomplete WS aggregate is considered, but only its central compact part, then as d not WS aggregate dimension, which is equal to ~2.5 [24] is accepted, but compact region dimension is considered, that is, $d \approx 3$.

Proceeding from the stated above considerations, the following equation for σ_p determination can be written [19]:

$$\sigma_p = 6\pi\eta_0 \left(\frac{n_{cl}N_{cl}^2}{\rho} \right)^{1/3}, \text{MPa}. \tag{6.12}$$

In Fig. 6.5, the dependence of experimental values σ_p on parameter $n_{cl}^{1/3}$ for PC is adduced. As one can see, this dependence is approximated well by a straight line, passing through coordinates origin and expressed by the following empirical equation [19]:

$$\sigma_p = \left(1, 4 \times 10^7 \, Pa \right) n_{cl}^{1/3}. \tag{6.13}$$

The Eqs. (6.12) and (6.13) comparison shows that the plot of Fig. 6.5 assumes the condition $\eta_0 = \text{const}$ and η_0 absolute value can be determined from the indicated equations combination. The calculation gave the following result: $\eta_0 = 0.69 \times 10^7$ Pa×s [19].

Alternatively the value η_0 can be estimated according to the relationship [25]:

$$\eta_0 = \frac{\sigma_p}{\dot{\varepsilon}}, \tag{6.14}$$

where $\dot{\varepsilon}$ is strain rate.

The estimation according to the Eq. (6.14) gave the following value: $\eta_0 = 0.82 \times 10^9$ Pa×s for $T = 293$ K. Besides, from the Eq. (6.14) it follows, that η_0 is not constant and will be reduced at T growth that contradicts to Fig. 6.5 experimental data.

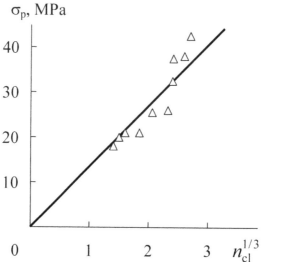

FIGURE 6.5 The dependence of forced high-elasticity stress σ_p on parameter $n_t^{1/3}$ value for PC [19].

The pointed above discrepancy of the η_o values, determined by the indicated methods, by two order of magnitude, can be explained easily within the frameworks of the model [26]. It is easy to see, that the calculation according to the Eq. (6.14) was fulfilled in the assumption that polymer is in glassy state, and the model [26] supposes, that clusters motion is realized in mechanically devitrificated loosely packed matrix. For the last case elasticity modulus E of devitrificated polymer can be estimated according to the rubber high-elasticity theory [27]:

$$E = kT\nu_e, \tag{6.15}$$

where k is Boltzmann constant, ν_e is macromolecular binary hooking's network density, since in devitrificated state cluster decay occurs [28]. The value ν_e for PC is accepted according to the data of Ref. [29] and in this case $E \approx 1.82$ GPa at $T = 293$ K [19].

The stress σ for such devitrificated loosely packed matrix can be estimated according to the following fractal relationship [30]:

$$\sigma = \frac{E}{2,5}\left(\lambda^2 - \lambda^{-2,5}\right), \tag{6.16}$$

where λ is draw ratio. The value λ is accepted equal to 1.2, that is, it corresponds to point on diagram $\sigma - \varepsilon$ immediately behind yield stress.

The calculation according to the Eq. (6.14) with $\sigma = 0.33$ MPa (the Eq. (6.16)) using gives the value $\eta_0 = 0.66 \times 10^7$ Pa×s, that corresponds excellently to the estimation according to the Eqs. (6.12) and (6.13). This correspondence confirms clearly, that polymer cold flow behind yield stress is possible only at loosely packed matrix devitrification condition. Actually calculation according to the Eq. (6.12) with the value $\eta_0 = 0.82 \times 10^9$ Pa×S for glassy loosely packed matrix using gives the value $\sigma_p \approx 1.75$ GPa, that is, without having physical significance. This σ_p value exceeds PC theoretical strength [32] and that is why polymer should fail up to cold flow realization that is observed in the brittle fracture case.

As it was noted above, the cluster model [18, 23] explains two more features of glassy polymers behavior on cold flow plateau. An experimentally observed high values σ_p are due to high values v_{cl}, which are about of order larger than v_l [23] and glassy polymer rubber-like behavior on the indicated plateau is due to loosely packed matrix rubber-like state.

The authors of Ref. [19] carried out the theoretical estimations σ_p for two polymers (HDPE and PAr) according to the Eq. (6.12) at the condition $\eta_0 = const = 0.69 \times 10^7$ Pa×s. The parameters v_{cl} and n_{cl} values for these polymers were accepted according to the data of Refs. [32] and [23], respectively. The calculation gave the following σ_p values: 18.0 MPa for HDPE and 47.8 MPa for PAr, that corresponds well (within the limits of 10%-th error) to experimental data.

Hence, the cluster model of polymers amorphous state structure and the model of WS aggregates friction at translational motion in viscous medium [24] combination allows to describe solid-phase polymers behavior on cold flow (forced high-elasticity) plateau not only qualitatively, but also quantitatively. In addition the cluster model explains these polymers behavior features on the indicated part of diagram $\sigma - \varepsilon$, which are not responded to explanation within the frameworks of other models [14].

As it is known [33], a macroscopic polymer sample is capable to bear large enough stress only at definite molecular weight MW (MW_{cr}) reaching. MW_{cr} is defined by macromolecular entanglements network formation, which is capable to spread load over entire polymer sample. Thus, entanglements network represents itself some bonds frame, by its physical significance similar to conductive bonds frame in mixture metal-insulator [34]. At present it is known, that in glassy polymers two types of macromolecular entanglements network are formed. The first from them represents itself a traditional macromolecular binary hooking's network [35] and the second – macromolecular entanglements cluster network [18, 23]. For the two indicated types of macromolecular networks the distinctions of their density temperature dependence are the most characteristic: if a macromolecular binary hooking's network density v_e is independent on temperature and is approximately the same above an below polymer glass transition temperature T_g [29], then a cluster network density v_{cl} decreases a testing temperature growth in virtue of its thermofluctuation origin [23] and at T_g $v_{cl} = 0$ [28]. As it has been shown in Refs. [36, 37], a cluster network at T_g forms percolation system, that is, it becomes capable to bear stresses. The authors of Ref. [38] were elucidated, which from the two indicated above macromolecular networks defined stress transfer in glass polymers macroscopic samples, using the mentioned above analogy with conductive bonds net in mixtures metal-insulator.

As it is known [39], the ability to conduct current with definite conductivity level g mixtures metal-insulator are acquired at percolation threshold reaching, that is, in the case, when conductive bonds form continuous percolation network. As it was noted above, macroscopic polymer samples are acquired ability to bear stress at formation in them of macromolecular entanglements continuous network. This obvious analogy allows to use modern physical models of conductivity in disordered systems for description of the dependence of cold flow plateau stress σ_p on macromolecular entanglements network density in amorphous polymers, As it is known [40], the dependent on length scale L conductivity $g(L)$ is described by the relationship:

$$g(L) \sim L^b, \tag{6.17}$$

where the exponent β is determined as follows [41]:

$$\beta = d_f - 2 - \theta. \tag{6.18}$$

In the Eq. (6.18) d_f is polymer structure fractal dimension, θ represents itself the exponent in the equation of dependent on distance r diffusivity D_r [41]:

$$D_r(r) \sim r^{-\theta}. \tag{6.19}$$

As it was noted above, an amorphous glassy polymers structure can be simulated as a WS clusters large number totality [21, 22], for which the following intercommunication of d_f and spectral (fraction) dimension d_s is valid [42]:

$$d_s = \frac{2d_f}{1+d_f}.$$ (6.20)

In its turn, in d_s terms the exponent β is determined as follows [41]:

$$\beta = d_f\left(1 - \frac{2}{d_s}\right).$$ (6.21)

The Eqs. (6.17) ÷ (6.21) combination allows to obtain the following simple relationship [38]:

$$g(L) \sim L^{-1}.$$ (6.22)

For polymers the natural choice is L definition as chain part length between its fixation points (chemical cross-linking nodes for cured polymers and physical entanglements – for linear ones) [41]. The chain part length between clusters L_{ch} can be determined according to the Eq. (2.13). Further, accounting for the indicated above analogy it can be written [38]:

$$\sigma_p \sim L_{ch}^{-1}.$$ (6.23)

In Fig. 6.6, the dependence $\sigma_p\left(L_{ch}^{-1}\right)$, corresponding to the Eq. (6.23) is adduced for PC and PAr. As one can see, the good linear correlation is received, that confirms the assumed above analogy. It is significant, that at L_{ch}^{-1} = 0 (that corresponds v_{cl} = 0 or rubber-like polymer [28]) the dependence $\sigma_p\left(L_{ch}^{-1}\right)$ is extrapolated to nonzero value σ_p = 5 MPa. As it is known [18, 23] at $T \geq T_g$ the sole entanglements network in polymer is macromolecular binary hooking's network, whose density $v_e < v_{cl}$. This explains the availability of small, but finite σ_p values at L_{ch}^{-1} = 0 (or v_{cl} = 0) [38].

Hence, the offered analogy between conductivity and stress transfer in disordered structures (amorphous polymers) allows to explain the σ_p temperature dependence. The dependence $\sigma_p(L_{ch})$ can be described analytically by the following empirical equation [38]:

$$\sigma_p = 5 + 44L_{ch}^{-1}, \text{ MPa},$$ (6.24)

where L_{ch} is given in nm.

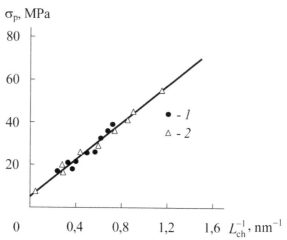

FIGURE 6.6   The dependence of cold flow plateau stress σ_p on reciprocal value of chain part between cluster length L_{ch} for PC (1) and PAr (2) [38].

Let us note that to obtain the similar dependence with macromolecular binary hooking network notions using is impossible, since its density ν_e (and, hence, chain part length between binary hooking's) is independent on testing temperature [14].

KEYWORDS

- cold flow
- forced high-elasticity plateau
- polymer
- rubber high-elasticity
- turbulency
- viscosity

REFERENCES

1. Balankin, A. S. (1991). Fractal Dynamics of Deformable Mediums. Pis ma v ZhTF, *17(6)*, 84–89.
2. Balankin, A. S. (1991). Synergetics of Deformable Body. Moscow, Publishers of Ministry Defence SSSR, 404 p.

3. Balankin, A. S. (1990). Self-Organization and Dissipative Structures in Deformable Body. Ris ma v ZhTF, *16(7)*, 14–20.
4. Belousov, V. N., Kozlov, G. V., Mashukov, N. I., & Lipatov, Yu. S. (1993). The Application of Dislocation Analogues for Yielding Process Description in Crystallizable Polymers. Doklady AN, *328(6)*, 706–708.
5. Honeycombe, R. W. K. (1968). The Plastic Deformation of Metals. London, Edward Arnold Publishers, Ltd, 398 p.
6. Flory, J. (1984). Conformations of Macromolecules in Condensed Phases. Pure Appl. Chem., *56(3)*, 305–312.
7. Gazaev, M. A., Kozlov, G. V., Mil man, L. D., & Mikitaev, A. K. (1996). The Turbulent Character of Amorphous Glassy Polymers Forced High-Elasticity. Fizika i Tekhnika Vysokikh Davlenii, *6(1)*, 76–81.
8. Goldstein, R. V., & Mosolov, A. B. (1992). Flow of Fractal-Broken Ice. Doklady AN, *324(36)* p. 576–581.
9. Sanditov, D. S., & Bartenev, G. M. (1982). Physical Properties of Disordered Structures. Novosibirsk, Nauka, 256 p.
10. Belousov, V. N., Kozlov, G. V., Mikitaev, A. K., & Lipatov, Yu. S. (1990). Entanglements if Glassy State of Linear Amorphous Polymers. Doklady AN SSSR, *313(3)*, 630–633.
11. Shogenov, V. N., Belousov, V. N., Potapov, V. V., Kozlov, G. V., & Prut, E. V. (1991). The Glassy Polyarylatesulfone Curves Stress-Strain Description within the Frameworks of High-Elastiticty Concepts. Vysokomolek. Soed. A, *33(1)*, 155–160.
12. Balankin, A. S., Bugrimov, A. L., Kozlov, G. V., Mikitaev, A. K., & Sanditov, D. S. (1992). The Fractal Structure and Physical-Mechanical Properties of Amorphous Glassy Polymers. Doklady AN, *326(3)*, 463–466.
13. Kozlov, G. V., Beloshenko, V. A., Gazaev, M. A., & Novikov, V. U. (1996). Mechanisms of Yielding and Forced High-Elasticity of Cross-Linked Polymers. Mekhanika Kompozitnykh Materialov, *32(2)*, 270–278.
14. Haward, R. N. (1993). Strain Hardening of Thermoplastics. Macromolecules, *26(22)*, 5860–5869.
15. Bekichev, V. I., & Bartenev, G. M. (1972). About the Method of Thermomechanical Curves of Solid Oriented Polymers Shrinkage. Vysokomolek. Soed., *14(3)*, 545–550.
16. Bekichev, V. I. (1974). To Question about Forced-Elastic Deformation Mechanism Vysokomolek. Soed. A, *16(8)*, 1745–1747.
17. Belyaev, O. F., & Belvatseva, E. M. (1974). Macromolecules Tension Influence on Polymers Crystallization from Solutions. Vysokomolek. Soed. A, *16(1)*, 141–145.
18. Kozlov, G. V., & Novikov, V. U. (2001). The Cluster Model of Polymers Amorphous State. Uspekhi Fizicheskikh Nauk, *171(7)*, 717–764.
19. Kozlov, G. V., Shystov, G. B., & Zaikov, G. E. (2003). Solid Polymers Deformation Mechanism on Forced High-Elasticity Plateau. Izvestiya VUZov, Severo-Kavkazsk. region, estestv. nauki, *2*, 58–60.
20. Witten, T. A., & Sander, L. M. (1983). Diffusion-Limited Aggregation. Phys. Rev. B, *27(9)*, 5687–5697.
21. Kozlov, G. V., Beloshenko, V. A., & Varyukhin, V. N. (1988). Simulation of Cross-Linked Polymers Structure as Diffusion-Limited Aggregate. Ukrainskii Fizicheskii Zhurnal, *43(3)*, 322–323.

22. Kozlov, G. V., Shogenov, V. N., & Mikitaev, A. K. (1998). Local Order in Polymers – the Description within the Frameworks of Irreversible Colloidal Aggregation. Inzhenerno-Fizicheskii Zhurnal, *71(6)*, 1012–1015.

23. Kozlov, G. V., Ovcharenko, E. N., & Mikiaev, A. K. (2009). Structure of the Polymers Amorphous State. Moscow, Publishers of the D.I. Mendeleev RKhTU, 392 p.

24. Chen, Z.-Y., Deutch, J. M., & Meakin, (1984). Translational Friction Coefficient of Diffusion Limited Aggregates. J. Chem. Phys., *80(6)*, 2982–2983.

25. Kalinchev, E. L., & Sakovtseva, M. B. (1983). Properties and Processing of Thermoplastics. Leningrad, Khimiya, 288 p.

26. Bekichev, V. I. (1974). About Poly(ethylene terephthalate) Crystallinity in its Cold Drawing Process. Vysokomolek. Soed. A, *16(7)*, 1479–1485.

27. Bartenev, G. M., & Frenkel, S. Ya. (1990). Physics of Polymers. Leningrad, Khimiya, 432 p.

28. Beloshenko, V. A., Kozlov, G. V., & Lipatov, Yu. S. (1994). Glass Transition Mechanism of Cross-Linked Polymers. Fizika Tverdogo Tela, *36(10)*, 2903–2906.

29. Wu S. (1989). Chain structure and entanglements. J. Polymer Sci.: Part B: Polymer Phys., *27(4)*, 723–741.

30. Balankin, A. S., Izotov, A. D., & Lazarev, V. B. (1993). Synergetics and Fractal Thermomechanics of Inorganic Materials. I. Thermomechanis of Multifractals. Neorganicheskie Materialy, *29(4)*, 451–457.

31. Shogenov, V. N., Kozlov, G. V.., & Mikitaev, A. K. (1986). The Physical Significance of the Notches Self-Blunting Model Parameters. Vysokomolek. Soed. A, *28(11)*, 2436–2440.

32. Malamatov, A. Kh., Serdyuk, V. D., & Kozlov, G. V. (1998). A Clusters Formation in Amorphous Phase of Modified High Density Polyethylene. Doklady Adygsk. (Cherkessk.) Internat. AN, *3(2)*, 74–77.

33. Kausch, H. H. (1978). Polymer Fracture. Berlin, Heidelberg, New York, Springer-Verlag, 435 p.

34. Stanley, H. E. (1986). Fractal Surfaces and "Termite" Model for Two-Component Random Materials. In: Fractals in Physics. Ed. Pietronero L., Tosatti E. Amsterdam, Oxford, New York, Tokyo, North-Holland, 463–477.

35. Kozlov, G. V., Sanditov, D. S., & Serdyuk, V. D. (1993). About Suprasegmental Formation Type in Polymers Amorphous State. Vysokomolek. Soed. B., 35*(12)*, 2067–2069.

36. Kozlov, G. V., Gazaev, M. A., Novikov, V. U., & Mikitaev, A. K. (1996). Simulation of Amorphous Polymers Structure as Percolation Cluster. Pis ma v ZhTF, *22(16)*, 31–38.

37. Kozlov, G. V., Novikov, V. U., Gazaev, M. A., & Mikitaev, A. K. (1998). Cross-Linked Polymers Structure as Percolation System. Inzhenerno-Fizicheskii Zhurnal, *71(2)*, 241–242.

38. Kozlov, G. V., Burya, A. I., Sviridenok, A. I., & Zaikov, G. E. (2003). The Analogy between the Dependences Conductivity – Deformations for Mixtures Metal-Insulator and Cold Flow Stress-Strain for Polymers. Doklady NAN Belarusi, *47(1)*, 62–64.

39. Gefen, Y., Aharony, A., & Alexander, S. (1983). Anomalous Diffusion on Percokating Clusters. Phys. Rev. Lett., *50(1)*, 77–80.

40. Rammal, R., & Toulouse, G. (1983). Random Walks of Fractal Structures and Percolation Clusters. J. Phys. Lett. (Paris), *44(1)*, L13–L22.

41. Alexander, S., Laermans, C., Orbach, R., & Rosenberg, H. H. (1983). Fracton Interpretation of Vibrational Properties of Gross-Linked Polymers, Glasses and Irradiated Quartz. Phys. Rev. B, *28(8)*, 4615–4619.
42. Meakin, P., Majid, I., Havlin, S., & Stanley, H. E. (1984). Topological Properties of Diffusion Limited Aggregation and Cluster-Cluster Aggregation. J. Phys. A., *17(18)*, L975–L981.

CHAPTER 7

FRACTURE

CONTENTS

As it is known [1, 2], the dilation concept of solid bodies fracture process assumes negative fluctuation density formation – dilaton, length of which is defined by phonons free run length Λ. In this case the overloading coefficient κ on breaking bonds can be expressed as follows [3]:

$$\kappa = \frac{\Delta}{a}, \qquad (7.1)$$

where a is the interatomic distance.

The values Λ of order of 100 Å were obtained for the oriented polymers [3]. However, in Refs. [4, 5] it was demonstrated for nonoriented polymers, subjecting to high-rate fracture, that the value $\Lambda \approx a$ (i.e., $\kappa \div 1$) and such dilaton, consisting of two neighboring atoms, was named "degenerated" one [4]. It was also found [6], that in this case experimentally determined values of athermic fracture stress turn out to be essentially ($2 \div 3$ times) smaller than theoretically calculated ones. A small values κ ($\div 0.2 \div 1.0$) is one more important feature of nonoriented polymers fracture in impact tests. This means, that the stress on breaking bonds is essentially lower than nominal fracture stress of bulk sample. And at last, it was found out [7], that the value κ reduces at testing temperature growth and the transition from brittle fracture to ductile (plastic) one. These effects explanation was proposed in Refs. [4–7], but development of fractal analysis ideas in respect to polymers lately and particularly, Alexander and Orbach work [8] appearance, which introduced the "fraction" notion, allows to offer the major treatment of polymer fracture process [9, 10], including the dilaton concept [1–3] as a constituent part.

The values κ can be calculated by two modes. In the first from them the value κ (κ_1) is calculated according to the equation [5]:

$$\gamma_s = \frac{C_a}{\alpha E}, \qquad (7.2)$$

where γ_s is structure-sensitive coefficient, C_a is atomic heat capacity, α is thermal expansion linear coefficient, E is elasticity modulus.

The second method is consists of the equation application [11]:

$$\sigma_f = \frac{\sigma_0}{e\kappa_2} \frac{T}{T_{tr}} \left[1 - \left(1 - \frac{3}{4e} \frac{T_{tr}}{T} \right) \right], \qquad (7.3)$$

where σ_f is the fracture stress, σ_0 is theoretical strength, equal to 0.1E, T is testing temperature, T_{tr} is transition temperature, which is equal to glass transition temperature T_g for amorphous glassy polymers and melting temperature T_m – for semicrystalline ones. The necessary for calculation parameters and final values κ_1 and κ_2 are adduced in Table 7.1.

TABLE 7.1 The Local Overloading Coefficients (κ_1 and κ_2) and Necessary For Their Calculation Parameters [5]

Polymer	Conventional sign	$\alpha \times 10^5$, K^{-1}	E, GPa	κ_1	σ_0, MPa	$T_g(T_m)$, K	κ_2
Polytetrafluoroethylene	PTFE	2.56	0.27	0.38	293	600	0.20
Poly(ethylene terephtalate)	PET	1.98	1.26	0.98	905	520	0.25
Polypropylene	PP	1.70	0.50	1.11	629	440	0.67
Polyethylene	PE	2.29	0.29	0.47	553	380	0.93
Polyamide-6	PA-6	2.38	1.64	0.86	651	500	0.20
Poly(vinyl chloride)	PVC	2.57	0.47	0.52	844	360	1.04
Polycarbonate	PC	2.35	1.29	0.79	833	425	0.44
Polystyrene	PS	1.96	1.44	1.08	725	373	0.44
Poly(methyl methacrylate)	PMMA	2.33	1.94	0.77	756	375	0.38
Polysulfone	PSF	1.56	1.70	1.50	909	465	0.25

Alexander and Orbach [8] showed that vibrational excitations on fractal were localized in virtue of medium fractal geometry and named such localized excitations fractons. It has been shown experimentally that for polymers the fracton regime is realized within the linear sizes scale from several Ångströms up to several tens Ångströms [12, 13] and the last value corresponds excellently with distances between clusters R_{cl} range for the indicated above polymers (Table 7.1). Let us remind, that chains parts between clusters breaking represents itself polymers fracture on molecular level. A localization of excited states (phonons) within the indicated above scales range means, that phonon cannot be moved over distances larger than interatomic ones, in other words, $\Lambda \approx a$ and $\kappa \approx 1$ according to the Eq. (7.1). This condition is fulfilled in virtue of chain part between its topological fixation points fractality [14, 15]. The transition to phonons usual behavior, that is,

$\Lambda > a$ and $\kappa > 1$, is realized in the case, when the indicated chain part loses its fractal properties and becomes Euclidean object, that is, when its fractal dimension D_{ch} becomes equal to topological dimension, in other words, at the condition $D_{ch} = 1.0$ fulfillment [9].

As it is known [14], the value D_{ch} can be determined according to the Eq. (2.12), from which it follows, that the condition $D_{ch} = 1.0$ means $R_{ch} = L_{ch}$, that is, it defines chain part, stretched completely between macromolecular entanglements. Such situation is typical for the oriented polymers [16] that predetermines dilation concept correct application for them [3]. Hence, the condition $\kappa \geq 1$ is due to chain part between macromolecular entanglements transition from fractal behavior to Euclidean one and corresponds to phonons delocalization ($\Lambda \geq a$) [10].

Let us consider the reasons of the adduced in Table 7.1 values $\kappa < 1$ for nonoriented polymers. As it has been shown in chapter two, the value D_{ch} characterizes molecular mobility (deformability) level of the indicated chain part [17]. At $D_{ch} = 1.0$ this mobility is suppressed completely and at $D_{ch} = 2.0$ it reaches the greatest possible level, typical for rubber-like state. Molecular mobility intensification results to corresponding stress relaxation intensification, applied to chain part between entanglements and, as consequence, to its reduction lower than macroscopic sample fracture stress σ_f [18]. Such treatment assumes availability of the correlation between parameters κ and D_{ch}. This assumption is confirmed by the plot of Fig. 7.1, where the dependence $\kappa_2(D_{ch})$ for 10 polymers, pointed out in table 7.1, is adduced and, besides, two points of the data for HDPE samples with a sharp notch, accepted according to [19], were shown. As it follows from the plot of Fig. 7.1, the dependence $\kappa(D_{ch})$ is stronger than the linear one, that is, the value κ_2 grows faster than D_{ch} reduces, pointing out to strong dependence of stress relaxation on molecular level on molecular mobility degree [10].

In insertion of Fig. 7.1 the dependence of overloading coefficient mean values $\kappa_m (\kappa_m = 0.5(\kappa_1 + \kappa_2))$ on value D_{ch}^{-1} is adduced. As one can see, a good linear correlation between the indicated parameters is observed, with the exception of only two polymers (PTFE and PSF) which have considerable, although symmetrical, scatter. This gives the possibility to express κ_m as D_{ch} function as follows [9]:

$$\kappa_m \approx D_{ch}^{-1}. \qquad (7.4)$$

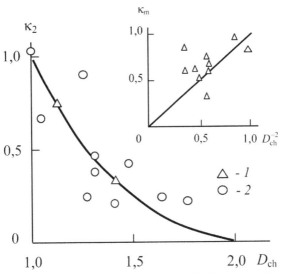

FIGURE 7.1 The dependence of local overloading coefficient κ_2 (the Eq. (7.3)) on fractal dimension of chain part between clusters D_{ch} for 10 polymers, pointed out in Table 7.1. 1 – results of impact tests of HDPE samples with sharp notch, where figures indicated notch length in mm. In insertion the dependence $\kappa_m(D_{ch})$ is shown for the same 10 polymers [9]

In Fig. 7.2, the dependence of κ_2 on molecular draw ratio λ_{mol}, determined according to the Eq. (5.8), is adduced for extrudates of ultra-high-molecular polyethylene (UHMPE) and polymerization-filled compositions on its basis (UHMPE/Al) [20]. The theoretical strength ($\sigma_0 \approx 0.1E$) for the latter was determined according to the obtained in Ref. [21] relationship $E(\lambda_{mol})$, since E reduction at large λ_{mol} for compositions UHMPE/Al is due to not molecular mechanisms, but macroscopic ones (interfacial boundaries polymer matrix-filler fracture) [16]. The linear dependence $\kappa_2(\lambda_{mol})$ is obtained, which at $\lambda_{mol} = 1$ is extrapolated to $\kappa_2 = 0$. This means, that polymer fracture is impossible without chain preliminary drawing. In other words, this assumes definite nonzero failure strain for nonoriented polymers samples. As it has been shown in Ref. [22], the greatest value λ_{mol} (λ_{mol}^{max}) can be determined according to the Eq. (5.8) and the authors of Ref. [14] demonstrated, that the ratio in right-hand side of this equation represented polymer structure automodellity coefficient, which was equal to characteristic ratio C, serving chain statistical flexibility indicator [23]. This gives the possibility to write for λ_{mol}^{max} [10]:

$$\lambda_{mol}^{max} \approx C_\infty. \tag{7.5}$$

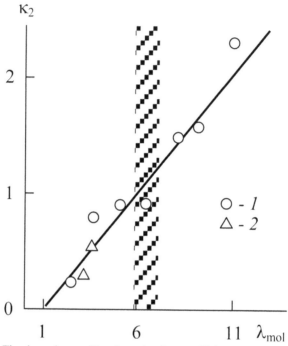

FIGURE 7.2   The dependence of local overloading coefficient κ_2, calculated according to the Eq. (7.3), on molecular draw ratio 1_{mol} for UHMPE (1) and UHMPE/Al (2). The shaded region indicates literary values C_∞ range [10].

The greatest molecular draw ratio λ_{mol}^{max} is reached at complete orientation of chain between entanglements nodes and, hence, it corresponds to the conditions $D_{ch} = 1.0$ and $\kappa = 1.0$, as it has been noted above. In Fig. 7.2, the shaded region indicates literary values C_∞ range for polyethylenes from $C_\infty = 5.4$ [24] up to $C_\infty = 7.0$ [25]. As one can see, exactly in this range of values $\lambda_{mol}^{max} \approx C_\infty$ the value κ_2 reaches the one, that means complete stretching of chain part between entanglements, this part transition from fractal behavior to Euclidean one and fracture regime change from fracton to dilaton one. Let us note, that even intuitively it is difficult to expect mechanical overloads on unstretched completely chain fragment.

Let us consider further reasons of polymer chains breaking at so small stresses, which can be on order lower than fracture macroscopic stress σ_f (i.e., at hypothetical $\kappa = 0.1$). The reasons were pointed for the first time in Refs. [1, 26]. Firstly, anharmonicity intensification in fracture center gives the effect, identical to mechanical overloading effect [26]. Quantitatively this effect is expressed by the ratio of thermal expansion coefficient in fracture center α_c and modal thermal expansion coefficient α_m [5]. The second reason is close inter communication of local yielding and fracture processes [1]. This allows to identify fracture center for nonoriented polymers as local plasticity zone [27, 28]. The ratio α_c/α_m in this case can be reached ~100 [5]. This effect compensates completely κ reduction lower than one. So, for PC $\alpha_c/\alpha_m \approx 70$, $\kappa_2 = 0.44$, $\sigma_0 = 0.1E_0 \approx 700$ MPa and then $\sigma_f = \sigma_0\alpha_m/\kappa_2\alpha_c \approx 23$ MPa, that by order of magnitude corresponds to experimental value σ_f for PC, which is equal approximately to 50 MPa at $T = 293$ K [7].

Thus, the stated above results demonstrated, that fractal analysis application for polymers fracture process description allowed to give more general fracture concept, than a dilation one. Let us note, that the dilaton model equations are still applicable in this more general case, at any rate formally. The fractal concept of polymers fracture includes dilaton theory as an individual case for nonfractal (Euclidean) parts of chains between topological fixation points, characterized by the excited states delocalization. The offered concept allows to revise the main factors role in nonoriented polymers fracture process. Local anharmonicity of intra and intermolecular bonds, local mechanical overloads on bonds and chains molecular mobility are such factors in the first place [9, 10].

The capability to bear large strains with the following complete recovery at stress removal is the property, displayed by actually all polymer substances, consisting of long chain-like macromolecules [29]. What is more, it is displayed by only such structure materials. This property is important over narrow boundaries of the term "rubber elasticity," which it is usually designated. The indicated property acts at polymer networks swelling, at deformation of substances, not included in semicrystalline polymers, at viscoelastic behavior of linear polymers in the case of flow in liquid or amorphous states. The assumption, that stress in rubbers is the consequence of covalent network chains deformation, whereas interchain intercommunication is negligible small, serves as main premise of rubber high-elasticity molecular theory. Strictly speaking, this is not quite correct even for true rubbers [29] and the more so for polymers in glassy state. The rubber high-elasticity theory gives the Eq. (5.7) for estimation of limiting draw ratio λ of elastomers [30].

An empirical assumptions number, taking into consideration considerable stronger intermolecular intercommunication in glassy polymers, usually is made for the considered concept application to such systems. Edwards and Vilgis [31] offered the slipping links concept, which assumes division of chain between macromolecular binary hooking's by into fragments, which are fixed, but have considerable internal freedom. This results to polymers limiting strain reduction in comparison with the estimated one according to the Eq. (5.7). The authors of Ref. [52] offered more general method of the considered above effects appreciation, based on fractal concepts application for analysis of polymer deformation behavior.

The correctness of fractal analysis methods application to chain part between entanglements (chemical cross-linking) is proved by the indicated parts fractality experimental confirmation [33–35]. In this case for the macromolecule, simulated by freely formed chain from statistical segments, the Eq. (2.12) was obtained, where $L_{ch} = L_e$ and $R_{ch} = R_e$ (L_e and R_e are chain part length between macromolecular binary hooking's, respectively) and dimension D_{ch} characterizes the mobility (deformability) of this chain part. The known scaling relationship [36]:

$$R_e^2 = C_\infty \frac{M_e l_0}{m_0},$$

(7.6)

where M_e is molecular weight of chain fragment between binary hooking's, l_0 and m_0 are length and molecular weight of real skeletal bond, respectively, can be transformed with the Eq. (2.15) appreciation and by division of its both parts by l_{st}^2 to the form [32]:

$$\frac{L_e}{l_{st}} = \left(\frac{R_e}{l_{st}}\right)^2,$$

(7.7)

which is the partial case of the fractal Eq. (2.12) at $D_{ch} = 2.0$, that presents itself, the limiting case of rubbers, for which, strictly speaking, the Eqs. (5.7) and (7.6) were derived. Dividing both parts of the Eq. (2.12) by the product $R_e l_{st}$ and taking into account, that $L_e/R_e = \lambda$ [37], let us obtain the equation fractal variant for λ determination [32]:

$$\lambda = \left(\frac{R_e}{l_{st}}\right)^{D_{ch}-1}$$

(7.8)

In the Eq. (7.8), polymer chain mobility (deformability) change for any reasons is taken into consideration by quite natural mode, namely, the fractal dimension D_{ch} variation. In Fig. 7.3 the relation of experimental λ and calculated according to the Eqs. (5.7) and (7.8) λ^T values of limiting draw ratio for PASF is adduced. If calculation according to the first from the indicated equations gives obviously overstated λ^T values, then the good enough theory and experiment correspondence are obtained in the case of the Eq. (7.8) using [32].

Let us consider the reasons, decreasing polymer chain mobility in glassy state and, hence, reducing D_{ch} value. In Ref. [31], it has been assumed that such effect can be consequence of chain denser environment in comparison with melt. As a matter of fact, the same treatment, but more precise and responding to quantitative estimation, gives the cluster model of polymers amorphous state structure [14]. A segment fixation in cluster excludes it from free-contacted chain that reduces the value L_e (at fixed R_e) and according to the Eq. (2.12) decreases dimension D_{ch}. Thus, it is to be expected, that cluster network density ν_{cl} enhancement (or increase of "frozen" in glassy state local order level φ_{cl}) will be resulted to chain mobility decrease, that is, to D_{ch} reduction. The correlation $D_{ch}(\nu_{cl})$, adduced in Fig. 7.4, confirms this supposition. It is significant, that the plot $D_{ch}(\nu_{cl})$ extrapolation to typical for rubbers values $D_{ch} \approx 2$ gives zero value ν_{cl}, that was to be expected within the frameworks of cluster model, which supposes $\nu_{cl} = 0$ at $T = T_g$ [38]. The Eqs. (2.5) and (2.16) are given the indicated intercommunication of structural characteristics (d_f or φ_{cl}) and D_{ch}, accordingly, in the analytical form

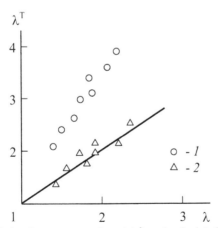

FIGURE 7.3 The relation between experimental λ and calculated according to the Eqs. (5.7) (1) and (7.8) (2) λ^T values of limiting draw ratio for PASF [32].

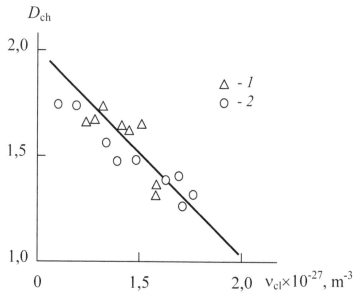

FIGURE 7.4 The dependence of fractal dimension of chain part between entanglements D_{ch} on entanglements cluster network density v_{cl} for PC(1) and PAr (2) [32].

In Fig. 7.5, the experimental and theoretical dependences of limiting draw ratio λ on testing temperature T were shown for PC and PAr. The Eq. (5.7) gives λ = const, approximately equal to 6.5, that obviously does not correspond to the experimental data. The Eq. (7.8) gives the results, corresponding to experiment by both dependence $\lambda(T)$ course and λ absolute values. Some systematic exceeding of λ calculated values in comparison with the experimental ones is quite natural and reflects all sorts of defects availability in real samples [39].

Since the ratio (R_e/l_{st}) in the Eq. (7.8) represents automodellity coefficient Λ [14], equal to C_∞, then the indicated equation can be simplified as follows [15]:

$$\lambda = C_\infty^{D_{th}-1},$$
                                                                                            (7.9)

that at D_{ch} = 2 gives the greatest draw ratio $\lambda = C_\infty$ according to the Eq. (7.5).

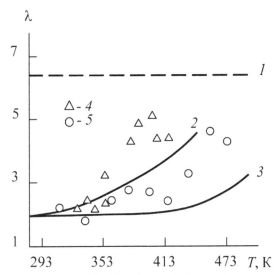

FIGURE 7.5 The dependences of limiting draw ratio λ on testing temperature T for PC (1, 2, 4) and PAr (3, 5). 1 – calculation according to the Eq. (5.7); 2, 3 – calculation according to the Eq. (7.8); 4, 5 – the experimental data [32].

Hence, the stated results demonstrated undoubted profit of fractal analysis application for polymer structure analytical description on molecular, topological and supramolecular (suprasegmental) levels. These results correspond completely to the made earlier assumptions (e.g., in Ref. [31]), but the offered treatment allows precise qualitative personification of slowing down of the chain in polymers in glassy state causes [32].

As it has been noted above, the main feature of polymers is that they consist of long chain macromolecules. Therefore, it is to be expected that polymer chains structure and their characteristics will be influenced essentially on bulk polymers properties. One of such polymer chain structural factors is availability in it of bulk side groups, which results to bulk polymers brittleness enhancement [40]. A side groups effect on plasticity level for heterochain polymers was considered in Ref. [41], where brittleness increase was explained by side groups nonparticipation in local or macroscopic plasticity processes.

In Ref. [42] for series of bromide-containing aromatic copolyethersulfones (B-PES) their high brittleness (the strain up to fracture $\varepsilon_f = 0.03 \div 0.04$) and low strength (fracture stress $\sigma_f = 29 \div 44$ MPa) were found. One of the causes of such low limiting characteristics for B-PES may be the buck side groups availability, caused by bromide-containing monomer using. Therefore, the authors of Ref. [43] attempted aromatic bromine-containing increased brittleness causes elucidation and their limiting characteristic quantitative prediction with the fractal analysis methods using.

The fractal dimension D of macromolecular coil in solution was determined with the help of the following relationship [44]:

$$D = \frac{5\left([\eta]/[\eta]_\theta\right)^{2/3} - 3}{3\left([\eta]/[\eta]_\theta\right)^{2/3} - 2}, \tag{7.10}$$

where $[\eta]$ and $[\eta]_\theta$ are intrinsic viscosities of copolymers in tetrachloroethane and 1,4-dioxane, accordingly.

The structure fractal dimension d_f for B-PES can be calculated according to the Eq. (2.3). Besides, the combination of the Eqs. (2.16) and (7.9) allows to obtain one more formula for limiting draw ratio λ_f estimation [43]:

$$\lambda_f = \frac{2}{\phi_{cl} C_\infty} \tag{7.11}$$

and the ε_f values can be estimated according to the Eq. (4.39).

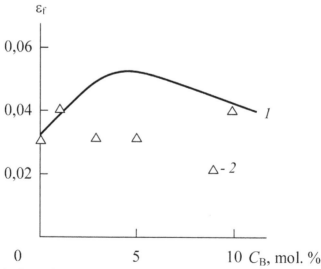

FIGURE 7.6 Dependence of fracture strain ε_f on bromide-containing monomer contents C_B for B-PES. 1 – calculation according to the Eq. (7.11), 2 – the experimental data [43].

In Fig. 7.6, the comparison of ε_f^T calculated by the indicated method and obtained experimentally ε_f values of fracture strain for B-PES is adduced. As follows from the adduced data, the good correspondence of ε_f^T and ε_f is obtained, particularly taking into account low precision of ε_f determination in this strains range. Besides, the theoretical calculation gives the greatest ε_f value for free-defects samples that in practice is achieved with difficulty, particularly for brittle polymers. Therefore, for the data of Fig. 7.6 the expected condition $\varepsilon_f^T \geq \varepsilon_f$ is carried out.

From the Eqs. (1.12), (2.3), (2.16), (4.39), (4.40), (7.9) and (7.11) comparison it is easy to see, that the value ε_f is determined in fact by two parameters, namely, D and S. The side groups availability in the given case plays a dual role. On the one hand, their availability results to D growth and, consequently, d_f according to the Eq. (2.3) that should result to ε_f increase owing to φ_{cl} reduction. On the other hand, the side groups availability results to S rising, that should result to an opposite effect. Thus, in order to obtain polymers, having side groups, with large deformability the two following conditions fulfillment is required. Side groups should increase D as much as possible and at the same time to rise S as little as possible. For copolymers B-PES S enhancement effect prevails and this defines their high brittleness. In Fig. 7.7, the dependences of ε_f on dicarbaminediphenyloxide (bromine-containing monomer [42]) contents C_B are shown for B-PES and three hypothetical polymers, having the same characteristics as B-PES, excluding S. As for these hypothetical polymers the following values were chosen: 30.7 Å^2 (polycarbonate), 22.0 Å^2 (polyformal) and 18.7 Å^2 (polyethylene) [45]. As follows from the data of Fig. 7.7, S reduction from ~41.5   60.4 Å^2 for B-PES up to 18.7 Å^2 for polyethylene results to ε_f increase in about 40 times. Therefore, in B-PES case the increase S effect prevails owing to the side groups introduction.

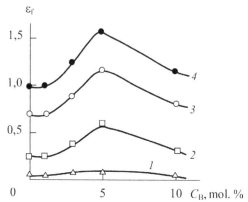

FIGURE 7.7 The dependence of fracture strain ε_f on bromine-containing monomer contents C_B for B-PES (1) and hypothetical polymers, having $S = 30.7$ (2), 22 (3) and 18.7 Å² (4) [43].

Let us consider further the strength (fracture stress) σ_f of copolymers B-PES. The value σ_f can be estimated theoretically according to the following relationship [46]:

$$\sigma_f = 1,4 \times 10^5 \left[\frac{\rho}{M_e} \left(1 - \frac{M_e}{\overline{M}_n} \right) \right]^{5/6}, \text{Pa,} \qquad (7.12)$$

where ρ is polymer density, M_e is molecular weight of polymer chain part between entanglement nodes, $\overline{M}_n$ is average number molecular weight of polymer.

The value M_e can be determined from the Eqs. (1.4), (1.5) and (1.11) combination. Since $\overline{M}_n \gg M_e$ [42], then in the first approximation the member $\left(1 - \frac{M_e}{\overline{M}_n} \right)$ in the Eq. (7.12) can be neglected and the Eqs. (1.4), (1.5), (1.11) and (7.12) gives (for high-molecular polymers):

$$\sigma_f = 1,4 \times 10^5 \left(\frac{\phi_{cl}}{2N_A SC_\infty l_0} \right)^{5/6}, \text{Pa,} \qquad (7.13)$$

Let us note, that the side groups influence on σ_f, in contrast to ε_f, has unequivocal effect, namely, σ_f reduction in virtue of simultaneous D growth and, hence, d_f increase and ϕ_{cl} reduction, and S enhancement. In Fig. 7.8, the comparison of the calculated according to the Eq. (7.13) σ_f^T and obtained experimentally σ_f fracture stress values for B-PES is adduced. As one can see, in general case $\sigma_f^T > \sigma_f$ and the condition $\sigma_f^T = \sigma_f$ is realized only for one copolymer. This occurs owing to the increased copolymers brittleness, that is, owing to very low values ε_f. The copolymers B-PES fracture is realized on an initial linear part of stress − strain $(\sigma - \varepsilon)$ diagram, on which σ_f is determined according to the relationship:

$$\sigma_f = \varepsilon_f E, \qquad (7.14)$$

where E is elasticity modulus.

Hence, the copolymers B-PES brittleness (the low values ε_f) does not allow to achieve their strength theoretical values. Thus, polymers can have two resources of limiting characteristics: strength resource and deformability resource. The last one for copolymers B-PES is exhausted faster than the first one and this defines very low values σ_f for them.

Hence, the offered techniques, using the methods of fractal analysis and cluster model of polymers amorphous state structure, allow the theoretical estimation of both deformability and strength of polymers. The side groups influence on these characteristics was considered in detail. It has been shown that the polymer' strength is defined by both strength resource and deformability resource.

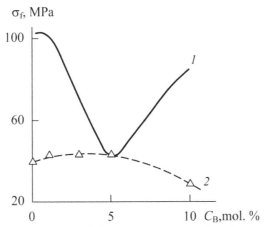

FIGURE 7.8 The dependences of fracture stress σ_f on bromine-containing monomer contents C_B for B-PES. 1 – calculation according to the Eq. (7.13); 2 – the experimental data [43].

At present in world scientific laboratories a new polymers large amount is synthesized from which small part only reaches industrial production stage [47]. It is naturally, that this work requires much time and means expenditure. These expenditures can be decreased essentially by new polymers properties prediction techniques development, proceeding from their chemical constitution.

Polymer mechanical properties are one from the most important ones, since even for polymers of different special-purpose function a definite level of these properties always requires [20]. Besides, in Ref. [48] it has been shown, that in epoxy polymers curing process formation of chemical network with its nodes different density results to final polymer molecular characteristics change, namely, characteristic ratio C, which is a polymer chain statistical flexibility indicator [23]. If such effect actually exists, then it should be reflected in the value of cross-linked epoxy polymers deformation-strength characteristics. Therefore, the authors of Ref. [49] offered limiting properties (properties at fracture) prediction techniques, based on a methods of fractal analysis and cluster model of polymers amorphous state structure in reference to series of sulfur-containing epoxy polymers [50].

In the Ref. [49] the two deformation-strength characteristics prediction was carried out: strain up to fracture ε_f and fracture stress σ_f. For the value ε_f theoretical estimation two methods can be used. The first from them does not include in the calculation of molecular characteristics and, hence, does not take into account their change, in any case, directly [51]. This method is based on the cluster model of polymers amorphous state structure notions [14], taking into account the order availability, and the limiting draw ratio λ_f value in this case is given as follows [51]:

$$\frac{1}{\lambda_f} = \frac{\phi_{cl}}{f} + \frac{\left(1 - \phi_{cl}\right)^{1/2}}{n_{st}^{1/2}}, \tag{7.15}$$

where ϕ_{cl} is clusters relative fraction, f is times number, for which macromolecule passes through cluster, n_{st} is statistical segments number between macromolecular entanglements nodes or clusters.

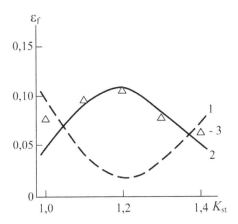

FIGURE 7.9 The dependence of strain up to fracture ε_f on the value K_{st} for sulfur – containing epoxy polymers. 1 – calculation according to the Eq. (7.15); 2 – calculation according to the Eq. (7.11); 3 – the experimental data [49].

The values φ_{cl} and n_{st} can be determined according to the Eqs. (4.66) and (2.16), respectively. The theoretical dependence of ε_f (where $\varepsilon_f = \lambda_f - 1$) on the ratio curing agent – oligomer K_{st}, obtained by the indicated mode, is adduced in Fig. 7.9 (the shaded line). Its comparison with the experimental data shows the Eq. (7.15) inadequacy for epoxy polymers considered series ε_f estimation. Since the same equation describes well the data for a linear polymers number [51], then the comparison of the data of Fig. 7.9 and the results of Ref. [51] assumes adequate usage of this method in the case of polymer molecular characteristics invariability only.

The second method is based on the fractal analysis notions, where the value λ_f is calculated according to the Eq. (7.11).

In Fig. 7.9, the comparison of values ε_f, calculated according to the Eq. (7.11) and obtained experimentally, is adduced. Now the good conformity of theory and experiment is observed, both quantitative and qualitative. This confirms postulate that molecular characteristics change in epoxy polymers curing process actually occurs and, if this factor is not taken into account, the ε_f prediction can give incorrect results.

It is curious to note, that sulfur-containing epoxy polymers the greatest deformability is reached at the greatest chemical cross-linking density realized at $K_{st} = 1.20$ [49] (see Fig. 7.9). This nontrivial observation can be explained within the frameworks of the fractal analysis. As it is known [17], the molecular mobility level can be characterized by the value of fractal dimension D_{ch} of the chain part between its fixation points (chemical cross-linkings, entanglements, clusters and so on). The value D_{ch} is determined with the help of the Eq. (2.16). The dependence of D_{ch} values, calculated by the indicated mode, on K_{st} is adduced in Fig. 7.10. As one can see, the value D_{ch} reaches maximum at $K_{st} = 1.20$, that is, at this K_{st} value epoxy polymer possesses the greatest molecular mobility. The molecular mobility intensi-fication results to the supplied mechanical energy dissipation growth and, as consequence, to polymer deformability enhancement [20]. From the Eq. (2.16) it follows, that the only cause of D_{ch} extreme growth is C decrease, that is, we return again to the necessity of cross-linking polymers molecular characteristics change in their curing process accounting for these materials properties correct description.

The Eq. (7.13) was used for the studied epoxy polymers strength (frac-ture stress) σ_f calculation. In Fig. 7.11 the comparison of strength values σ_f, obtained experimentally and calculated according to the Eq. (7.13), for sulfur-containing epoxy polymers is adduced. As the adduced comparison shows, the good conformity of theory and experiment is obtained.

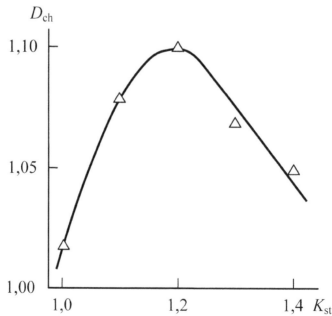

FIGURE 7.10   The dependence of fractal dimension of chain part between clusters D_{ch} on K_{st} value for sulfur-containing epoxy polymers [49].

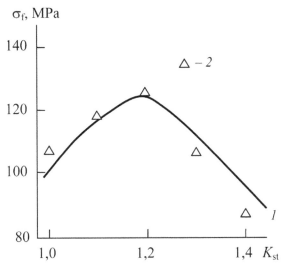

FIGURE 7.11   The dependences of fracture stress σ_f on K_{st} value for sulfur-containing epoxy polymers. 1 – calculation according to the Eq. (7.13); 2 – the experimental data [49].

Hence, the offered techniques of limiting mechanical characteristics calculation allow stress and strain of fracture (strength and deformability) estimation on the basis of two independent parameters only: T_g and S. Let us remind, that value T_g is determined by the ratio of two molecular characteristics [48]:

$$T_g \approx 147 \left(\frac{S}{C_\infty} \right)^{1/2}, \text{K.} \qquad (7.16)$$

The value T_g can be estimated theoretically as well (for example according to the group contributions method [52]). A cross-linked polymers molecular characteristics change in curing process is necessary to account for polymers with different cross-linking density (or K_{st} variation) at their deformation-strength properties prediction.

Let us consider in the present chapter conclusion the fractal analysis of amorphous glassy polyamide (phenylone) and particulate-filled nanocomposites on its basis fracture in compression tests. The experimental data show that elastic bodies can be destroyed by brittle mode at compression. In addition the fracture has often columnar character and body division on vertical columns is formed by cracks, propagating in uniaxial compression

direction [53]. This phenomenon does not correspond to theoretical postulates of classical fracture mechanics, since from the point of view of traditional theory the stresses intensity factor at crack (one-dimensional cut),
oriented along compression direction, is equal to zero. Since the indicated
factor is the main parameter, characterizing fracture, then in accordance with
classical postulates such crack cannot be propagated, that contradicts to the
experimental results [54]. Therefore, the authors of Ref. [55] used the fracture fractal model [54] for phenylone and particulate-filled nanocomposites
on its basis fracture process in compression tests analysis.

The fracture of both matrix phenylone and nanocomposites on its basis in compression tests occurs by the mode of samples division to vertical
columns (Fig. 7.12), as it was mentioned above. The indicated observation
makes necessarily the fractal model [54] application for the given process
description. For this task solution it is necessary firstly to determine the fracture surface fractal dimension d_{surf}. As it is known [56], the dimension d_{surf}
is defined by the fracture type and the Poison's ratio ν value (see the Eqs.
(4.50) and (4.51)). The value ν calculation according to the Eq. (2.20) with
diagrams stress-strain, adduced in Fig. 7.13, usage shows, that $\nu = 0.208 =$
const for all nanocomposites phenylone/β-sialone. At the same time from
the diagrams $\sigma - \varepsilon$, shown in Fig. 7.13, it follows that the fracture stress σ_f
increases with nanofiller contents φ_n growth and its increment within the
range $\varphi_n = 0 \div 10$ mas.% makes up 21%. With this appreciation it can be assumed that the fracture stress σ_f value is determined by the effective dimension d_{surf}^{ef}, which can be estimated, proceeding from the following consideration. First of all the value $\nu < 0.25$ assumes ideally brittle fracture type, for
which dimension d_{surf} is determined as follows [56]:

$$d_{surf} = 2(1 + \nu). \tag{7.17}$$

FIGURE 7.12   The general view after fracture in compression tests of samples of phenylone
(a) and nanocomposite phenylone/β-sialone with nanofiller content 1.0 mas.% (b) [55].

Let us note, that for the indicated fracture type d_{surf} coincides with fractal dimension d_f of nanocomposites structure for three-dimensional Euclidean space (see the Eq. (1.9) for $d = 3$). In this case the value $d_{surf} = d_f$ is determined according to the Eq. (1.12). For the initial phenylone the value φ_{cl} can be determined according to the Eqs. (1.9) and (1.12) and in this case φ_{cl} = 0.458. Since both polymer nanocomposites and matrix polymer fracture occurs identically (see Fig. 7.12), then it should be expected, that a crack passes through polymer matrix, without touching nanofiller particles and interfacial regions. The latter also can be considered as polymer nanocomposite densely packed regions in virtue of the definition: such regions are all structure parts, in which molecular mobility is suppressed [57]. Between relative volume fractions of nanofiller φ_n and interfacial regions φ_{if} the intercommunication exists, expressed by the equation [58].

$$\varphi_{if} = c\varphi_n, \tag{7.18}$$

where coefficient c can be changed within wide enough limits depending on nanofiller particle geometry, its surface structure and polymer matrix molecular characteristics.

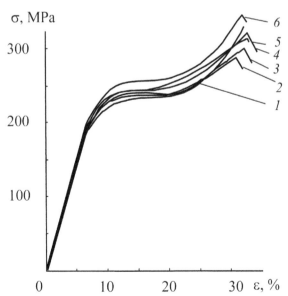

FIGURE 7.13   The diagrams stress – strain $\sigma - \varepsilon$ in compression tests for phenylone (1) and nanocomposites phenylone/β-sialone with nanofiller constants 0.2 (2), 1.0 (3), 5.0 (4), 7.5 (5) and 10 mas. % (6) [55].

Further the fracture surface effective dimension d_{surf}^{ef} can be calculated by analogy with the Eq. (1.12) as follows [55]:

$$d_{surf}^{ef} = d_f^{ef} = 3 - 6,44 \times 10^{-10} \left(\frac{\phi_{cl} + \phi_n + c\phi_n}{SC_\infty} \right)^{1/2} \tag{7.19}$$

Since the authors of Ref. [54] consider fracture process in Euclidean space with dimension $d = 2$ (flat crack), then in assumption of crack propagation direction independence it can be written [59]:

$$d_{surf}^{ef} = \frac{d_f^{ef}}{2}. \tag{7.20}$$

The authors of Ref. [55] considered two main case of crack presentation: by a stochastic self-affine fractal (coordinates quasihomogeneous tension occurs at scaling) and by usual isotropic fractal. For the indicated cases it can be written, respectively [54]:

$$K_{Ic} \sim \left(\frac{d_{surf}^{ef} - 1}{2 - d_{surf}^{ef}} \right)^{1/2} \tag{7.21}$$

and

$$K_{Ic} \sim \left(\frac{d_{surf}^{ef} - 1}{2 - d_{surf}^{ef}} \right)^{1/2}. \tag{7.22}$$

where K_{Ic} is stress intensity critical factor, determined as follows [20]:

$$K_{Ic} = \pi a \sigma_f^{1/2}, \tag{7.23}$$

where a is critical defect length, σ_f is fracture stress.

With the Eq. (7.23) appreciation the Eqs. (7.21) and (7.22) can be rewritten as follows [55]:

$$\sigma_f \sim \left(\frac{d_{surf}^{ef} - 1}{2 - d_{surf}^{ef}} \right) = B_1 \tag{7.24}$$

and

$$\sigma_f \sim d_{surf}^{ef} - 1 = B_2 \qquad (7.25)$$

for the cases of crack presentation by self-affine and isotropic fractals, respectively.

The dependences of σ_f on parameters B_1 and B_2 are presented in Fig. 7.14. As one can see, in both cases these dependences are linear, but the correlation $\sigma_f(B_1)$ does not pass through coordinates origin and, hence, at d_{surf}^{ef} =1.0 (nonfractal crack) $\sigma_f \neq 0$, that does not correspond to the known at present notions [54]. At the same time the dependence $\sigma_f(B_2)$ can be presented as follows [55]:

$$\sigma_f = 1110B_2 = 1110(d_{surf}^{ef} - 1), \text{MPa}, \qquad (7.26)$$

that corresponds completely to the main postulates, stated above. This means, that a crack in fracture process of phenylone and particulate-filled nanocomposites in its basis can be presented only by isotropic fractal (the Eqs. (7.22) and (7.25), that was to be expected in virtue of fracture statistical nature [60].

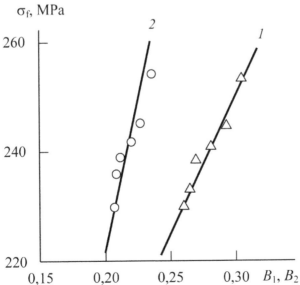

FIGURE 7.14 Dependences of fracture stress σ_f on parameters B_1 (1) and B_2 (2) for phenylone and nanocomposites/β-sialone [55].

Hence, the stated above results have shown, that the correct description of fracture process of phenylone and particulate-filled nanocomposites on its basis can be obtained within the frameworks of fractal model only [54]. In addition the fracture crack should be simulated by an isotropic fractal [55].

KEYWORDS

- characteristic ratio
- fraction
- fracture
- limiting draw ratio
- overloading coefficient
- polymer

REFERENCES

1. Zhurkov, S. N. (1983). The Dilaton Mechanism of Solid Bodies Strength. Fizika Tverdogo Tela, *25(10)*, 3119–3123.
2. Petrov, V. A. (1983). The Dilaton Model of Crack Thermofluctuation Origination. Fizika Tverdogo Tela, *25(10)*, 3124–3127.
3. Vettegren, V. I. (1984). About Physical Nature of Polymers Fracture Thermofluctuation Mechanism. Fizika Tverdogo Tela, 26, 1699–1704.
4. Kozlov, G. V., & Mikitaev, A. K. (1988). A Brittle and Ductile Fracture of Polyarylate and Polyarylatesulfone in Impact Tests Conditions. In: Polycondensation Processes and Polymers. Ed. Mikitaev, A. K. Nal'chik, KBSU, 3–8.
5. Kozlov, G. V., & Sanditov, D. S. (1994). Anharmonic Effects and Physical-Mechanical Properties of Polymers. Novosibirsk, Nauka, 261 p.
6. Kozlov, G. V., & Mikitaev, A. K. (1988). Athermic Fracture of Monoriented Polymers in Impact Tests Conditions. Vysokomolek. Soed. B, *30(7)*, 520–522.
7. Shogenov, V. N., Kozlov, G. V., Gazaev, M. A., Kardanova, M. Sh., & Mikitaev, A. K. (1986). Fracture of Rigid-Chain Polymers at Impact Loading. Plast. Massy, *8*, 32–33.
8. Alexander, S., & Orbach, R. (1982). Density of States on Fractals: "Fractons". J.Phys. Lett. (Paris), 43(17), L 625–L 631.
9. Kozlov, G. V., & Novikov, V. U. The Physical Principles of Polymer. Particulate-Filled Composites Brittlenes Enhancement. Prikladnaya Fizika, *1*, 94–100.
10. Kozlov, G. V., & Novikov, V. U. (1997). The Fracton Concept of Polymers Fracture. Materialovedenie, *8–9*, 3–6.
11. Askadskii, A. A., & Matveev, Yu. I. (1983). The Chemical Constitution and Physical Properties of Polymers. Moscow, Khimiya, 248 p.

12. Bagraynskii, V. A., Malinovskii, V. K., Novikov, V. N., Pushchaeva, L. M., & Sokolov, A. P. (1988). Inelastic Light Diffusion on Fractal Vibrational Modes in Polymers. Fizika Tverdogo Tela, *30(8)*, 2360–2366.

13. Zemlyanov, M. G., Malinovskii, V. K., Novikov, V. N., Parshin, P., & Sokolov, A. P. (1992). The Study of Fractons in Polymers. Zhurnal Eksperiment. i Teoretich Fiziki, *101(1)*, 284–293.

14. Kozlov, G. V., Ovcharenko, E. N., & Mikitaev, A. K. (2009). Structure of the Polymers Amorphous State. Moscow, Publishers of the D.I. Mendeleev RKhTU, 392 p.

15. Mikitaev, A. K., & Kozlov, G. V. (2008). The Fractal Mechanics of Polymeric Materials. Nal'chik, Publishers KBSU, 312 p.

16. Aloev, V. Z., & Kozlov, G. V. (2002). The Physics of Orientation Phenomena in Polymeric Materials. Nal'chik, Polygraphservice and T, 288 p.

17. Kozlov, G. V., Temiraev, K. B., Shetov, R. A., & Mikitaev, A. K. (1999). A Structural and Molecular Characteristics Influence on Molecular Mobility in Diblock-Copolymers Oligoformal *2,2-di-(4-oxyphenil)*-propane-oligosulfone Phenolphthaleine. Materialovedenie, *2*, 34–39.

18. Kauch, H. H. (1978). Polymers Fracture. Berlin, Heidelberg, New York, Springer-Verlag, 435 p.

19. Novikov, V. U., & Kozlov, G. V. (2001). Polymer Fracture Analysis within the Frameworks of Fractal Concept. Moscow, Publishers MSOU, 136 p.

20. Belochenko, V. A., Kozlov, G. V., Slobodina, V. G., Prut, E. V., & Grinev, V. G. (1995). Thermal Shrinkage of Ultra-High-Molecular Polyethylene and Polymerization-Filled Compositions on its Basis. Vysokomolek. Soed. B, *37(6)*, 1089–1092.

21. Kozlov, G. V., Gazaev, M. A., Bloshenko, V. A., Varyukhin, V. N., & Slobodina, V. G. (1995). Intercommunication of Molecular Characteristics and Molecular Draw Ratio for Oriented Polyethylene and Compositions on its Basis. Ukrainskii Fizicheskii Zhurnal, *40(8)*, 883–886.

22. Kramer, E. J. (1984). Craze Fibril Formation and Breakdown. Polymer Engng. Sci., *24(10)*, 761–769.

23. Budtov, V. P. (1992). Physical Chemictru of Polymer Solutions. Sankt-Peterburg, Khimiya, 384 p.

24. Wu, S. (1992). Secondary Relaxation, Brittle-Ductile Transition Temperature and Chain Structure. J. Appl. Polymer Sci., *46(4)*, 619–624.

25. Aharoni, S. M. (1983). On Entanglements of Flexible and Rod-Like Polymers. Macromolecules, *16(9)*, 1722–1728.

26. Zhurkov, S. N. (1980). To the Question about Strength Physical Basis. Fizika Tverdogo Tela, *22(11)*, 3344–3349.

27. Kozlov, G. V., & Mikitaev, A. K. (1988). The Estimation of Heat Expansion Coefficient at Nonoriented Polymers Fracture. Izvestiya VUZov, Severo-Kavkazsk. Region, Estestv. Nauki, *1*, 94–96.

28. Shogenov, V. M., Kozlov, G. V., & Mikitaev, A. K. (1989). Prediction if Rigid-Chain Polymers Fracture Process Parameters. Vysokomolek. Soed. B, *31(11)*, 809–811.

29. Flory, J. (1985). Molecular Theory of Rubber Elasticity. Polymer J., *17(1)*, 1–12.

30. Haward, R. N., & Thackray, G. (1968). The Use of a Mathematical Model to Describe Isothermal Stress-Strain Curves in Glassy Thermoplastics. Proc. Roy. Soc. London, A *302(14716)*, 453–472.

31. Edwards, S. F., & Vilgis, T. (1987). The Stress-Strain Relationship in Polymer Glasses. Polymer, *28(3)*, 375–378.

32. Kozlov, G. V., Serdyuk, V. D., & Dolbin, I. V. (2000). Chain Fractal Geometry and Amorphous Glassy Polymers Deformability. Materialovedenie, *12,* 2–5.

33. Halvin, S., & Ben-Avraham, D. (1982). New Approach to Selfavoiding Walks as Critical Phenomenon. Phys. Rev. A, *26(3)*, 1728–1734.

34. Muthukumar, M., & Winter, H. H. (1986). Fractal Dimension of a Cross-Linking Polymer at the Gel Point. Macromolecules, *19(4)*, 1284–1285.

35. Chu, B., Wu, Ch., Wu, D. Q., & Phillips, J. C. (1987). Fractal Geometry in Branched Epoxy Polymer Kinetics. Macromolecules, *20(10),* 2642–2644.

36. Lin Y. H. (1987). Number of Entanglement Strands per Gubed Tube Diameter, a Fundamental Aspect of Topological Universality in Polymer Viscoelasticity. Macromolecules, 20(12), 3080–3083.

37. Donald, A. M., & Kramer, E. J. (1982). Effect of Molecular Entanglements on Craze Microstructure in Glassy Polymers. J. Polymer Sci.: Polymer Phys. Ed., *20(4)*, 899–909.

38. Beloshenko, V. A., Kozlov, G. V., & Lipatov, Yu. S. (1994). The Vitrification Mechanism of Cross-Linked Polymers. Fizika Tverdogo Tela, *36(10)*, 2903–2906.

39. Belousov, V. N., Kozlov, G. V., & Mikitaev, A. K. (1983). The Experimental Detemination of Critical Defect in Polymer in Impact Fracture Conditions. Doklady AN SSSR, *270(5)*, 1120–1123.

40. Kozlov, G. V., Kekharsaeva, E. R., Shogenov, V. N., Beriketov, A. S., Kharaev, A. M., & Mikitaev, A. K. (1986). The Correlation between Polymer Films Strain and Polymers Molecular Parameters. Vysokomolek. Soed. B, *28(1)*, 3–4.

41. Shogenov, V. N., Kozlov, G. V., & Mikitaev, A. K. (1989). A Figid-Chain Polymers Forced Elasticity Prediction. Vysokomolek. Soed. A, *31(8)*, 1766–1770.

42. Temiraev, K. B., Shystov, G. B., & Mikitaev, A. K. (1993). Bromine-Containing Aromatic Copolyethersulfones. Vysokomolek. Soed. B, *35(12)*, 2057–2059.

43. Kozlov, G. V., Mikitaev, A. K., & Zaikov, G. E. (2008). The Side Groups Influence on Polymers Limiting Mechanical Characteristics: Fractal Analysis. Polymers Research J., *2(4)*, 381–388.

44. Kozlov, G. V., & Dolbin, I. V. (2002), Fractal Variant of the Mark-Kuhn-Houwink Equation. Vysokomolek. Soed. B, *44(1)*, 115–118.

45. Aharoni, S. M. (1985). Correlations between Chain Parameters and Failure Characteristics of Polymers below their Glass Transition Temperatire. Macromolecules, *18(12)*, 2624–2630.

46. Kozlov, G. V., Belousov, V. N., & Lipatov, Yu. S. (1990). The Molecular Entanglements Fluctuation Network and Amorphous Glassy Polymers Strength. Doklady AN USSR, *6,* 50–53.

47. Moshey, A., McGrath, J. E. (1977). Block Copolymers. New York, San Francisco, London, Academic Press, 478 p.

48. Kozlov, G. V., Beloshenko, V. A., Kuznetsov, E. N., & Lipatov, Yu. S. (1994). An Epoxy Polymer Molecular Parameters Change in their Curing Process. Doklady NAN Ukraine, *12,* 126–128.

49. Kozlov, G. V., Mil'man, L. D., & Mikitaev, A. K. (2008). A Densely Cross-Linked Epoxy Polymers Defoprmation-Strength Properties Prediction: The Fractal Analysis. Mater. IV

Internat. Sci.-Part. Conf. "New Polymer Composite Materials". Nal'chik, KBSU, 127–133.

50. Kozlov, G. V., Burmistr, M. V., Korenyako, V. A., & Zaikov, G. E. (2002). The Kinetics of Dissipative Microstructures Formation in Cross-Linked Polymers Curing Process. Voprosy Khimii I Khimicheskoi Teknologii, *6*, 77–81.

51. Kozlov, G. V., Sanditov, D. S., & Serdyuk, V. D. (1993). On Suprasegmental Formations Type in Polymers Amorphous State. Vysokomolek. Soed. B, *35(12)*, 2067–2069.

52. Askadskii, A. A. (1981). Structure and Properties of Thermostable Polymers. Moscow, Khimiya, 320 p.

53. Obert, L. (1972). The Brittle Fracture of Rocks. In: Fracture, 7, part. I. Ed. Liebowitz. New York, London, Academic Press, 59–128.

54. Mosolov, A. B., & Borodich, F. M. (1992). Fractal Fracture of Brittle Bodies at Compression. Doklady AN, *324(3)*, 546–549.

55. Mikitaev, A. K., Kozlov, G. V., & Zaikov, G. E. (2008). Polymer Nanocomposites: Variety of Structural Forms and Applications. New York, Nova Science Publishers Inc., 319 p.

56. Balankin, A. S. (1991). Synergetics of Deformable Body. Moscow, Publishers of Ministry Defence SSSR, 404 p.

57. Kozlov, G. V., & Novikov, V. U. (2001). The Cluster Model of Polymers Amorphous State. Uspekhi Fizicheskikh Nauk, *171(7)*, 717–764.

58. Kozlov, G. V., Malamatov, A. Kh., Burya, A. I., & Lipatov, Yu. S. (2006). A Polymer Nanocomposites Strengthening Mechanisms. Doklady NAN Ukraine, *7*, 148–152.

59. Long, Q. Y., Suqin, L., & Lung, C. W. (1991). Studies of the Fractal Dimension of a Fracture Surface Formed by Slow Stable Crack Propagation. J. Phys. D: Appl. Phys., *24(4)*, 602–607.

60. Bessendorf, M. H. (1987). Stochastic and Fractal Analysis of Fracture Trajectories. Int. J. Engng. Sci., *25(6)*, 667–672.

CHAPTER 8

FRACTAL CRACKS

CONTENTS

The studies carried out earlier have shown that polymer film samples strength to a considerable extent is defined by growth parameters of stable crack in local deformation zone (ZD) at a notch tip [1–3]. As it has been shown in Refs. [4, 5], the fractal concept can be used successfully for the similar processes analysis. This concept is used particularly successfully for the relationships between fracture processes on different levels and subjecting fracture material microstructure derivation [5]. This problem is of the interest in one more respect. As it has been shown earlier, both amorphous polymers structure [7] and Griffith crack [4] are fractals. Therefore, the possibility to establish these objects fractal characteristics intercommunication appears. The authors of Refs. [8, 9] consider stable cracks in polyarylatesulfone (PASF) film samples treatment as fractals and obtain intercommunication of this polymer structure characteristics with samples with sharp notch fracture parameters.

In Fig. 5.2, ZD and stable crack development succession at PASF sample with sharp notch deformation (the sample was obtained from the solution in chloroform) is shown. This figure demonstrates, that sample macroscopic strain increase results to both ZD size raising and advancement of stable crack, having the form of triangle with a sharp tip. Maslow [4] showed that the Eq. (5.9) was valid for a fractal crack. This relationship allows to verify correctness of a crack presentation as stochastic fractal with dimension D_{cr}. In Fig. 8.1, the dependences $2\ln\delta_{cr}(\ln l_{cr})$ of cracks parameters, measured by photographs, for PASF three samples (solvents − dichloroethane, chloroform and N, N-dimethylformamide) are shown in double logarithmic coordinates. The indicated dependences are approximated well by straight lines, passing through coordinates origin, and this means, that the Eq. (5.9) conditions are fulfilled and a stable crack in PASF film samples can be considered as stochastic self-similar fractal with dimension D_{cr}, determined from the plots, adduced in Fig. 8.1, slope.

The electron microscopy data confirm the made conclusion. In Fig. 8.2, the microphotograph of stable crack boundary in PASF sample (solvent − chloroform) is adduced. As one can see, the fracture surface has microroughnesses at any rate of two levels (~1 mcm and ~20 nm) that allows to apply fractal models for PASF samples fracture process [4, 5].

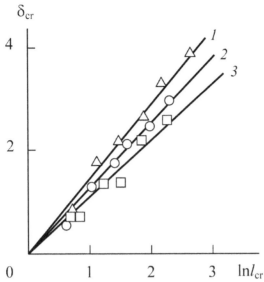

FIGURE 8.1 The dependences of stable crack opening δ_{cr} on its length l_{cr} in double logarithmic coordinates for PASF samples, obtained from solutions in dichloroethane (1) chloroform (2) and N, N-dimethylformamide (3) [8].

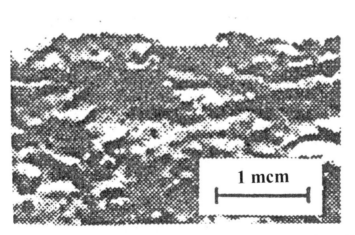

FIGURE 8.2 The electron microphotograph of stable crack boundary for PASF sample (solvent – chloroform). Enlargement 15,000 [10],

In Fig. 8.3, the relation between fractal dimensions of structure d_f and stable crack D_{cr} for PASF samples is adduced. As it was to be expected from the most general considerations, the intercommunication existed between dimensions D_{cr} and d_f, one of possible causes of which can be the influence of d_f on local plasticity level (see chapter five) and, hence, on crack boundaries (fracture surface) roughness degree. In Fig. 8.3, two straight lines are drawn, giving possible theoretical relation $D_{cr}(d_f)$. The straight line 1 is drawn in assumption, that the brittle fracture, to which the criterion $\nu = 0.25$ (or $d_f = 2.5$, see the Eq. (1.9)) corresponds [11], is realized at $D_{cr} = 1$, that is carried out, and ductile fracture – at $\nu = 0.5$, that is, for true rubbers. The straight line 2 as a matter of fact is similar to straight line 1, but it is drawn in assumption of ductile fracture achievement (sample general yielding) at $\nu = 0.475$ [11]. However as the data of Fig. 8.3 show, the value $D_{cr} = 2.0$ (limiting value D_{cr}) is reached at $d_f \approx 2.72$, that is, at the transition from quasibrittle (quasiductile) fracture to ductile one [11].

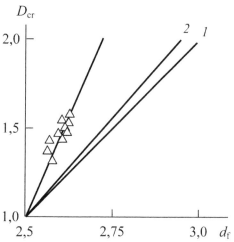

FIGURE 8.3   The relation between fractal dimensions of stable crack D_{cr} and polymer structure d_f for PASF samples. The straight lines drawing mode explanations are given in the text [8].

In Ref. [4], it has been shown that between stress intensity factors (resistance to crack propagation) for fractal crack $K_I(D_{cr})$ and smooth one-dimensional cut (i.e., idealized crack with smooth boundaries) K_{Io} the following relationship exists with precision to a multiplicative constant of order one:

$$K_I(D_{cr}) \sim K_{Io} l_{cr}^{(1-D_{cr})/2},$$

(8.1)

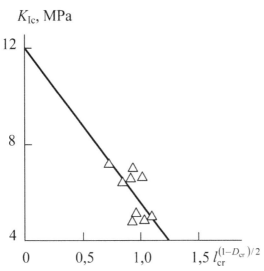

FIGURE 8.4   The dependence of critical stress intensity factor K_{Ic} on fractal parameter $l_{\mathrm{cr}}^{(1-D_{\mathrm{cr}})/2}$ in logarithmic coordinates for PASF [8].

In Fig. 8.4, the dependence of K_{Ic} on $l_{\mathrm{cr}}^{(1-D_{\mathrm{cr}})/2}$ corresponding to the Eq. (8.1), in logarithmic coordinates is shown, which is approximated satisfactorily by a straight line. As the Eq. (8.1) shows, the fractal resistance to crack propagation $K_{\mathrm{I}}(D_{\mathrm{cr}})$ is approximately equal to K_{Io} at the condition $D_{\mathrm{cr}} = 1$ or $l_{\mathrm{cr}}^{(1-D_{\mathrm{cr}})/2} = 1$. From the plot of Fig. 8.4 it follows, that absolute value $K_{\mathrm{I}}(D_{\mathrm{cr}})$ can be both greater and smaller than K_{Io}, which can be assumed as material constant. The relation between $K_{\mathrm{I}}(D_{\mathrm{cr}})$ and K_{Io} is defined mainly by stable crack advancement value through ZD (l_{cr}). The value l_{cr} itself is the function of G_{ch}, as it follows from the plot of Fig. 8.5. The stable crack is not propagated (i.e., samples fracture occurs by instable Griffith crack propagation) at $D_{\mathrm{cr}} = 1.0$ or, as follows from the plot of Fig. 8.3 at $d_{\mathrm{f}} = 2.5$ ($v = 0.25$). In other words, from the point of view of fractal analysis the condition of transition from stable to instable crack can be expressed as follows [8]:

$$\delta_{\mathrm{cr}} \sim l_{\mathrm{cr}}^{1/2} \,. \tag{8.2}$$

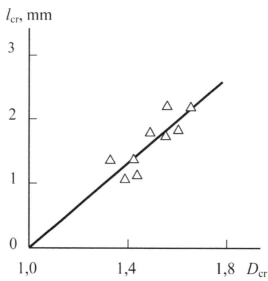

FIGURE 8.5   The dependence of stable crack length l_{cr} on its fractal dimension D_{cr} for PASF [8].

Let us consider in conclusion the physical significance of stable crack fractal dimension D_{cr} in PASF samples. As it has been shown in Ref. [12], the stress concentration coefficient K_S of triangular crack is given by the Eq. (5.10). In Fig. 8.6 the relation between parameters D_{cr} and K_S for PASF sample (solvent – chloroform) is adduced, from which K_S linear reduction follows at D_{cr} growth. Thus, the dimension D_{cr} for stable crack has the simple physical significance – it is the value, reciprocal to stress concentration level at crack tip.

Hence, the stated above results have shown, that the stable crack in PASF samples, obtained from different solvents, can be treated as stochastic fractal, whose dimension D_{cr} has clear physical significance – its value is reciprocal to stress concentration degree at a crack tip. The value D_{cr} is determined unequivocally by polymer structure fractal dimension d_f and in its turn, influences on PASF samples fracture process parameters [8].

Since by their physical significance the dimensions of fracture surface d_{surf} and stable crack D_{cr} are identical, then this circumstance allows to reveal the dependence of d_{surf} on polymer structure characteristics and its fracture mechanism. The authors of Ref. [9] accepted structure fractal dimension d_f as its characteristic and Fig. 8.7 the dependences $d_{fr}^{br}(d_f)$ and $d_{fr}^{duc}(d_f)$, are adduced where d_{fr}^{br} and d_{fr}^{duc} were calculated according to the Eqs. (4.50) and (4.51), respectively.

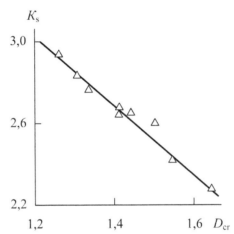

FIGURE 8.6 The dependence of stress concentration coefficient K_s of stable crack on its fractal dimension D_{cr} for PASF [8].

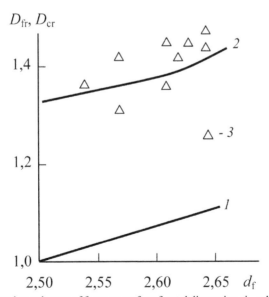

FIGURE 8.7 The dependences of fracture surface fractal dimension d_{fr}, calculated according to the Eqs. (4.50) (1) and (4.51) (2), and stable crack fractal dimension D_{cr} (3) on structure fractal dimension d_f for PASF [9].

Besides, in the same figure the points represent the values D_{cr} for PASF, calculated according to the Eq. (5.9) and the plots of Fig. 8.1 slope. As one can see, the dimensions D_{cr} and d_{fr}^{duc} good correspondence is observed, from which the conclusion can be made, that PASF film samples fracture by stable crack mechanism is related to ductile fracture mechanism. These results demonstrate clearly, that the d_{fr} value for polymer samples can be predicted, proceeding from the two indicated above factors: polymer structure and its fracture mechanism [13].

In Ref. [1], the stationary crack (notches) self-braking effect for some film samples of blends polyarylate arylene sulfone oxide/polycarbonate (PAASO/PC) was described. The indicated effect consists of monotonous increase of fracture stress σ_f at sharp notch length a growth. In Refs. [2, 3], this effect was explained within the frameworks of mechanics of continua by mechanism of notch blunting by local plastic deformation zone. Methods of fractal analysis development in respect to fracture process [4, 5] allows to apply them for the indicated effect explanation [14]. Let us note that the found dependence $\sigma_f(a)$ contradicts to the known Griffith equation. It has been assumed, that this contradiction is due to crack boundaries fractality (roughly speaking, roughness, see, Fig. 8.2) influence underestimation and the fractal Griffith crack model [4] using will allow to explain the indicated discrepancy the more so, the mobile cracks self-braking possibility was shown in Ref. [15] within the frameworks of fractal analysis.

In Fig. 8.8, the optical microphotographs block is shown, illustrating a stable crack and local plastic deformation zone development in sample of blend PAASO/PC (PC content $C_{PC} = 20$ mas. %), which is very similar to analogous process for PASF samples (Fig. 5.2). As one can see, (the stable crack is developed from a sharp notch ($a = 1$ mm), has a triangular form and a sharp tip. The most important feature of this crack is its self-similarity, which is expressed in the constant ratio of its opening δ_{cr} to length l_{cr}. As and earlier, this circumstance allows to simulate stable crack by stochastic fractal and determine its dimension D_{cr} according to the Eq. (5.9). As the estimations show, the stable crack fractal dimension D_{cr} is monotonously increasing C_{PC} function and it changes within the limits of $1.38 \div 1.81$ (Fig. 8.9).

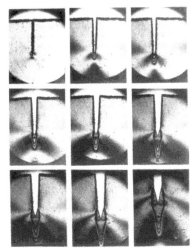

FIGURE 8.8 The optical microphotographs block, showing ZD and stable crack evolution in blend PAASO/PC with PC content 20 mas. % sample. Enlargement 22 [14].

Now the equation for the fracture stress σ_f estimation as a function of D_{cr} and sharp notch length a can be obtained. In Ref. [4], the following definition of stress intensity factor K_I as a function of D_{cr} was given:

$$K_I\left(D_{cr}\right)=C\frac{a^\alpha}{\sqrt{D_{cr}}}, \quad \alpha=\frac{2-D_{cr}}{2}, \tag{8.3}$$

where C is coefficient, dependent on crack form details and external loads concrete distribution.

One more estimation $K_I(D_{cr})$ method the Eq. (8.1) gives. Assuming $K_I(D_{cr})$ as polymer constant [16] and equating the Eqs. (8.1) and (8.3), let us obtain [14]:

$$\sigma_f \sim a^{D_{cr}-3/2}\left(D_{cr}\right)^{1/2}. \tag{8.4}$$

The Eq. (8.4) has three characteristic features. Firstly, it shows, that at $D_{cr} < 1.5$ σ_f will be decreased at a enhancement and at $D_{cr} > 1.5$ will raise. If to compare these predictions with the data of Fig. 8.9 and Refs. [1, 2], then one can see, that they are fulfilled completely. Secondly, at $D_{cr} = 1.5$ σ_f must not depended on notch length, that is also confirmed experimentally [1, 2]. And thirdly, at $a = 1$ mm the condition $\sigma_f \sim \sqrt{D_{cr}}$ should be fulfilled. As the dependence $\sigma_f\left(\sqrt{D_{cr}}\right)$, adduced in Fig. 8.9 insertion, shows, this condition is also fulfilled. In other words, the Eq. (8.4) gives qualitative correct description of the dependence $\sigma_f(a)$ for the indicated polymer blends [14].

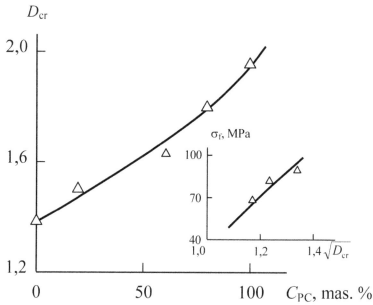

FIGURE 8.9   The dependence of stable crack fractal dimension D_{cr} on PC content C_{PC} for blends PAASO/PC. In insertion: the dependence $\sigma_f\left(\sqrt{D_{cr}}\right)$ for samples with $a = 1$ mm [14].

In Ref. [4] it has been shown that D_{cr} increase from 1 up to 2 by its very nature represents itself the transition from ideally sharp notch ($D_{cr} = 1$) to cavity ($D_{cr} = 2$), that is, the crack with blunted tip. Thus, the physical significance of fractal model corresponds completely to the offered in Ref. [2] treatment, namely, to the mechanism of notch blunting by local plastic deformation zone.

Let us carry out the quantitative estimations according to the Eq. (8.4) and compare them to the experimental dependences $\sigma_f(a)$ by a fitting method. The comparison of the experimental and theoretical dependences of fracture stress on the notch length (fitting was made by σ_f values at $a = 1.5$ mm [14]) is adduced in Fig. 8.10 for PAASO and two blends of PAASO/PC. As one can see, the Eq. (8.4) describes well the experimental results and the observed discrepancies (particularly at $a = 0.5$ mm) are due to the two following reasons. Firstly, the calculation simplicity was not taken into account, a some D_{cr} reduction at sharp notch length growth, observed experimentally. Secondly, strictly speaking, at calculation one should not use sharp notch length a, but values a and stable crack length sum should be used, since it is a matter of fracture process, which is realized by instable crack propagation after reaching by a stable crack of a definite length (Fig. 8.8). Nevertheless, the adduced in Fig. 8.10 comparison shows, that the fractal Griffith crack model gives exhaustive explanation of crack self-braking effect.

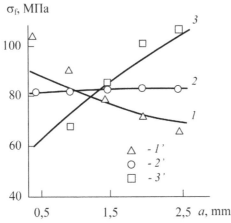

FIGURE 8.10 The dependences of fracture stress σ_f on notch length a for PAASO (1, 1') and blends PAASO/PC with PC contents 20 (2, 2') and 80 (3, 3') mas. %. 1, 2, 3 – calculation according to the Eq. (8.4); 1', 2', 3' – the experimental data [14].

Let us note in conclusion, that the adduced above results have shown that the fractal Griffith crack model application can not only improve quantitative conformity of theoretical and experimental data, but also obtain qualitatively new picture of fracture processes.

The term "cracking" at simultaneous action of stress and environment" was introduced for polymers (mainly polyethylenes) being in stressed state at mobile polar liquids presence. It has been shown [17], that in the end the most weak amorphous part of semicrystalline polymer is responsible for materials strength at this fracture kind. This allows to connect occurring at cracking phenomena to polar liquid diffusion in amorphous regions.

In Ref. [18], it has been shown that in case of HDPE, modified by high-disperse mixture Fe/FeO (Z), strong extreme growth of stability to cracking, expressed by time up to fracture τ_{50}, is observed. So, if for the initial HDPE the normative value τ_{50} makes up 10 h, then for composition HDPE + Z with

Z content $C_z = 0.05$ mas. % the value τ_{50} reaches 250 h. The authors of Refs. [19, 20] explained this important from the theoretical and practical points of view effect within the frameworks of the transport processes fractal concept [21, 22].

For theoretical prediction of the value τ_{50}, characterizing stability to cracking, two main assumptions were made [19]. Firstly, it has been supposed, that samples fracture occurs then, when active medium in diffusion process reaches their median plane. This assumption is based on the analysis of cracking under stress process [17]. Then the theoretical value τ_{50}^T is given by the main equation of stationary diffusion [23]:

$$\tau_{50}^T = \frac{l^2}{6D_{dif}}, \tag{8.5}$$

where l is one half of sample thickness. D_{dif} is active medium diffusivity in HDPE.

The second assumption consists in the fact, that water molecules clusters diffusion is considered without appreciation of detergent OP-7 molecular size. In essence this means that in H_2O molecules cluster the replacement of one from these molecules by OP-7 molecule is supposed. Water molecules cluster formation at interaction with polymer is a well-known fact [24, 25]. The estimations showed, that in this case cluster consisted of H_2O three molecules [25]. The water cluster schematic presentation according to the data of Ref. [24] is adduced in Fig. 8.11. This scheme allows to calculate the greatest cluster size $d_m = 7.8$ Å with appreciation of the fact, that water molecule diameter is equal to 3.08 Å [24].

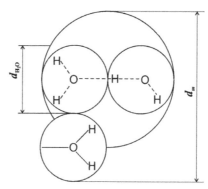

FIGURE 8.11   The model of adsorbed by polymers cluster [24]. d_{H_2O} is water molecule diameter, d_m is water cluster rated diameter [20].

For the diffusivity D_{dif} calculation the transport processes fractal model [21] was used, according to which the D_{dif} value is equal to:

$$D_{dif} = D_0' f_g \left(d_h / d_m\right)^{2(D_t - d_s)/d_s},$$ (8.6)

where D_0' is the universal constant, equal to 3.7×10^{-7} cm²/s, f_g is relative free volume, d_h is this volume microvoid diameter, D_t is polymer structure dimension, controlling transport process, d_s is spectral dimension, accepted equal to 1.0 for linear HDPE [26].

Dimension D_t choice depends on the ratio d_h/d_m value [27]. At $d_m < 0.6 d_h$ the interaction of diffusate molecules with free volume microvoid walls is small and transport process is controlled by structure fractal dimension d_f (the structural transport). At $d_m \geq d_h$ the interaction of diffusate molecules with free volume microvoids walls, which presented itself polymer macromolecules surface with dimension D_f, influences strongly on transport processes [28]. In this case $D_t = D_f$ is accepted, where D_f is excess energy localization regions dimension (molecular transport) [27].

Let us consider the parameters, included in the Eq. (8.6), estimation methods. The value f_g can be determined according to the Eq. (2.19). The dimension D_f is calculated as follows [29]:

$$D_f = \frac{2(1-v)}{1-2v},$$ (8.7)

where v is Poisson's ratio.

The volume of free volume microvoid v_h within the frameworks of the fractal model [28] can be calculated according to the equation:

$$v_h = \frac{D_f (1-2v) k T_m}{f_g E},$$ (8.8)

where T_m is polymer melting temperature ($T_m = 403$ K for HDPE [30]), k is Boltzmann constant, E is elasticity modulus and then it can be estimated this microvoid diameter d_h from geometrical considerations.

The values D_f and d_h for compositions HDPE + Z are adduced in Table 8.1. The values d_h can be compared with the corresponding experimental data, obtained by the positrons annihilation method. For HDPE at $T = 323$ K the experimental value $d_h \approx 6.8$Å [31]. This value d_h corresponds well enough to corresponding calculated values d_h, adduced in Table 8.1.

TABLE 8.1 The Active Medium Diffusion Process Characteristics For Compositions HDPE+Z [19]

Composition	D_f	D_h, Å	$D_{dif} \times 10^8$, Cm²/s	τ_{50}, hours	τ_{50}^T
HDPE	5.17	8.65	4.76	10 (standard)	38.6
HDPE + 0.01 Z	5.0	8.56	4.11	36	44.1
HDPE + 0.05 Z	4.33	6.65	0.69	250	268.0
HDPE + 0.10 Z	5.0	8.56	4.11	38	44.1
HDPE + 0.15 Z	5.17	8.65	4.76	37	38.6
HDPE + 1.0 Z	5.17	8.65	4.76	39	38.6

The figures in conventional signs of compositions indicate Z contents in mas. %.

Now the values D_{dif} can be calculated according to the Eq. (8.6) (Table 8.1) and then – the theoretical values of stability to cracking τ_{50}^T according to the Eq. (8.5). The comparison of experimental values τ_{50} and τ_{50}^T is adduced in Table 8.1. As one can see, the good correspondence of theory and experiment is obtained – the overage discrepancy of τ_{50} and τ_{50}^T makes up ~9%. Taking into account statistical character of tests on cracking in active mediums, one can say, that this discrepancy does not exceed an experimental error. It is important to note, that at τ_{50}^T calculation the fitted empirical coefficients were not used and τ_{50} extreme change is explained completely by structural changes, which are due to Z introduction and characterized by dimension D_f. Besides the theory and experiment conformity confirms correctness of the made above assumptions.

Let us consider structural aspect of τ_{50} change, which is due to Z introduction. As it is known [32], for the composition HDPE + 0.05Z clusters relative fraction φ_{cl} extreme increasing is observed, that results to structure fractal dimension d_f decreasing according to the Eq. (1.12). In its turn, the dimensions d_f and D_f are connected with each other by the relationship [29]:

$$D_f = 1 + \frac{1}{3 - d_f}. \qquad (8.9)$$

Hence, d_f decreasing results to D_f reduction that is the main factor, defining diffusivity Ddif reduction (the Eqs. (8.6) and (8.8)) and, as consequence, τ_{50} growth (the Eq. (8.5)). Let us also note, that comparatively small D_f variation (on about 20%) defines strong (eightfold) D_{dif} change and, as consequence, τ_{50} in virtue of the Eq. (8.6) power dependence, typical for a fractal relationships.

Hence, the adduced above results have shown that polyethylenes samples fracture criterion in tests on cracking under stress in active mediums is polar liquid reaching of sample median plane. The stability to cracking is described correctly within the frameworks of fractal model of transport in polymers. The stability to cracking extreme growth cause is structural changes, which are due to high-disperse mixture Fe/FeO introduction and characterized by dimension D_f.

KEYWORDS

- **active medium**
- **crack**
- **deformation zone**
- **polymer**
- **stochastic fractal**
- **stress intensity factor**

REFERENCES

1. Shogenov, V. N., Kozlov, G. V., Gazaev, M. A., Shustov, G. B., & Mikitaev, A. K. (1985). The Correlation Strength-Notch Length in Polycarbonate-Polyarylate Blend Films. Vysokomolek. Soed. B, *27(4)*, 244–245.
2. Shogenov, V. N., Kozlov, G. V., Gazaev, M. A., & Mikitaev, A. K. (1986). A Notches Self-Blunting in Blends Polycarbonate-Polyarylatearylenessulfonoxide Block-Copolymers Film Samples. Vysokomolek. Soed. A, *28(11)*, 2430–2435.
3. Shogenov, V. N., Kozlov, G. V., & Mikitaev, A. K. (1986). The Physical Significance of Notches Self-Blunting Model Parameters. Vysokomolek. Soed. A, *28(11)*, 2436–2440.
4. Mosolov, A. B. (1991). The Fractal Griffith Crack. Zhurnal Tekhnicheskoi Fiziki, *61(7)*, 57–60.
5. Ivanova, V. S. (1982). From the Griffith Theory to Fracture Fractal Mechanics. Fiziko-Khimicheckaya Mekhanika Materialov, 1993, *29(3)* 432 p.

6. Mandelbrot, B. B. (1982). The Fractal Geometry of Nature. San-Francisco, Freeman and Co. 432 p.
7. Kozlov, G. V., & Zaikov, G. E. (2004). Structure of the Polymer Amorphous State. Utrecht, Boston, Brill Academic Publishers, 465 p.
8. Shogenov, V. N., Byrya, A. I., Shustov, G. B., & Kozlov, G. V. (2000). The Stable Cracks Fractality in Deformation Zones of Glassy Polyarylatesulfone. Fiziko-Khimicheskaya Mekhanika Materialov, *36(1)*, 51–55.
9. Novikov, V. U., & Kozlov, G. V. (2001). Polymers Fracture Analysis within the Frameworks of Fractal Concept. Moscow, Publishers MSOU, 136 p.
10. Kozlov, G. V., Beloshenko, V. A., & Shogenov, V. N. (1995). Fractality of Polyarylatesulfone Film Samples Fracture. Fiziko-Khimicheskaya Mekhanika Materialov, *31(4)*, 116–119.
11. Balankin, A. S. (1991). A Fractals Elastic Properties, Transverse Strain Effect and Solid Bodies Free Fracture Dynamics. Doklady AN SSSR, *319(5)*, 1098–1101.
12. Kaieda, Y., & Pae, K. D. (1982). Fracture Stress Difference of Notched Polycarbonate between Atmospheric Pressure and a Hydrostatic Pressure. Y. Matter. Sci., *17(2)*, 369–376.
13. Izotov, A. D., Balankin, A. S., & Lazarev, V. B. (1993). Synergetics and Fractal Thermomechanics of Inorganic Materials. II. Fractal Geometry of Solids Fracture. Neorganicheskie Materialy, *29(7)*, 883–893.
14. Kozlov, G. V., Shogenov, V. N., & Zaikov, G. E. (2002). The Fractal Analysis of Effect of Auto-Stopping of Stationary Cracks. In: Perspectives on Chemical and Biochemical Physics. Ed. Zaikov, G. E. New York, Nova Science Publishers Inc., 153–158.
15. Mosolov, A. B. (1992). The Fractal Decay of Elastic Fields at Fracture. Zhurnal Tekhnicheskoi Fiziki, *62(6)*, 23–32.
16. Kausch, H. H. (1978). Polymer Fracture. Berlin, Heidelberg, New York, Springer-Verlag, 435 p.
17. Howard, J. B. (1966). The Cracking at Stress Action. In: Engineering Design for Plastics. Ed. Baer E. New York, Reinhold Publishing Corporation, 331–378.
18. Mashukov, N. I., Krupin, V. A., Mikitaev, A. K., & Malamatov, A. Kh. (1990). The Stability to Cracking of Modified Polyethylene. Plast. Massy, *1*, 31–33.
19. Kozlov, G. V., Ovcharenko, E. N., & Zaikov, G. E. (2008). The Fractal Model of Modified Polyethylene Stability to Cracking. Teoreticheskie Ocnovy Khimicheskoi Tekhnologii, *42(4)*, 453–456.
20. Malamatov, A. Kh., & Kozlov, G. V. (2006). Fractal Model of Stability to Cracking of Modified Polyethylene. J. Balkan Tribologic. Assoc., *12(3)*, 328–333.
21. Kozlov, G. V., Zaikov, G. E., & Mikitaev, A. K. The Fractal Analysis of Gas Transport Process in Polymers. Moscow, Nauka, 2009, 199 p.
22. Kozlov, G. V., Zaikov, G. E., & Mikitaev, A. K. (2009). The Fractal Analysis of Gas Transport in Polymers. The Theory and Practical Applications. New York, Nova Science Publishers Inc., 238 p.
23. Rogers, C. E. (1966). Permeability and Chemical Stability. In: Engineering Design for Plastics. Ed. Baer E. New York, Reinhold Publishing Corporation, 193–273.
24. Belfort, G., & Sinai, N. (1980). The Study of Adsorbed Water Relaxation in Porous Glasses. In: Water in Polymers. Rowland, S. P. Ed.; Washington DC, 314–335.

25. Brown, G. L. (1980). Water Clusters Formation in Polymers. In: Water in Polymers. Ed. Rowland, S. P. Washington, D.C., 419–428.
26. Alexander S., & Orbach R. (1982). Density of states on Fractals: "Fractons". J. Phys. Lett. (Paris). *43(17)*, L625–L631.
27. Kozlov, G. V., Afaunov, V. V., Mashukov, N. I., Lipatov, Yu. S. (2000). Fractal Analysis of Polyethylenes Permeability to Gas. Doklady NAN Ukraine, *10*, 140–145.
28. Kozlov, G. V., Sanditov, D. S., & Lipatov, Yu. S. (2004). Structure Analysis of Fluctuation Free Volume in Amorphous State of Polymers. In: Achievements in Polymers Physics-Chemistry. Ed. Zaikov, G. E., Berlin, A. A., Slotskii, S. S., Minsker, K. S., Monakov, Yu. B. Moscow, Khimiya, 412–474.
29. Balankin, A. S. (1991). Synergetics of Deformable Body. Moscow, Publishers of Ministry Defence SSSR, 404 p.
30. Kalinchev, E. L., & Sakovtseva, M. B. (1983). Properties and Processing of Thermoplastics. Leningrad, Khimiya, 288 p.
31. Lin, D., & Wang, S. J. (1992). Structure Transitions of Polyethylene Studied by Position Annihilation. J. Phys.: Condens. Matter, *4(16)*, 3331–3336.
32. Mashukov, N. I., Gladyshev, G. P., & Kozlov, G. V. (1991). Structure and Properties of High Density Polyethylene Modified by High-Disperse Mixture Fe and FeO. Vysokomolek. Soed. A, *33(12)*, 2538–2546.

CHAPTER 9

CRAZING

CONTENTS

The well-known heterogeneity of amorphous glassy polymers plastic deformation [1, 2] allows to assume them nonhomogeneous systems. The same affirmation is valid in respect of semicrystalline polymers amorphous phase [3, 4]. As well nevertheless, both models of continues (let us remind, that the known Dugdale model, often used for crazes characteristics, was developed originally for a metals [5]) and molecular concepts, are applied successfully for both classes polymers behavior description. In this connection the question arises about scale, which can be considered as lower boundary of models of continua applicability.

The one more problem consists what in with probability of heterogeneous on molecular level system (which is unambiguously the amorphous glassy polymer) can be considered as a two-phase system. If there is such a probability, this will allow the application of so-called composite models for description of the amorphous polymers behavior. These models are well developed and successfully used, for example, for description of artificially created two-phase systems, which also include the filled ones. These two problems were discussed in the Ref. [6].

Fellers and Huang [7] have applied the fluctuation statistical theory to describe crazing process in amorphous polymers. They have deduced the expression for estimation of the polymer volume V_0, in which the fluctuation probability is equal to one:

$$\frac{\sigma_c V_0^{1/2}}{\left(2kT_0 B\right)^{1/2}} = 3,87 \, , \tag{9.1}$$

where σ_c is the crazing stress, T_0 is a equilibrium temperature, the lower boundary of which range is the glass transition temperature T_g, B is a bulk modulus, which is connected with Young's modulus by the following relationship [8]:

$$B = \frac{E}{3(1 - v)} \, , \tag{9.2}$$

where v is the Poisson's ratio.

The distance between cluster R_{cl} can be assessed according to the Eq. (4.63). In Table 9.1, the comparison of R_{cl} and linear scale L_0, at which the fluctuation probability is equal to one $\left(L_0 = V_0^{1/3}\right)$ for five amorphous glassy polymers. As it follows from the data of this table, parameters R_{cl} and L_0 are close by both the absolute values and variation tendencies. This means that in the scales of the cluster structure characteristic sizes amorphous polymer (or semicrystalline polymer amorphous phase) may be considered as inhomogeneous system [6].

Katsnelson [9] has given the following definition of the matter phases: they are "…states of the matter able, being in touch, to exist simultaneously in equilibrium with one another. Obviously different properties correspond to different phases. In addition, it should be taken into account that by different phases… parts of a body are meant, related to the solid phase, but possessing different structure and properties." Clusters and loosely packed matrix which in accordance with the cluster model [10] are the main structural elements of the polymer amorphous state meet the above definition, at least, partly. It is known that these elements possess different mechanical properties [11] and different glass transition temperatures [12]. All these facts together allow to consider the polymer amorphous state to be a quasi-two-phase state, disclaiming full strength of line definition [6].

TABLE 9.1 The Comparison of Characteristic Sizes of the Fluctuation Theory L_0 and the Cluster Model R_{cl} For Amorphous Glassy Polymers [6]

Polymer	L_0, Å	R_{cl}, Å
Polystyrene	76.4	36.1
Poly(methyl methacrylate)	31.7	31.6
Poly(vinyl chloride	54.0	27.1
Polycarbonate	39.7	31.1
Polysulfone	36.4	25.0

As it has been noted in chapter five, the local plasticity zone type defines the fracture type: if a craze forms at critical defect tip, then polymer failed quasibrittle and if deformation zone (ZD) or local shear yielding zone ("shear lips") – then quasiductile [13]. The inelastic deformation mechanism change is considered as brittle-ductile transition [14]. The treatment of the indicated transition will be considered below within the frameworks of both cluster model and solid body synergetics.

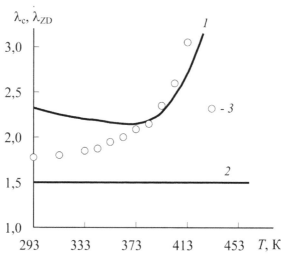

FIGURE 9.1   The experimental dependences of limiting draw ratio in craze λ_c (1) and ZD λ_{ZD} (2) on testing temperature T for PC according to the data of Ref. [15]. 3 – calculation according to the Eq. (5.8) [16].

Plammer and Donald [15] have measured the draw ratio value λ_c and λ_{ZD} for crazes and ZD, respectively, in PC, the temperature dependences of which are adduced in Fig. 9.1 (the curves 1 and 2). The value λ_c begins to increase sharply at $T \approx 383$ K (compare with T_g' for PC) and the value λ_{ZD} remains approximately constant. The limiting draw ratio theoretical value in crazes can be determined according to the Eq. (5.8). The results of the carried out by such mode calculation of λ_c and λ_{ZD} for PC are presented in Fig. 9.1 by the points and their comparison with the experimental data of Ref. [15] is very interesting. Let us indicate first of all, that at calculation the cluster structure corresponding characteristics (L_{cl} and R_{cl}) were used as L_e and d_e. In the crazing region for PC (as it was noted above, this region beginning corresponds to T_g') the calculated and experimental values λ_c are coordinated excellently, that is unexpectedly accounting for the fact that the results were obtained in different laboratories and for PC different marks. In the realization ZD region ($T \le 343$ K) the values λ_{ZD} are coordinated well enough by both temperature dependences course and absolute values (the discrepancy does not exceed 15%). But the theory and experiment comparison result in transitional region from ZD to crazes ($T = 333 \div 373$ K) is the most interesting, from which it follows that necessary for the indicated transition macromolecular entanglements network density decreasing and, hence, λ

increasing is defined only by clusters thermofluctuation decay at T growth. Let us remind, that the concepts [15, 17, 18] attract macromolecules slip through entanglements nodes intensification for this transition explanation.

Thus, the cluster model allows the alternative treatment of plastic deformation mechanisms and controlling them parameters changes at testing temperature variation for amorphous glassy polymers. Both qualitative and quantitative conformity of the offered treatment to the obtained earlier experimental data is shown [16].

The authors of the Ref. [19] studied the plastic deformation mechanisms for polymers within the temperatures wide range. They showed that for PC and polyphenyleneoxide (PPO) the transition from shear to crazing was observed at testing temperature approach to the glass transition temperature of those polymers. The indicated transition was observed at temperatures 373   393 K for PC (compare with the data of Fig. 9.1) and ~413 K for PPO [19]. The authors of Ref. [20] considered the transition "shear-crazing" as nonequilibrium phase transition and obtained its universal criterion within the frameworks of deformable solid body synergetics [21, 22].

Let us consider for the transition "shear-crazing" analysis on the basis of synergetics principles the intercommunication of critical characteristics of structure adaptability to deformation and governing parameter on the basis of the data [22], on entanglements cluster network density change before and after yielding for epoxy polymers on the basis of diglycidyl ether of bisphenol A of anhydride (EP-1) and amine (EP-2) curing. The ratio of curing agent to epoxy oligomer reactive groups K_{st} was accepted as governing parameter and critical values of cluster network density v_{cl} as order parameter. The data of Ref. [22] shoed that in the undeformed state curing agent type influences on coupling form between K_{st} and v_{cl}: the transition from amine curing agent to anhydride one results to governing parameter K_{st} critical value increase from 1.0 up to 1.25 (Table 9.2). After yielding point reaching the value v_{cl} is independent on K_{st}. Besides, the linear correlation between yield strain ε_Y and the values v_{cl} difference before and after yielding is observed and within the experimental error limits the curing agent type does not influence on this correlation form [20]. The adaptability to deformation A_m calculation at $K_{st} = K_{st}^*$ was fulfilled as the cluster network critical densities before $(v_{cl}^*)_{in}$ and after $(v_{cl}^*)_{def}$ deformation ratio determination [21]:

$$A_m = \frac{\left(v_{cl}^*\right)_{def}}{\left(v_{cl}^*\right)_{in}} = \Delta_i^{1/m}, \qquad (9.3)$$

where Δ_i is structure stability measure, m is a possible reorganizations number.

TABLE 9.2 The Adaptability Critical Characteristics of Structure of Epoxy Polymers with Different Curing Agents Before and After Yielding [20]

Parameter	Epoxy Polymer			
	EP-1		EP-2	
	The initial state	After yielding	The initial state	After yielding
K_{st}^*	1.25	—	1.0	—
$V_{cl}^* \times 10^{-27}$, m^{-3}	3.0	0.90	2.85	0.90
$\left(V_{cl}^*\right)_{def}/\left(V_{cl}^*\right)_{in}$	—	0.30	—	0.32
$A_m^* = A_m$	—	0.29	—	0.32
m	—	1.0	—	1.0
Δ_i	—	0.285	—	0.324

From the data of Table 9.2 it follows, that anhydride curing agent usage reduces polymer adaptability to deformation in comparison with amine curing agent from $A_m^* = 0.324$ up to 0.285, but in both indicated cases adaptability is controlled by entanglements cluster network stability. The value $m = 1$ indicates linear feedback between parameters A_m and Δ_i [21].

Thus, the data of Ref. [20] analysis within the frameworks of synergetics shows that the order parameter at transition from shear to crazing is as before the ratio $\left(V_{cl}^*\right)_{def}/\left(V_{cl}^*\right)_{in}$, characterizing stability loss by cluster structure at deformation. This ratio value is close to Poisson's ratio magnitude v_* of medium at nonequilibrium phase transitions ($v_* = 0.33$). It is known [23], that Poisson's ratio is one of the most important characteristics of various mediums, including polymers [24]. The value ε_Y can be determined as follows [24]:

$$\varepsilon_Y = \frac{3\sigma_Y}{E}, \tag{9.4}$$

where σ_Y is yield stress, E is elasticity modulus and critical shear strain, corresponding to solid-phase polymer stability loss, is given as follows [24]:

$$\gamma_* = \frac{\sigma_Y}{E}. \qquad (9.5)$$

From the Eqs. (9.4) and (9.5) the condition of stability loss by cluster structure follows [20]:

$$\frac{\gamma_*}{\sigma_Y} = 0.33 \qquad (9.6)$$

The criterion (9.6) shows that amorphous glassy polymers yielding process is controlled by cluster structure stability loss (for more details see chapter four) and the condition of transition from shear to crazing can be written as follows [20]:

$$\frac{\left(v_{cl}^*\right)_{def}}{\left(v_{cl}^*\right)_{in}} = 0.33 = v_* \qquad (9.7)$$

where v_* is the invariant value of Poisson's ratio for medium at nonequilibrium phase transitions.

Let us consider experimental proofs of the criterion (9.7) correctness. In Fig. 9.2, the temperature dependences of Poisson's ratio v, determined according to the Eq. (2.20) are adduced for three amorphous glassy polymers and the value v_* is indicated by the shaded horizontal line. As it follows from the data of this figure, for polystyrene (PS) at the used T the value v is always larger than v_*, that is, crazing is always main plastic deformation mechanism for PS. This observation is confirmed experimentally [13, 25] and the indicated fact explains high brittleness of PS samples [25]. For plastic PC and PPO the transition from shear to crazing is realized at temperatures $T_{sc} = 373$ and 423 K, accordingly, that corresponds excellently to the experimental results of Refs. [16, 19] ($T_{sc} = 373 \div 393$ K for PC and 413 K for PPO).

Let us note, that the indicated temperatures T_{sc} also correspond well to loosely packed matrix glass transition temperature of PS and PPO T_g', which is approximately by 50 K below than macroscopic T_g [26]. This assumes, that the crazing process at high temperatures for PC and PPO is due to mechanically devitrificated loosely packed matrix drawing facility (in which instable clusters, restricting this process, are absent) in craze fibrils. The Eqs. (1.9), (1.12) and (4.66) indicate unequivocally the cause of PS tendency to crazing: a large S and C_∞ values (69.8 $Å^2$ [27] and 10 [28], accordingly), that defines high values d_f and, hence, v [20].

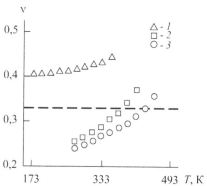

FIGURE 9.2   The dependence of Poisson's ratio v on testing temperature T for PS (1), PC (2) and PP (3). The horizontal shaded line shows the value v_* [20].

In Fig. 9.3, the dependences $v(K_{st})$ for the considered epoxy polymers are adduced where the value v_* is indicated again by a horizontal shaded line. As one can see, crazing is possible only for epoxy polymer samples with K_{st} = 0.50 and 1.50, having the smallest values of both v_{cl} and chemical cross-linking nodes density v_c [29], that corresponds completely to the concepts of Refs. [16, 30]. Exactly these epoxy polymers in compression testing show brittle fracture (up to yield stress) that confirms experimentally the plots of Fig. 9.3 correctness.

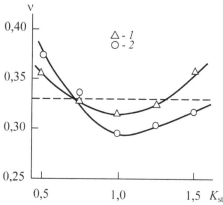

FIGURE 9.3   The dependence of Poisson's ratio n on the value K_{st} for EP-1 (1) and EP-2 (2). The horizontal shaded line shows the value n$_*$ [20].

Hence, the adduced above synergetic analysis of nonequilibrium phase transition "shear-crazing" with the experimental and theoretical data for glassy linear and cross-linked amorphous polymers at different temperatures allows to establish, that:

1. the cluster network critical density v_{cl}^{*}, corresponding to the governing parameter critical value K_{st}, is the order parameter, controlling amorphous polymers structure stability to deformation;

2. the nonequilibrium phase transition "shear-crazing" is connected with self-organizing criticality effect, corresponding to the universal order parameter, which is equal to Poisson's ratio $v_{*} = 0.33$ for mediums with self-organizing criticality [20].

KEYWORDS

- craze
- fluctuation theory
- Poisson's ratio
- polymer
- self-organizing criticality
- synergetics

REFERENCES

1. Kambour, R. P. (1973). A Review of Crazing and Fracture in Thermoplastics. J. Polymer Sci.: Macromol. Rev., 7(1), 1–154.
2. Ishikawa, M., Ogawa, H., & Narisawa, I. (1981). Brittle Fracture in Glassy Polymers. J. Macromol. Sci. Phys., B 19(3), 421–443.
3. G'Sell, C., Aly-Helal, N. A., Semiatin, S. L., & Jonas, J. J. (1992). Influence of Deformation Defects on the Development of Strain Gradients during the Tensile Deformation of Polyethylene. Polymer, 33(6), 1244–1254.
4. Liu, Y., & Truss, R. W. (1994). A Study of Tensile Yielding of Isotactic Polypropylene. J. Polymer Sci.: Part B: Polymer Phys., 32(12), 2037–2047.
5. Dugdale, D. (1960). Yielding of Steel Sheets Containing Slits. J. Mech. Phys. Solids, 8(2), 100–104.
6. Kozlov, G. V., Belousov, V. N., & Mikitaev, A. K. (1998). A Solid Polymers Description as Quasi-Two-Phase Bodies. Fuzuka i Tekhnika Vysokikh Davlenii, 8(1), 101–107.
7. Fellers, J. F., & Huang, D. C. (1979). Crazing Studies of Polystyrene. II. Application of Fluctuation Theory. J. Appl. Polymer Sci., 23(8), 2315–2326.

8. Kozlov, G. V., & Sanditov, D. S. (1994). Anharmonic Effects and Physical-Mechanical Properties of Polymers. Novosibirsk, Nauka, 261 p.

9. Katsnelson, A. A. (1984). Introduction is Solid Body Physics. Moscow, Publishers MSU, 293 p.

10. Belousov, V. N., Kozlov, G. V., Mikitaev, A. K., & Lipatov, Yu. S. (1990). Entanglements if Glassy State of Linear Amorphous Polymers. Doklady AN SSSR, 313(3), 630–633.

11. Shogenov, V. N., Belousov, V. N., Potapov, V. V., Kozlov, G. V., & Prut, E. V. (1991). The Glassy Polyarylatesulfone Curves Stress-Strain Decription within the Frameworks of High-Elasticity Concepts. Vysokomolek. Soed. A, 33(1), 155–160.

12. Startsev, O. V., Abeliov, Ya. A., Kirillov, V. N., & Voronkov, M. G. (1987). Two-Stage Character of Amorphous Mixed Polyorganilsiloxanes α-Relaxation. Doklady AN SSSR, 293(6), 1419–1422.

13. Bucknall, C. B. (1977). Toughened Plastics, London, Applied Science, 318 p.

14. Narisawa, Y. (1987). The Strength of Polymeric Materials. Moscow, Khimia, 400 p.

15. Plummer, C. J. G., & Donald, A. M. (1989). The Deformation Behavior of Polyethersulfone and Polycarbonate. J. Polymer Sci.: Polymer Phys. Ed. 27(2), 325–336.

16. Kozlov, G. V., Beloshenko, V. A., & Lipatov, Yu. S. (1998). Temperature Dependence of the Mechanisms of Crazing and Shear in Amorphous Glassy Polymers: a Current Review and a New Approach. Intern., J. Polymer Mater., 39(3–4), 201–212.

17. Plummer, C. J. G., & Donald, A. M. (1990). Disentanglement and Crazing in Glassy Polymers. Macromolecules, 23(12), 3929–3937.

18. Berger, L. L., & Kramer, E. J. (1987). Chain Disentanglement during High-Temperature Crazing of Polystyrene. Macromolecules, 20(6), 1980–1985.

19. Wellinghoff, S. T., & Baer, E. (1978). Microstructure and its Relationship to Deformation Processes in Amorphous Polymer Glasses. J. Appl. Polymer Sci., 22(7), 2025–2045.

20. Bashorov, M. T., Kozlov, G. V., Ovcharenko, E. N., & Mikitaev, A. K. (2008). The Nanostructures in Polymers: Synergetics if the Nonequilibrium Phase Transition "Shear-Crazing". Nano-i Microsistemnaya Tekhnika, 11, 5–7.

21. Ivanova, V. S., Kuzeev, I. R., & Zakirnichnaya, M. M. (1998). Synergetics and Fractals. Universality of Materials Mechanical Behavior. Ufa, Publishers USNTU, 366 p.

22. Kozlov, G. V., Beloshenko, V. A., & Varyukhin, V. N. (1996). Evolution of Dissipative Structures in Yielding Process of Cross-Linked Polymers. Prikladnaya Mekhanika i Teknicheskaya Fizika, 37(3), 115–119.

23. Balankin, A. S. (1991). Synergetics of Deformable Body. Moscow, Publishers of Ministry of Defence SSSR, 404 p.

24. Kozlov, G. V., & Novikov, V. U. (2001). The Cluster Model of Polymers Amorphous State. Uspekhi Fizicheskikh Nauk, 171(7), 717–764.

25. Kozlov, G. V., & Mikitaev, A. K. (1987). The dependence of Crazing Stress on Testing Temperature at Polystyrene Impact Fracture. Izvestiya VUZov, Severo-Kavkazsk. Region, estestv. Nauki, 3, 66–69.

26. Belousov, V. N., Kotsev, B. Kh., & Mikitaev, A. K. (1983). Two-Step of Amorphous Polymers Glass Transition. Doklady AN SSSR, 270(5), 1145–1147.

27. Aharoni, S. M. (1985). Correlations between Chain Parameters and Fracture Characteristics of Polymer below their Glass Transition Temperature. Macromolecules, 18(12), 2624–2630.

28. Aharoni, S. M. (1983). On Entanglements of Flexible and Rodlike Polymers. Macromolecules, *16(9)*, 1722–1728.
29. Kozlov, G. V., Beloshenko, V. A., Varyukhin, V. N., & Lipatov, Yu. S. (1999). Application of Cluster Model for he Description of Epoxy Polymer Structure and Properties. Polymer, *40(4)*, 1045–1051.
30. Donald, A. M., & Kramer, E. J. (1982). Effect of Molecular Entanglements on Craze Microstructure in Glassy Polymers. J. Polymer Sci.: Polymer Phys. Ed., *20*, 899–909.

CHAPTER 10

IMPACT TOUGHNESS

CONTENTS

At present it is accepted to consider [1], that devitrificated at testing temperature amorphous phase of semicrystalline polymers is the main source of such polymers high impact toughness. However, this conclusion has speculative enough grounds and is not supported by any quantitative estimations, which is due to corresponding models absence. The cluster model and fractal analysis allow to fill a gap in our knowledge in respect to semicrystalline polymers behavior at mechanical loading. In Refs. [2, 3], this is carried out on the example of HDPE samples, which is typical representative of a polymers considered class.

The parameter χ in the Eq. (4.9) characterizes a polymer fraction, which does not participated in plastic deformation process, but subjects to elastic deformation. For semicrystalline polymer this fraction consists of devitrificated amorphous phase and crystalline phase part, which was subjected to partial mechanical disordering [4]. In other words, the parameter χ characterizes he deformed polymer structural state. For the considered in Refs. [2, 3] HDPE crystallinity degree $K = 0.687$ and, hence, amorphous phase fraction φ_{am}, makes up $1 - K = 0.313$. As estimations according to the Eq. (4.9), have shown the value χ for HDPE depending on notch length and testing temperature changes within the limits of $0.40 \div 0.83$, that is, exceeds φ_{am}. Since the yielding process in polymers is realized in densely packed regions, then this means the necessity of crystalline phase some part disordering, the fraction of which χ_{cr} can be determined according to the Eq. (4.68). Such disordered component of HDPE structure may also be an effective impact energy dissipative element and therefore, as candidates on a structural component role, defining HDPE high plasticity in impact tests, one can assume the following: devitrificated loosely packed matrix of amorphous phase, which fraction is equal to $\varphi_{l.m.}$; crystalline phase disordered part, which fraction is equal to χ_{cr} or their sum with fraction $(\varphi_{l.m.} + \chi_{cr})$.

The fraction of dissipated in impact loading process energy η_d can be estimated within the frameworks of solid body synergetics according to the equation [5]:

$$\eta_d = 1 - \Delta_i^{-\beta_{fr}}, \tag{10.1}$$

where Λ_i is a polymer structure automodelity coefficient, β_{fr} is a fractional part of fracture surface fractal dimension d_{fr}, which, in its turn, is defined according to the Eqs. (4.50) and (4.51) for quasibrittle and quasiductile fracture types, accordingly. The indicated fracture types boundary is defined by the condition $v = 0.35$ [6]. As it has been shown in Ref. [7], for a polymers structure automodelity coefficient is equal to characteristic ratio C (for HDPE $C_\infty = 6.8$ [8, 9]).

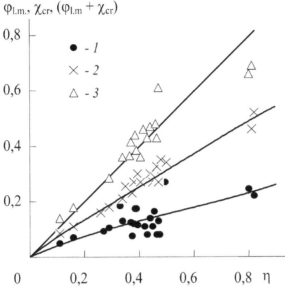

FIGURE 10.1   The relation between fractions of loosely packed matrix $\varphi_{l.m.}$ (1), disordered crystallites χ_{cr} (2) and their sum $(\varphi_{l.m.} + \chi_{cr})$ (3) and dissipated energy fraction η_d for HDPE samples with sharp notch within the range of $T = 293 \div 353$ K [2].

In Fig. 10.1, the relations of the values η_d and $\varphi_{l.m.}$, χ_{cr}, $(\chi_{cr} + \varphi_{l.m.})$ are adduced for HDPE samples within the range of $T = 293 \div 353$ K. As one can see, the increase of any from the indicated structural components results to η_d growth, but the relation 1:1 was obtained only for $(\chi_{cr} + \varphi_{l.m.})$. This relation of complete energy dissipation is typical for rubbers, which is both devitrificated loosely packed matrix with fraction $\varphi_{l.m.}$ and mechanically disordered crystallites part with fraction χ_{cr}. Hence, the indicated structural components of deformed state define energy dissipation in HDPE, but not amorphous phase itself, the part of which (clusters) does not participated at all in impact energy dissipation process [3].

It can be expected, that the parameter η_d is defined in the long run by polymer structure, the state of which can be characterized by its fractal dimension d_f. In Fig. 10.2, the dependence $\eta_d(d_f)$ for HDPE is adduced, which turns out to be a linear one and shows η_d increase at d_f growth. Such dependence type was expected, since the condition $d_f \to 3$ means the approach to the rubber – like state [10], for which $\eta_d = 1.0$. It is significant, that at $\eta_d = 0$ the value $d_f = 2.5$, which gives boundary of the transition to brittle fracture [6]. The value $\eta_d = 1.0$ it reached at $d_f = 2.95$, that corresponds to limiting possible for solid bodies Poisson's ratio $v = 0.475$ [6]. The adduced in Fig. 10.2 relation can be written analytically as follows [3]:

$$\eta_d \approx 2.1(d_f - 2.5). \tag{10.2}$$

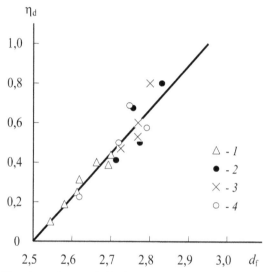

FIGURE 10.2 The dependence of dissipated energy fraction η_d on structure fractal dimension d_f for HDPE samples with sharp notch at $T = 293$ (1), 313 (2), 333 (3) and 353 K (4) [2].

In this case the fracture surface fractal dimension d_{fr} can be written from the Eq. (10.1) and with the condition $\Lambda_1 = C_\infty$ appreciation as follows [2]:

$$d_{fr} = 2 - \frac{\ln(6 - d_f)}{\ln C_\infty}. \tag{10.3}$$

From the Eq. (10.3), it follows that the structural (d_f) and molecular (C_∞) characteristics of HDPE define the value d_{fr} (and, hence, impact toughness A_p) [2].

As it is known [11], the main mechanism of impact energy dissipation for HDPE is a local shear deformation mechanism, that follows from A_p linear increasing at enhancement of the indicated deformation zones size η_p ("shear lips"). Proceeding from these considerations, definite correlation between η_p and η_d. Should be expected the adduced in Fig. 10.3 plot $\eta_p(\eta_d)$ confirms this supposition. As it was expected, η_d increase results to η_p growth. At η_d 1 the value η_p makes up approximately 1.8 mm. The brown model [12] assumes, that brittle-ductile transition is realized at the condition:

$$r_p = \frac{B}{2},$$ (10.4)

where B is a sample width.

For HDPE samples the brittle-ductile transition is realized at $v \approx 0.35$ or $\eta_d \approx 0.42$ (the Eqs. (1.9) and (10.2)), corresponding to $r_p \approx 0.85$ mm (Fig. 10.3), that is much smaller than the value, assumed by the criterion (10.4) (at $B = 6$ mm). The relation between r_p and η_d for HDPE can be written as follows [3]:

$$r_p \approx 2\eta_d, \text{ mm},$$ (10.5)

or, with the Eq. (10.2) appreciation:

$$r_p \approx 4.2(d_f - 2.5), \text{ mm}.$$ (10.6)

It is known [13], that molecular mobility intensification results to energy dissipation increase and enhancement of polymer toughness at failure. In chapter two, it has been shown that the molecular mobility level can be characterized by the value of fractal dimension of chain part between macromolecular entanglements nodes $D_{ch}(1 < D_{ch} \leq 2$ [14]). Fractional part D_{ch} change (i.e., $D_{ch} - 1$) from 0 up to 1 gives all possible spectrum of molecular mobility of this chain part. In Fig. 10.4, the relation between η_d and $(D_{ch} - 1)$ is adduced, which has the expected character and is written analytically as follows [3]:

$$\eta_d \approx D_{ch} - 1.$$ (10.7)

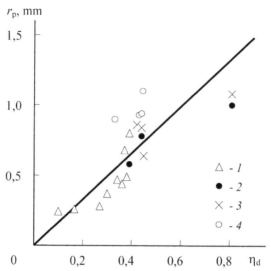

FIGURE 10.3   The dependence of "shear lips" size r_p on dissipated energy fraction η_d for HDPE samples with sharp notch at $T = 293$ (1), 313 (2), 333 (3) and 353 K (4) [2].

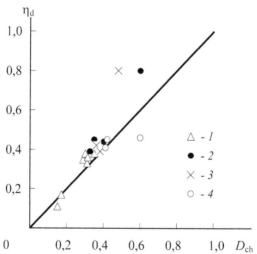

FIGURE 10.4   The relation between dissipated energy fraction η_d and fractional part of fractal dimension of chain part between entanglements D_{ch} for HDPE samples with sharp notch at $T = 293$ (1), 313 (2), 333 (3) and 353 K (4) [2].

Let us obtain with the Eq. (10.2) appreciation [3]:

$$D_{ch} = 2.1\, d_f - 4.25, \qquad\qquad (10.8)$$

that in two limiting cases gives: at $d_f = 2.5$ (brittle fracture) $D_{ch} = 1.0$ and at $d_f = 2.95$ $D_{ch} = 1.0$ and at $d_f = 2.95$ $D_{ch} = 1.945$ [3].

The Eq. (10.1) for polymers with the relations $\Lambda_i = C_\infty$ and $\eta_d = D_{ch} - 1$ can be modified as follows [2]:

$$d_{fr} = 2 - \frac{\ln(2 - D_{ch})}{\ln C_\infty}. \qquad (10.9)$$

This equation is outstanding in virtue of the fact, that it couples the fracture surface fractal dimension d_{fr} with two specific (and the most important) for polymers factors-chain mobility and flexibility. Besides, the adduced relations allow to connect microscopic (D_{ch} and d_f) and macroscopic (η_p) structure and fracture process parameters with polymer properties. So, the Eqs. (10.5) and (10.7) can be generalized as follows [3]:

$$r_p = 2(D_{ch} - 1), \text{ mm.} \qquad (10.10)$$

The Eq. (10.10) assumes, that the fractal dimension d_{fr}, determined on microlevel (the Eq. (10.9)), can be connected with the value d_{fr} on macrolevel by the value r_p corresponding usage.

Let us consider further the intercommunication of the initial and deformed HDPE structure. As it has been shown in Fig. 10.5, the values χ_{cr} and K_{η_d} coincide practically, as it has been supposed earlier, that is, the ration χ_{cr}/K defines the value η_d. From the data of Figs. 10.1 and 10.5 it can be written [2]:

$$\chi_{cr} + \phi_{l.m.} = \frac{\chi_{cr}}{K}, \qquad (10.11)$$

or

$$\chi_{cr} = \frac{\phi_{l.m.} K}{1 - K}. \qquad (10.12)$$

Hence, a notch length increase results to HDPE amorphous phase mechanical vitrification, decreasing the value $\phi_{l.m.}$. The value χ_{cr} at the condition $K = $ cons decreases simultaneously. Both indicated effects result to dissipated energy fraction decrease, "shear lips" size reduction and impact toughness A_p decreasing. The Eq. (10.12) demonstrates as a matter of fact the intercommunication of HDPE structure in the initial state and after deformation.

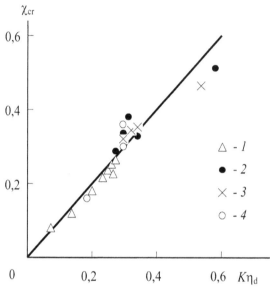

FIGURE 10.5   The relation between disordered crystallites fraction χ_{cr} and product $K\eta_{d}$ for HDPE samples with sharp notch at $T = 293$ (1), 313 (2), 333 (3) and 353 K (4) [2].

The adduced results have shown that loosely packed matrix of devitrificated amorphous phase and disordered in deformation process crystalline phase part are structural components, defining impact energy dissipation and hence, impact toughness of semicrystalline polymers. The fractal analysis allows correct quantitative description of processes, occurring at HDPE impact loading. It is important, that the intercommunication exists between polymer initial structure characteristics and its changes in deformation process [2, 3].

The dependence of impact toughness A_p on macromolecular entanglements cluster network density ν_{cl} was considered briefly by the authors of Ref. [15]. In Fig. 10.6, the dependence of impact toughness A_p of PASF aged samples on the value ν_{cl} in double logarithmic coordinates is adduced. Its linearity allows to obtain the following empirical relationship between A_p and ν_{cl} [15]:

$$A_p \sim \nu_{cl}^{3,7}, \tag{10.13}$$

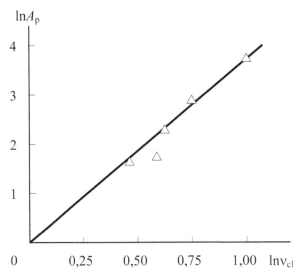

FIGURE 10.6 The dependence of impact toughness A_p on macromolecular entanglements cluster network density ν_{cl} in double logarithmic coordinates for PASF aged samples [15].

It is easy to see, that the Eq. (10.13) power correlation supposes very strong dependence of A_p on ν_{cl}. This observation was confirmed experimentally. The ν_{cl} decrease approximately in 1.5 times results to A_p reduction more than on order of magnitude [16]. The cause of such sharp A_p decrease is weakening of shear mechanism (which is the most effective impact energy dissipation mechanism) at ν_{cl} reduction. The similar relation between A_p and ν_{cl} was obtained and for polyethylenes. The Eq. (10.13) together with the dependence of ν_{cl} on molecular weight, adduced in Ref. [15], also explains the strong dependence of PASF impact toughness on samples molecular weight [19].

As it is known [20], the strain energy release critical rate G_{Ic}, which characterizes value and its dependence on notch length a can be expressed as follows:

$$G_{Ic} = \frac{K_{Ic}^2}{E} \sim G_{I0} \cdot a^{d_r - 1}, \tag{10.14}$$

where d_r is also fractal dimension and the reasons of its special designation will be considered below.

Attempts of the parameter G_{Ic} representation as a function $a^{d_r - 1}$ were unsuccessful, since through this correlation is given by a straight line, it does not pass through coordinates origin [21]. This discrepancy reasons are obvious. G_{Ic} value is controlled by plastic strain in "shear lips" or crazes in the case of instable crack. Therefore, the value G_{Ic} should depend not on fracture surface, but on local deformation zone, which was designated as d_r. If to present such zone as triangle with height of r_p and basis of δ_p (where δ_p is the indicated zone opening, Fig. 10.7), as it was often made [22], then by analogy with crack (Fig. 8.2) the value d_r can be determined as follows [10]:

$$d_r = \frac{2 \ln \delta_p}{\ln r_p}. \qquad (10.15)$$

The obvious physical significance of local plasticity zone fractal dimension d_r follows from the adduced analysis: the greater δ_p at fixed r_p the larger strain in the indicated zone (see the Eq. (5.6)) and the larger its fractal dimension [21].

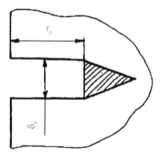

FIGURE 10.7   The schematic picture of crack and local plasticity zone (which is shaded) within the frameworks of the Dugdale-Barenblatt model [22].

In Fig. 10.8, the dependence of G_{Ic} on a^{d_r-1} is presented, corresponding to the Eq. (10.14), for HDPE and PS. Let us note, that within the frameworks of linear fracture mechanics the value G_{Ic} is assumed as material property, that is, at fixed testing temperature and strain rate it should be constant and independent on a value [23]. However, in impact tests HDPE and PC samples the value G_{Ic} changes in $2 \div 3$ times. This is usually explained by the circumstance, that the condition $G_{Ic} = \text{const}$ is valid only for plane-deformed state, which is not carried out in usual impact tests, since the samples of large sizes, particularly for such ductile polymer as HDPE, are required for its fulfillment, that happens technically difficult or even impossible [24].

The Eq. (5.16), which both qualitatively and quantitatively corresponds to experimentally obtained dependence $A_p(B)$ [13, 24], allows to confirm this postulate. So, at testing temperature $T = 293$ K the value $r_p \approx 0.8$ mm for HDPE [21]. The calculated at this condition according to the Eq. (5.16) correlation $d_{frl}(B)$ is adduced in Fig. 10.9. It is obvious, that at the condition $2r_p = B$, corresponding to the brittle-ductile transition [12], the greatest possible value $d_{fr} = 2$ is reached.

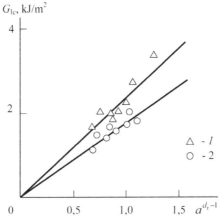

FIGURE 10.8   The dependence of critical strain energy release rate G_{Ic} on parameter a^{d_r-1} for HDPE (1) and PS (2) [21].

In addition the obtained value $B = 1.6$ mm corresponds well to the literary data for samples with the greatest plasticity thickness [13]. Then the fast d_{frl} reduction at B growth is observed, that corresponds completely to similar fracture toughness decrease within the frameworks of polymer fracture two-component model [25, 26] and also corresponds to the stressed state transition from plane-stressed to plane-strained one [13]. At $B = 20 \div 25$ mm the values d_{frl} reach asymptotic magnitude, that also corresponds to the experimental B value, obtained for transition to plane-strained state [24].

The plots of Fig. 10.8 are of interest from two points of view. Firstly, they confirm the intercommunication of G_{Ic} and local plastic deformation zones fractality. Secondly, such plot allows to determine the value G_{I0} in the Eq. (10.14), which is actually material property. From the data of Fig. 10.8 it follows, that for HDPE $G_{I0} = 2.35$ kJ/m^2 and for PS – 1.70 kJ/m^2, that is very close to the cited earlier results [22].

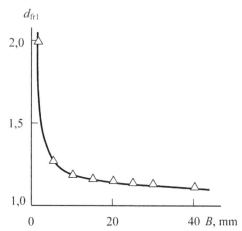

FIGURE 10.9 The dependence of fracture surface fractal dimension d_{fr1}, calculated according to the Eq. (5.16), on sample width B for HDPE [21].

At polymers deformation the brittle-ductile transition can be treated as type fracture change from low-energetic brittle one (without noticeable plastic strain) to failure with large plastic strains and corresponding to the applied from outside energy expenditures for samples failure [22]. This effect importance is mentioned repeatedly above. The authors of Ref. [27] offered the percolation model of brittle-ductile transition for toughened by rubber polymers, in which this transition is realized at percolation network (cluster) formation on the entire extent of sample cross-section and this network consists of glassy matrix parts between rubber particles, subjected to shear strain. It is easy to see, that the Brown model [12] as a matter of fact is percolation model variant, postulating brittle-ductile transition as shear deformation bands continuous network formation. On this notion basis the authors of Ref. [28, 29] carried out the brittle-ductile transition description within the frameworks of percolation models on the example of HDPE impact tests.

As it is known [11, 30], at HDPE samples with sharp notch failure in impact tests two zones of local shear yielding with width r_p are formed at sample edges. The greater r_p is the higher samples fracture energy U or their impact toughness A_p is. These zones merging in sample center will be considered as percolation network formation with the probability P_∞ of shear bands belonging to this network (cluster) $P_\infty = 1$ and the transition to sample plastic fracture with macroscopic yielding formation [29]. The last effect in HDPE impact tests is expressed by the fact, that samples do not break, but bend only. In this case the impact energy U dissipation probability is proportional to P_∞ or local shear zone relative width r_p^{rel} ($r_p^{rel} = 2r_p/B$, where B is a sample width, which is equal to 6 mm), that can be written in the percolation relationship form [30]:

$$U \sim \left(r_p^{rel}\right)^{v_p},\qquad(10.16)$$

where v_p is the critical percolation index.

Thus, the offered in Refs. [29, 30] treatment assumes percolation not by a sample cross-section area, but by a line – sample width B. Such treatment reason will be considered below. Besides it is supposed [30], that percolation cluster formation beginning (percolation threshold) corresponds to any small amount of local shear zone (shear bands network) formation in deformation process and therefore, to the first approximation the percolation threshold r_p^c is accepted equal to zero [29].

In Fig. 10.10, the dependence of U on r_p^{rel} in double logarithmic coordinates for HDPE impact tests, corresponding to the Eq. (10.16), is adduced. As one can see, the indicated dependence is linear and from its slope the critical index v_p can be calculated, which is equal to ~0.77. This value is close to the corresponding classical v_p value, which is equal to ~0.80 [31].

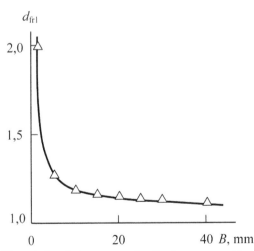

FIGURE 10.10 The relation between samples fracture energy U and local shear zone relative length r_p^{rel} for HDPE [29].

Let us consider the reason of usage in the Eq. (10.16) namely the index v_p, but not other any classical or nonclassical critical index [31]. Let us appeal for this to the percolation cluster fractal model, offered in Ref. [32] for particulate-filled composites. In this model the structure of infinite cluster, consisting of disperse particles, represents itself a rare spatial distorted network, composed from a filler particles set. For simplification a particles chains can be replaced by lines, assuming, that along the line particles density remains constant and the greatest one and sharply decreases on the side of the line [32]. The completely similar network macromolecular entanglements cluster network forms in HDPE amorphous phase [18]. This allows to consider the indicated phase as quasi-two-phase system and apply for its description a composite models [33]. In this treatment the critical index β, which is equal to ~0.40, characterizes macromolecular entanglements cluster network (the first subset [32]). Since impact energy dissipation and, hence, local shear zones formation is associated with loosely packed matrix [2] (the second subset [32]), then in the Eq. (10.16) exponent equality to index v_p, is expected which characterizes this subset, that confirms the plot of Fig. 10.10. Let us note, that within the frameworks of treatment [32] the value v_p is defined by polymer structure and connected with fractal dimension d_f of the latter by the Eq. (5.5). The estimation d_f according to the indicated equation at $v_p \approx 0.77$ gives the value ~2.60. The calculation d_f by the mechanical tests results [3] for HDPE samples with notch lengths $a = 0.5 \div 1.5$ mm was gives d_f range of 2,688 and 2,546 with mean value $d_f \approx 2,617$, that corresponds excellently to the estimation according to the Eq. (5.5).

Now the value U can be calculated theoretically, proceeding from the known r_p values, estimating the constant coefficient in the Eq. (10.16) from the best conformity of theory and experiment. The comparison of experimental U and calculated by the indicated mode U^T HDPE samples with different notch length and at different testing temperatures fracture energy is adduced in Fig. 10.11. The data of this figure has shown the good correspondence of theory and experiment, that the percolation relationship (the Eq. (10.16)) correctness confirms.

Let us consider now the reason for percolation linear model choice. As it has been shown in Ref. [34], for samples of HDPE, modified by high-disperse mixture Fe/FeO(Z) at Z content 0.05 mas.% the extreme U increase up to ~0.78 J is observed, accompanied by shear transvers band at notch tip appearance. Thus, two side and one transverse bands of shear are formed closed arch-like construction, forming actually the linear percolation in the indicated above sense by sample width. Despite small r_p increase for side bands of shear and small value r_p for transverse band of shear (~0.5 mm), the sharp increase U in the indicated compositions HDPE+Z is observed.

Calculation r_p according to the Eq. (10.16) shows, that the similar U increase will give the value $r_p = 3$ mm or $r_p^{rel} \approx 1$, that is, local shear zones percolation by the entire area of sample cross-section. Thus, the adduced example demonstrates local shear side bands merging in sample center importance, irrespective of the percolation model (linear or plane one) choice [28]. Hence, the adduced above results have shown the percolation model applicability for the brittle-ductile transition description in HDPE.

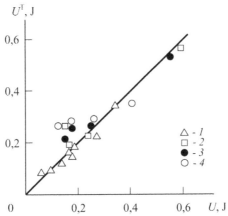

FIGURE 10.11 The relation between theoretical U^T and experimental U fracture energy of HDPE samples at testing temperature: 293 (1), 313 (2), 333 (3) and 353 K (4) [29].

Lately great attention it paid to the intercommunication of fractal characteristics and mechanical properties of materials [35]. Particularly, in the first from this trend Refs. [36] it has been found out that metals fracture energy U at impact loading displays linear correlations with fracture surface fractal dimension d_{fr}, decreasing at the latter growth. The value d_{fr} in Ref. [36] was determined by slit islands analysis (SIA), in which the relation of perimeter P and area S of roughnesses on fracture surface at its consecutive polishing is measured. In addition in microscope field of vision one can see such roughnesses of two types contours – "islands" (ledges) and "lakes" (hollows). As it has been shown in Ref. [37], P and S values of "islands using" for d_{fr} determination gives impact toughness A_p and d_{fr} linear correlation with A_p increase at d_{fr} growth and the same parameters for "lakes" using $-A_p$ reduction at d_{fr} growth. It is quite obvious, that the correlation of the first type has much more physical grounds than of the second one.

Another aspect of the indicated relationships is prevalent usage in them of the fracture surface fractal dimension, but not the structure, although it is quite obvious, that d_{fr} is defined by the structure fractal dimension d_f [21] and the fracture mechanism, which in its turn depends on d_f [6]. This is more so important that d_{fr} can have an extremely indirect relation to the value U (or A_p) from the point of view of physical significance. So, in polymers impact tests at quasibrittle fracture the value U is defined by the dynamical competition of two parameters – critical defect stiffness and local plastic deformation zone ability to dissipation of impact energy [38]. At a definite strain reaching a sample breaks down in such tests, as a rule, by an instable crack, the propagation of which does not required energetic expenditures and which gives almost no contribution in U (or A_p) for this reason [39]. Hence, if the connection between U and d_{fr} exists, then it is defined by more general correlation $d_{fr}(d_f)$ or $U(d_f)$. Williford [40] shows within the frameworks of fracture process multifractal analysis, that the material structure is supporting fractal (having the greatest dimension from possible fractal dimensions) in the indicated process and sample surface or fracture surface is the first subfractal. The authors of Ref. [42] obtained the direct analytical intercommunication of the values U and fractal dimension and elucidated, which from the latter (d_{fr} or d_f) defined fracture energy on the example of HDPE samples with varied length of sharp notch.

As Williford has shown [42], the fractal dimension D (not obligatory equal to d_{fr}), connected with fracture energy U, can be determined from the relationship:

$$D = \frac{\ln(U/C)}{\ln L},$$ (10.17)

where C is the material constant, L is observation characteristic scale.

The value L is coupled with damage volume V_f by the following relationship [42]:

$$V_f = L^D.$$ (10.18)

It is obvious, that in impact tests the volume V_f represents itself that sample part, where local plastic deformation and microcracks formation are occurred [43]. In HDPE impact tests such zone is formed at a notch tip, has in section approximately triangular form (see Fig. 10.7) and its volume can be estimated as follows [41]:

$$V_f = \frac{1}{2} r_p \delta_p B, \qquad (10.19)$$

where r_p is local plastic deformation zone length, δ_p its critical opening, B is sample thickness. In its turn, the value δ_p is determined according to the Eq. (5.15).

Believing for the first approximation $D = 3$, the value L can be estimated from the Eqs. (10.18), (10.19) and (5.15). The estimations showed, that D change within the limits of $2 \div 3$ at L calculation gave the greatest error of ~0.3%, that was not reflected practically on the value D, calculated according to the Eq. (10.17).

The material constant C is determined as follows [42]:

$$C = \frac{G_{Ic}}{a}, \qquad (10.20)$$

where a is sharp notch length.

The fracture surface fractal dimension at quasibrittle fracture d_{fr}^{br} is determined according to the Eq. (4.50) and at quasiductile fracture – according to the Eq. (4.51). And at last the value d_{fr} for fractal Griffith crack d_{fr}^{br} can be estimated according to the Eq. (5.17) at the condition $m = 2$.

In Fig. 10.12 the dependences of fracture and fracture surface fractal dimensions, calculated according to the Eqs. (1.9), (4.50), (4.51) and (5.17) and the values D, calculated according to the Eq. (10.17) are adduced as a notch length function for HDPE samples in Sharpy impact tests. The values D and d_f very good correspondence is quite obvious by both the dependence on the course and absolute values. This means, that the fracture energy U value for HDPE is defined by polymer structural state, which is characterized by fractal dimension d_f. The coupling of U (or A_p) with the value d_{fr} at fracture type (mechanism) correct choice is determined by the dimensions d_f and d_{fr} intercommunication, expressed through Poison's ratio value [41].

Fracture of a polymer samples with notch in impact tests conditions is usually analyzed usually within the frameworks of linear fracture mechanics [13, 22]. If samples are broken down by brittle mode and local plastic deformation zone at notch tip is small, then for critical strain energy release rate G_{Ic}, which is material plasticity measure, receiving the following expression is used (without a kinetic effects appreciation) [23]:

$$U = G_{Ic} BD\emptyset, \qquad (10.21)$$

where U is sample fracture energy, B and D are sample width and thickness, respectively, $\varnothing$ is the correction coefficient, which depends on tests geometry.

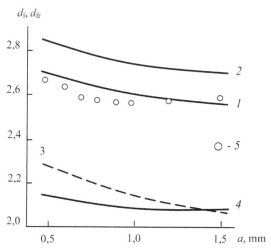

FIGURE 10.12   The dependences of structure d_f (1) and fracture surface dimensions on sharp notch length a for HDPE. The values d_{fr} were calculated according to: (2) the Eq. (4.51), (3) the Eq. (4.50), (4) the Eq. (5.17). The fractal dimension D (5) was calculated according to the Eq. (10.17) [41].

For plastic polymers samples, for which macroscopic yielding is realized in impact loading process in front of a notch, the Eq. (10.21) is not linear already and therefore, for plasticity estimation the equation is used [44]:

$$2U = J_{Ic}B(D - a), \tag{10.22}$$

where J_{Ic} is plasticity measure (J – integral), which is similar to G_{Ic}, a is notch length.

Depending on plasticity degree a polymers behavior in impact tests is described by either the Eq. (10.21) or the Eq. (10.22). Let us note, that the Eq. (10.21) was derived from the conditions of elastic energy in sample accumulation and dissipation and therefore, fracture process itself in the obvious form (i.e., new surfaces formation) is not taken into consideration at the derivation [23]. The Eq. (10.22) was obtained from the modified energetic Griffith criterion, in which a new surfaces formation at crack front advancement is directly taken into consideration [22]. At such treatment it is supposed, that crack edges are absolutely flat, though experimental observations show that polymers fracture surfaces are characterized by roughnesses of the very various scale availability (see Fig. 8.2). Therefore, a real crack in mesoscale is not much resembling ideal cracks with the flat edges, considered usually in the linear fracture mechanics. There was established statistical self-similarity of fracture surface microrelief was established and this means, that cracks surface structure can be simulated by fractal surfaces. In this case a crack true length l_{cr} is coupled with its current mean size R_{cr} according to the fractal law [20]:

$$l_{cr} = R_{cr}^{d_{fr}}, \qquad (10.23)$$

where d_{fr} is crack surface fractal dimension, presented by a stochastic fractal.

In the case of sample complete breakdown $l_{cr} = (D - a)$ and the effective area of rough fracture S_{ef} is determined as follows [45]:

$$S_{ef} = (D-a)^{fr} B^{fr}. \qquad (10.24)$$

Further the dependences of effective fracture energy U_{ef} on sample fracture surface effective area S_{ef} can be plotted and determined the values G_{Ic} and J_{Ic} can be determining, characterizing a polymer plasticity. It is clear, that in the Eq. (10.21) $U_{ef} = U$, $S_{ef} = BDØ$ and in the Eq. (10.22) $U_{ef} = 2U$, $S_{ef} = (D-a)^{fr} B^{fr}$. Let us note, that the parameters G_{Ic} and J_{Ic} are polymer property and in virtue of this they should not depended neither on testing geometry nor on their determination mode [23, 44].

The value d_{fr} can be determined according to the Eq. (4.51). As it was noted in chapter seven, the value d_{fr} was determined for fracture surface and to find the corresponding value d_1, for a curve (fracture surface section), that the Eq. (10.23) assumes, is necessary to divide the value d_{fr} by 2 [46] (see the Eq. (7.20)).

In Fig. 10.13, the dependences of U_{ef} on S_{ef} for HDPE are adduced for three testing temperatures. Despite a definite data scattering (typical for impact tests), it is obvious, that the Eqs. (10.21) and (10.22) in its fractal treatment are approximated by one straight line that allows to determine the only parameter $G_{Ic}(J_{Ic})$ for HDPE, characterizing its plasticity. Such correspondence was to be expected, since the values G_{Ic} and J_{Ic} characterize the same polymer property. The obtained at $T = 293$ K G_{Ic} values are typical for polyethylenes [22, 44].

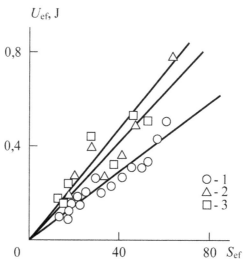

FIGURE 10.13 The dependences of effective fracture energy U_{ef} on samples effective cross-sectional area S_{ef} for HDPE at testing temperatures 293 (1), 313 (2), 333 (3) and 353 K (4). The black symbols correspond to the Eq. (10.21), bright ones – to the Eq. (10.22) fractal treatment [45]

Hence, the HDPE samples fracture surfaces real structure within the frameworks of fractal analysis allows to obtain the only plasticity characteristics by two different methods. Thus, the Eq. (10.22) is an individual case for cracks with flat edges, that corresponds to $d_{fr} = 1.0$ or $v = 0.25$ for brittle fracture and $v = 0$ – for ductile one [45].

The authors of Ref. [47] found out that the mentioned above discrepancies of the experimental correlations $A_p(d_{fr})$, where the value d_{fr} was determined by "slit islands" method (SIA), were due not to distinctions at d_{fr} measurement or other occasional factors, but had key character. With this purpose impact tests are conducted for HDPE samples two series. The first from these series included tests of samples with varied within the limits of $0.5 \div 1.5$ mm sharp notch length a and this series tests were conducted at temperature $T = \text{const} = 293$ K. The second series consisted of tests of samples with sharp notch of length $a = 0.5$ mm and tests were conducted within the range of $T = 213 \div 333$K. For critical strain energy release rate calculation the Eq. (10.21) was used and the value d_{fr} was calculated according to the Eq. (4.50).

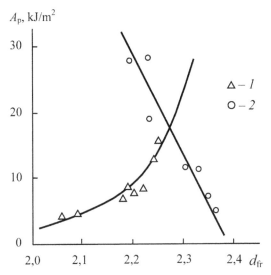

FIGURE 10.14 The dependence of impact toughness A_p on fracture dimension d_{fr} for HDPE samples in tests with varied sharp notch length a (1) and testing temperature T (2) [47].

In Fig. 10.14, the dependences $A_p(d_{fr})$ for both indicated HDPE samples series are adduced. As one can see, for the same polymer and at tests identical conditions the correlations $A_p(d_{fr})$ with opposite tendencies are obtained – the samples with varied notch length series shows A_p increase at d_{fr} growth and at T variation A_p reduction at d_{fr} enhancement is observed. Such discrepancy has several key reasons, the part of which is noted above. Firstly, the impact toughness A_p is the integrated characteristics, which is defined by polymer plasticity (G_{lc}) and critical defect stiffness (for a sharp notch – its length a) [22]. As it is known [48], a polymers, as and all real (physical) fractals possess statistical self-similarity and are fractals only in a definite linear scales range, which was determined experimentally and equal to ~3 50 Å [49]. The processes, defining plasticity (both local and macroscopic) for polymers, are controlled by exactly this structure level [50], whereas a sharp notch has the size on about 5 order more and in virtue of this is considered as Euclidean object within the frameworks of mechanics of continua. Secondly, as it was noted above, the polymers fracture process by instable crack (exactly such mechanism is realized in the considered tests) does not practically require energy expenditures and that is why the fracture energy U value is defined not by fracture process (main crack propagation, dividing the initial sample into two parts), but by local plastic deformation processes, in addition the indicated processes are separated on tests temporal scale. Therefore, the fracture surface (and, hence, its dimension d_{fr}) reflects only to a certain extent processes, controlling the value Ap. Thirdly, the Eq. (4.50) was received from the views of material dilatation in fracture process [6], where as in polymers dilatation process proceeds before and during plastic deformation [51]. Proceeding from the said above, one can make the conclusion, that correlations $A_p(d_{fr})$ have no key physical basis and can have only individual exclusively empirical character. At the same time the adduced above opinions show that the correlation between G_{Ic}, which is the material plasticity measure, and d_{fr} should have physical significance [21].

Let us consider this question in detail. As Meakin [52] has shown, the relation exists between specific surface energy γ value and parameter α, which within the frameworks of multifractal formalism is the scaling index, characterizing singularities concentration:

$$\ln\gamma = (\alpha - 2)\ln L_m, \tag{10.25}$$

where L_m is sample averaged size.

Since the polymers structure is multifractal [53], then, following to Williford [42], the fracture surface can be considered as the first subfractal, having dimension d_I (information dimension, see the Eq. (4.49)) [48]. In this case within the frameworks of the indicated above formalism [42] $\alpha = f$, where f is dimension of singularities α, equal to [47]:

$$f = d_I - 2 = d_{fr} - 2. \tag{10.26}$$

Further from the Eqs. (10.25) and (10.26) combination it can be written [47]:

$$\ln\gamma = (d_p - 4)\ln L_m. \tag{10.27}$$

In Fig. 10.15, the dependence of $\ln(G_{Ic}/2)$ is adduced, since $G_{Ic} = 2\gamma$ [42], on $(d_{fr} - 4)$, from which linear correlation follows, expected according to the Eq. (10.27). Thus, the Eq. (10.27) gives as matter of fact the analytical intercommunication of polymer plasticity (γ or G_{Ic}) and its fracture surface fractal dimension d_{fr}. The value L_m, calculated from the plot $[\ln(G_{Ic}/2)](d_{fr} - 4)$ slope, makes up $\sim$14.5 mm, which is approximately equal to arithmetical mean of sample three sizes: B, D and distance between device supports (span) L_{dev} ($\sim$13.3 mm). In such treatment the results for HDPE samples both series are described by the only straight line.

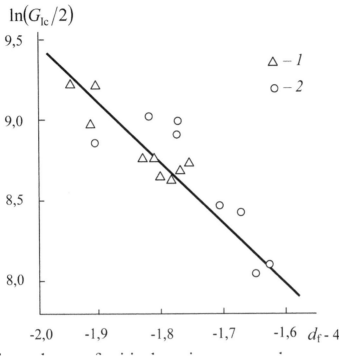

FIGURE 10.15   The dependence of critical strain energy release rate G_{Ic} on parameter $(d_f - 4)$ in logarithmic coordinates for HDPE samples. The designations are the same, that in Fig. 10.14 [47]

Hence, the adduced above data have shown the physical grounds for often cited in literature correlations of impact toughness A_p and fracture surface fractal dimension d_{fr}, at any rate for polymers. And on the contrary, such grounds exist for correlations of plasticity characteristics (G_{Ic} or γ) and d_{fr}. The Eq. (10.27) gives the possibility of the value G_{Ic} prediction, since d_{fr}, as it follows from the Eq. (4.50), is a function of structure characteristic – Poisson's ratio ν. In its turn, G_{Ic} and critical defect characteristics knowledge gives the possibility of A_p prediction [47].

As it is known [54], a polymers number exists, which do not break down in impact tests at samples without notch usage. As the conducted experiments have shown [55], impact energy increase in four times (from 1 up to 4 J) does not change the situation. The two main groups, to which the polymers of such type belong, can be distinguished. The semicrystalline polymers, which have devitrificated amorphous phase at testing temperature (polyethylenes, polypropylene), should be attributed to the first group, to the second one a rigid-chain polymers with high glass transition temperature T_g 425 K, typical representative of which is polycarbonate. Polycarbonate (PC) was called earlier even a polymer with anomalously high toughness [22]. However, it was clarified later, that polysulfone, polyarylates of industrially produced DV (PAr) and other polymers number possess the similar properties [56]. Such behavior of the two indicated groups of polymers can be named superplasticity, that is, the ability to sustain large plastic strain without fracture [6]. The authors of Ref. [57] considered the physical causes of a polymers superplasticity within the frameworks of fractal analysis.

The dissipated impact energy fraction η_d can be estimated according to the Eq. (10.1) and the value d_{fr} – according to the Eq. (4.50). As it is known [58], the fracture surface fractal dimension d_{fr} can be expressed alternatively with the help of η_d and polymer structure automodelity coefficient Λ with the condition $\Lambda_i = C_\infty$ [7] appreciation as follows [57]:

$$d_{fr} = 2 - \frac{\ln(1 - \eta_d)}{\ln C_\infty},$$ (10.28)

and the condition of solid bodies ductile fracture has the following form [6]:

$$d_{fr} \geq 2.70.$$ (10.29)

From the Eq. (10.28) it follows, that large values d_{fr} reaching, that is, ductile fracture can be realized either by η_d enhancement or by C reduction. For polymers the value η_d is connected with molecular mobility level, which can be characterized by fractal dimension of the chain part between entanglements D_{ch} [14]. In Fig. 10.16 the relation between $(D_{ch} - 1)$ and η_d for the 10 studied in Ref. [57] polymers is adduced, from which follows their approximate equality (deviation from the linear dependence for PAr, PSF and PASF is due to the condition $\beta_{fr} \leq 1$ in the Eq. (10.1) nonfulfillment [58]. The dependence $\eta_d(D_{ch} - 1)$ for the 10 different polymers, adduced in Fig. 10.16, is identical to the similar dependence for HDPE samples with various length of notch, tested at different temperatures (Fig. 10.4), that indicates high community degree of the relation between η_d and D_{ch}.

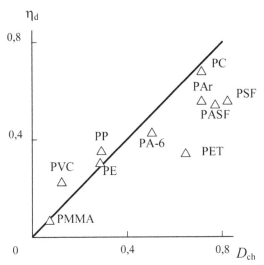

FIGURE 10.16   The relation between dissipated energy fraction part of fractal dimension of chain part between entanglements D_{ch} for 10 polymers at $T = 293$ K. The straight line shows the relation 1: 1 [57].

The Eq. (10.28) demonstrates clearly the parameters, defining the value d_{fr} or polymer fracture type and plasticity degree. In Figs. 10.17 and 10.18 the schematic dependences $d_{fr}(C_\infty)$ at $\eta_d = const$ and $d_{fr}(\eta_d)$ at $C_\infty = const$, accordingly, are adduced, which illustrate d_p dependence degree (and, hence, fracture type) on the indicated parameters. So, from the data of Fig. 10.17 it follows, that at small η_d values (the curve 1) ductile fracture and the more so superplasticity cannot be reached at all, at intermediate η_d (the curve 2) the indicated states are possible at small C_∞ only (within the range of $C_\infty \approx$ 2.0 ÷ 2.5) and at large η_d (the curve 3) ductile fracture is realized at $C_\infty < 10$ and superplasticity ($d_{fr} \approx 3$ [6] – at $C_\infty < 5$. C_∞ influence on the value d_{fr} as a function of η_d is pronounced particularly at small C_∞ (Fig. 10.18, the curve 3) and at large enough C_∞ the dependences $d_{fr}(\eta_d)$ are close to one another at different C_∞ (Fig. 10.18, the curves 2 and 3).

Five typical polymer samples were selected for the considered effects quantitative estimation: PS and HDPE, showing superplasticity availability; PVS, showing intermediate behavior (it breaks down immediately beyond yield stress); typical brittle PMMA and HDPE samples with sharp notch on length 1.5 mm, showing also brittle fracture. The necessary rated parameters for them in elastic and viscoelastic (up to yield stress) and plastic (beyond yield stress) regions are adduced in Table 10.1.

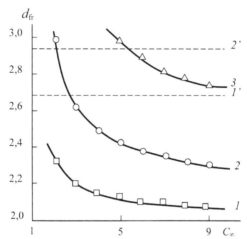

FIGURE 10.17 The dependences of fracture surface fractal dimension d_{fr}, calculated according to the Eq. (10.28), on characteristic ratio C at fixed dissipated energy fraction η_d: 0.2 (1), 0.5 (2) and 0.8 (3). The shaded lines indicate d_{fr} level for ductile fracture (1') and super plasticity (2') [57].

As one can see, both PC and HDPE in the first from the indicated regions have the values d_{fr}, calculated according to the Eq. (10.28) larger than 2.7, that is, they are subjected to ductile fracture according to the criterion (10.29). For PC this threshold d_{fr} value is reached at the expense of C_∞ small values η_d (or D_{ch}), that is the consequence of devitrificated amorphous phase availability for the latter at testing temperature $T = 293$ K. For PVC the value ν in principle allows ductile fracture, proceeding from the Eq. (4.51), but η_d (or D_{ch}) small values result to the fact, that samples are broken down either by brittle mode or immediately beyond yield stress. For PMMA the small values η_d (or D_{ch}) are typical and, as consequence, small values d_{fr}, indicating quasibrittle fracture, that is, $2.0 < d_{fr} < 2.5$ [6].

The results of comparison for HDPE samples with sharp notch and without it show, that notch (stiff macroscopic defect) introduction results to ν reduction, that is, local order level enhancement, which is expressed in mechanical vitrification of devitrificated amorphous phase part [59]. This effect defines molecular mobility considerable reduction, expressed by D_{ch} decrease, and η_d reduction. The last factor results to d_{fr} decrease up to the values typical for quasibrittle fracture, that is, $d_{fr} < 2.5$ [6].

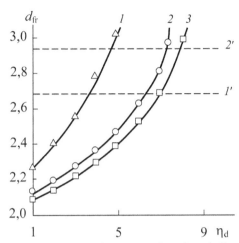

FIGURE 10.18 The dependences of fracture surface fractal dimension d_{fr}, calculated according to the Eq. (10.28), on dissipated energy fraction η_d at fixed characteristic ratio C_∞: 2.4 (1), 6.8 (2) and 10 (3). The shaded lines indicate d_{fr} level for ductile fracture (1') and super plasticity (2') [57].

As it is well-known [60], in deformation process up to yielding point in polymers the dilatation effect is observed, characterized by relative free volume increase by the value Δf_g. As the estimations shown [7], the increment Δf_g usually makes up the value ~0.06 ÷ 0.08. This effect induces a corresponding structural changes, which can be described with the aid of the fractal dimension d_f^Y in yielding point [6]:

$$d_f^Y = \frac{3d_f^{cl}}{3 - d_f^{cl} \Delta f_g},$$ (10.30)

where d_f^{cl} is the value of polymer structure fractal dimension in the elastic deformation region.

TABLE 10.1 The Results of Fractal Model Parameters Calculation for Five Polymer Samples [57]

Polymer sample	C_∞	Δf_g	Elastic and viscoelastic deformation				
			ν	d_f^{cl}	D_{ch}	η_d	d_{fr}
HDPE, without notch	6.8	0.06	0.335	2.670	1.80	0.80	2.84
PC	2.4	0.08	0.310	2.620	1.48	0.48	2.75
PVC	6.7	0.07	0.339	2.678	1.11	0.11	2.81

Polymer sample	Plastic deformation					A_p, kJ/m^2
PMMA	7.0	—	0.420	2.840	1.10	0.10	2.05
HDPE, sharp notch with length of 1.5 mm	6.8	—	0.280	2.560	1.40	0.40	2.27
	v	d_f^T	D_{ch}	η_d	d_{fr}	
HDPE, without notch	0.410	2.820	1.85	0.85	2.99	>167
PC	0.400	2.800	1.58	0.58	2.99	>167
PVC	0.428	2.856	1.45	0.45	2.31	14.5
PMMA	—	—	—	—	—	6.2
HDPE, sharp notch with length of 1.5 mm	—	—	—	—	—	4.7

In the general case d_f value is connected with Poisson's ratio v by the Eq. (1.9). Hence, using the values v, obtained according to the Eq. (2.20), the values d_f^{cl} can be calculated according to the Eq. (1.9) and then, using the adduced in Table 10.1 values Δf_g, the values d_f^Y according to the Eq. (10.30). Further recalculation of the parameters D_{ch} (or η_d) for the indicated samples plastic deformation region was conducted, excluding HDPE samples with a notch and PMMA, which break down by quasibrittle mode, that is, earlier yield stress reaching.

The calculation according to the Eq. (10.30) is correct only for those samples, for which yielding zone in impact tests occupies the entire cross-section of sample [55]. As it follows from the data of Table 10.1, a values d_{fr}, close to 3 and characterizing superplasticity state (determined according to the Eq. (10.28)), were obtained for PC and HDPE. As before, this effect reason for PC is C_∞ small value and for HDPE – high values η_d (or D_{ch}). For PVC, having molecular mobility level comparable with such one for PC, the higher C_∞ value defines smaller d_{fr} value, without allowing superplasticity effect realization. In the Table 10.1, last column the impact toughness A_p values for the considered samples are cited, in addition the A_p values for PC and HDPE were estimated by the division of device the greatest impact energy (4 J) by sample cross-section area. These data demonstrate clearly fracture type influence to A_p value, which is the most important operating characteristic of engineering polymers [54].

The transition from superplasticity to intermediate behavior (even at ductile fracture possibility) reduces A_p more than an order and some even less the value A_p in brittle fracture case.

Thus, the stated above results have shown that impact energy fraction, dissipated in sample deformation process (or molecular mobility level) and polymer chain statistical flexibility are the two main factors, defining polymers fracture type in impact tests and in the end their impact toughness. Although molecular mobility level influence on the A_p value was known for a long time [13], but this influence degree quantitative estimation is conducted for the first time, which becomes possible at fractal analysis methods usage. And at last, the comparison of results for HDPE samples with sharp notch and without it demonstrates the causes of macroscopic defects string influence on the A_p value [57].

The polymers physical aging represents itself the structure and properties change in time and is the reflection of the indicated materials thermodynamically nonequilibrium nature [61, 62]. As a rule, the physical aging results to polymer materials brittleness enhancement and therefore, the ability of structural characteristics in due course prediction is important for the period of estimation of polymer products safe exploitation. For cross-linked polymers the quantitative estimation of structure and properties changes in physical aging process was conducted in Refs. [63, 64] within the frameworks of fracture analysis [65] and cluster model of polymers amorphous state structure [7, 66]. The authors of Ref. [67] use the indicated theoretical models for the description of PC physical aging. Besides, for PC behavior closer definition in the indicated process such theoretical notions were drawn as structure quasiequilibrium state [68] and the thermal cluster model [69], which is one from variants of percolation theory.

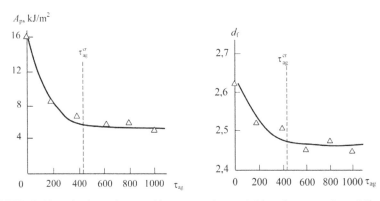

FIGURE 10.19 The dependences of impact toughness A_p (a) and structure fractal dimension d_f (b) on aging duration τ_{ag} for PC. The vertical shaded line indicates the value τ_{ag}^{cr}, calculated according to the Eq. (10.33) [67].

As it was to be expected from the very general considerations [13], in physical aging process the PC plasticity reduction is observed, that is displayed in its impact toughness A_p decreasing with aging duration τ_{ag} (Fig. 10.19). As it follows from the data of Fig. 10.19a, this A_p decreasing occurs during first 500 h of aging and then the values A_p achieve the dependence asymptotic branch with small absolute values A_p (~4.5 kJ/m²). Such the dependence $A_p(\tau_{ag})$ course can be explained by some limiting state of structure achievement, which is defined as quasiequilibrium state [68]. The polymer nonequilibrium structure tendency at physical aging to thermodynamical equilibrium within the frameworks of the model [7, 66] is expressed by local (short-range) order degree φ_{cl} enhancement [64]. However, on a definite aging stage this tendency is balanced by chains entropic tightness and φ_{cl} increase is ceased [68]. The value of excess energy localization regions dimension D_f^{cr} for this state is determined from the Eq. (3.5).

It has been shown earlier [70], that the polymer structure can be simulated as the thermal cluster, at which formation both geometrical and thermal interactions are important. The order parameter φ_{cl} for such cluster is determined according to the equation [69]:

$$\phi_{cl} = \left(\frac{T_g - T}{T_g} \right)^{\beta_T},$$

(10.31)

where T_g is polymer glass transition temperature T is testing temperature, β_T is critical index of thermal cluster.

As it has been shown in Ref. [70], the value β_T in the general case is a function molecular mobility level of polymer and $\beta_T \geq \beta_p$, where β_p is the corresponding critical index of percolation cluster, the formation of which is controlled by geometrical interactions only [31]. The equality $\beta_T = \beta_p$ is reached only in the case of completely inhibited molecular mobility., that is, in the case of quasiequilibrium state.

The calculated according to the Eqs. (1.9) and (2.20) and illustrated PC structure evolution in physical aging process dependence $d_f(\tau_{ag})$ is shown in Fig. 10.19b. As one can see, at $\tau_{ag} > 500$ h the value d_f reaches its asymptotic value, which is approximately equal to 2.46. The value f_g for this case can be estimated from the Eq. (2.19) and further the value D_f^{cr} can be calculated according to the Eq. (3.5) and then the corresponding value d_f^{cr} can be estimated, using the Eq. (8.9). Let us obtain $D_f^{cr} = 2.68$ and $d_f^{cr} = 2.41$ according to the Eqs. (3.5) and (8.9). The last value corresponds well to the adduced above value d_f for asymptotic branch of the dependence $d_f(\tau_{ag})$ (Fig. 10.20b). Thus, reaching by impact toughness values A_p on its asymptotic branch (Fig. 10.20a) can be explained by PC structure reaching of its quasiequilibrium state at physical aging [67].

Let us estimate further the value φ_{cl} at quasiequilibrium state reaching. This can be conducted according to the Eq. (10.31), determining the value $\beta_T = \beta_p$ as follows [32]:

$$\beta_T = \beta_p = \frac{1}{d_f}. \qquad (10.32)$$

Calculation according to the Eq. (10.31) gives $\phi_{cl}^{cr} = 0.619$. Further the value d_f^{cr} can be estimated according to the Eq. (1.12), that gives $d_f^{cr} = 2.45$, which corresponds well to the adduced above estimations. Thus, the thermal cluster model is adequate at polymers structure description, particularly, at quasiequilibrium state realization condition description. The estimations of D_f^{cr} value according to the Eqs. (3.5), (8.7) and (8.9) gave accordingly 2.68, 2.85 and 2.82, that also demonstrates the indicated techniques good correspondence.

Let us estimate physical aging duration τ_{ag}^{cr}, necessary for the quasiequilibrium state reaching. This can be conducted, using the equation, obtained in Ref. [64]:

$$\phi_{cl}^{cr} = C_1 \phi_{cl} \tau_{ag}^{(d_f-2)(d_f-d_f^{cr})}, \qquad (10.33)$$

where C_1 is the constant, which is equal to 0.14 [64], φ_{cl} is clusters relative fraction for initial polymer, and the value τ_{ag} is given in seconds.

The estimation of including in the Eq. (10.33) parameters can be conducted as follows. The value ϕ_{cl}^{cr} was calculated at the aging temperature $T_{ag} = 403$ K according to the Eq. (10.31) and the value φ_{cl} – at T_{ag} as well according to the Eq. (1.9). Then the values d_f and d_f^{cr} at T_{ag} were calculated according to the Eq. (1.12) at corresponding, values φ_{cl} and ϕ_{cl}^{cr}. The value τ_{ag}^{cr}, calculated at the mentioned conditions according to the Eq. (10.33) and equal approximately to 435 h, is shown in Fig. 10.19(a, b) by the vertical shaded line. As one can see, τ_{ag}^{cr} corresponds well to value τ_{ag}, conforming to the transition of the dependences $A_p(\tau_{ag})$ and $d_f(\tau_{ag})$ to asymptotic branch. This allows to determine transitional (nonstationary) period duration τ_{ag}^{cr}

, during which polymer changes structure and properties in physical aging process $\left(\tau_{ag}^{tr} \leq \tau_{ag}^{cr}\right)$.

As it is known [13], the value A_p increases at molecular mobility intensification and the latter is associated with chains mobility in polymer structure loosely packed regions [71]. Within the frameworks of model [7] loosely packed matrix, surrounding clusters is such region. Its relative fraction $\varphi_{l.m.}$ can be determined according to the Eq. (2.4). In Fig. 10.20, the dependence $A_p(\varphi_{l.m.})$ is shown, which has the expected character. In conformity with the data of Ref. [13] A_p linear growth at $\varphi_{l.m.}$ increase is observed and the condition $A_p = 0$ is realized not at $\varphi_{l.m.} = 0$, what was to be expected, but at $\varphi_{l.m.} \approx$ 0.352. This value $\varphi_{l.m.}$ according to the Eq. (2.4) corresponds to $\varphi_{cl} \approx 0.648$, that is close enough to the value ϕ_{cl}^{cr} at $T = 293$ K ($\sim$0.619). This corresponds to the postulate of completely stretched chains at structure quasiequilibrium state, restricting φ_{cl} growth, and completely inhibited owing to this molecular mobility, that defines the condition $A_p = 0$, that is, complete brittleness of polymer [67].

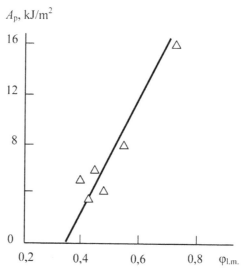

FIGURE 10.20   The dependence of impact toughness A_p on loosely packed matrix relatively fraction $\varphi_{l.m.}$ for PC [67].

Thus, the theoretical description of physical aging process of amorphous polymers (on the example of their typical representative-polycarbonate) within the frameworks of fractal analysis and thermal cluster model. It has been shown that the structure quasiequilibrium state is reached in physical aging process and polymers structure and properties quantitative description possibility at this process course is demonstrated.

KEYWORDS

- fracture surface
- impact toughness
- polymer
- scaling
- strain energy release rate
- superplasticity

REFERENCES

1. Yamamoto, I., Miyata, H., & Kobayashi, T. (1990). Instrumented Impact Test and Evaluation of Fracture Toughness in Polymers. Proceedings "Benibana" Intern. Symposium October 8–11, 1990. Yamagata, Japan, 184–189.
2. Novikov, V. U., & Kozlov, G. V. (1997). The Energy Dissipation Processes in Polymers in Impact Tests Conditions. Materialovedenie, 4, 6–9.
3. Kozlov, G. V., & Novikov, V. U. (1997). The Physical Significance of Dissipation Processes in Impact Tests of Semicrystalline Polymers. Prikladnaya Fizika, 1, 77–84.
4. Kozlov, G. V., Serdyuk, V. D., & Beloshenko, V. A. (1994). Plastic Constraint Factor and Mechanical Properties of High Density Polyethylene at Impact Loading. Mekhanika Kompozithykh Materialov, 30(5), 691–695.
5. Balankin, A. S., Ivanova, V. S., & Breusov, V. P. (1992). A Collective Effects in Metals Fracture Kinetics and Dissipative Structure Fractal Dimension Spontaneous Change at Ductile-Brittle Transition. Doklady AN, 322(6), 1080–1086.
6. Balankin, A. S. (1991). Synergetics of Deformable Body. Moscow, Publishers of Ministry Defence SSSR, 404 p.
7. Kozlov, G. V., Ovcharenko, E. N., & Mikitaev, A. K. (2009). Structure of the Polymers Amorphous State. Moscow, Publishers of the D.I. Mendeleev PKhTU, 392 p.
8. Graessley, W. W., & Edwards, S. F. (1981). Entanglements interaction in polymers and the chain contour concentration. Polymer, 22(10), 1329–1334.
9. Aharoni, S. M. (1983). On Entanglements of Flexible and Rodlike Polymers. Macromolecules, 16(9), 1722–1728.
10. Kozlov, G. V., & Novikov, V. U. (1998). Synergetics and Fractal Analysis of Cross-Linked Polymers. Moscow, Klassika, 112 p.

11. Mashukov, N. I., Serdyuk, V. D., Kozlov, G. V., Ovcharenko, E. N., Gladyshev, G. P., Vodakhov, A. B. (1990). Stabilization and Modification of Polyethylene by Oxygen Acceptors. (Preprint). Moscow, Publishers of IKhF AN SSSR, 64 p.

12. Brown, H. R. (1982). A Model for Brittle-Ductile Transitions in Polymers. J. Mater. Sci., *17(3)*, 469–476.

13. Kausch, H. H. (1978). Polymer Fracture. Berlin, Heidelberg, New York, Springer-Verlag, 435 p.

14. Kozlov, G. V., Temiraev, K. B., Shetov, R. A., & Mikitaev, A. K. (1999). A structural and Molecular Characteristics Influence on Molecular Mobility in Diblock-Copolymers Oligoformal 2,2-di-(4-oxyphenil)-propane-oligosulfone Phenolphthaleine. Materialovedenie, 2, 34–39.

15. Kozlov, G. V., Belousov, V. N., Mokaeva, K. Z., Gazaev, M. A., & Mikitaev, A. K. (1995). The Molecular Weight Influence on Polyarylatesulfone Forced Elasticity Stress. Manuscript deposited to VINITI RAN, Moscow, August 09, 2407–V95.

16. Shogenov, V. N., Kharaev, A. M., & Guchinov, V. A. (1988). The Polyarylatesulfone Mechanical Properties Change in Aging Process. In: Polycondensation Processes and Polymers. Ed. Mikitaev A.K. Nal chik, KBSU, 14–21.

17. Henkel, C. S., & Kramer, E. J. (1984). Crazing and Shear Deformation in Cross-Linked Polystyrene. J. Polymer Sci.: Polymer Phys. Ed, *22(4)*, 721–737.

18. Mashukov, N. I., Gladyshev, G. P., & Kozlov, G. V. (1991). Structure and Properties of High Density Polyethylene Modified by High-Disperse Mixture Fe and FeO. Vysokomolek. Soed. A, *33(12)*, 2538–2546.

19. Belousov, V. N., Kozlov, G. V., & Mikitaev, A. K. (1984). The Correlation of Impact Toughness and Molecular weight of Polyarylatesulfone. Plast. Massy, 6, 62.

20. Mosolov, A. B. (1991). Fractal Griffith Crack. Zhurnal Tekhnicheskoii Fiziki, *67(7)*, 57–60.

21. Novikov, V. U., & Kozlov, G. V. (2001). Polymer Fracture Analysis within the Frameworks of Fracture Concept. Moscow, Publishers MSOU, 136 p.

22. Bucknall, C. B. (1977). Toughened Plastics. London, Applied Science, 318 p.

23. Marshall, G. P., Williams, Y. G., & Turner, C. F. (1973). Fracture Toughness and Absorbed Energy Measurement in Impact Tests on Brittle Materials. J. Mater. Sci., *8(7)*, 949–956.

24. Mai, Y. W., & Williams, J. G. (1977). The Effect of Temperature on the Fracture of two Partially Crystalline Polymers; Polypropylene and Nylon. J. Mater. Sci., *12(11)*, 1376–1382.

25. Plati, E., & Williams, J. G. (1975). Effect of Temperature on the Impact Fracture Toughness of Polymers Polymer, *16(12)*, 915–920.

26. Hine, J., Duckett, R. A., & Ward, I. M. (1981). A Study of the Fracture Behaviour of Polyethyersulfone. Polymer, *22(12)*, 1745–1753.

27. Margolina, A., & Wu, S. (1988). Percolation Model for Brittle-Ductile Transition in Nylon/Rubber Blends. Polymer, *29(12)*, 2170–2173.

28. Kozlov, G. V., Shustov, G. B., & Zaikov, G. E. (2003). A Percolation Model of Brittle-Ductile Transition for Polyethylene. In: Fractal Analysis of Polymers: From Synthesis to Composites. Ed. Kozlov, G. V., Zaikov, G. E., Novikov, V. U. New York, Nova Science Publishers Inc., 99–105.

29. Kozlov, G. V., Shustov, G. B., & Zaikov, G. E. (2002). Percolation Model of Brittle-Ductile Transition for Polyethylene. J. Balkan Tribologic. Assoc. 9(3), 388–393.
30. Kozlov, G. V., Serdyuk, V. D., Mil man, L. D. (1993). The Determination of Polyethylene Forced Elasticity Local Stress in Impact Loading Conditions. Vysokomolek. Soed. B, 35(12), 2049–2050.
31. Sokolov, I. M. (1986). Dimensions and Other Geometrical Critical Exponents in Percolation Theory. Uspekhi Fizicheskikh Nauk, 150(2), 221–256.
32. Bobrychev, A. N., Kozomazov, V. N., Babin, L. O., & Solomatov, V. I. (1994). Synergetics of Composite Materials Lipetsk, NPO ORIUS, 153 p.
33. Bashorov, M. I., Kozlov, G. V., & Mikitaev, A. K. (2009). Nanostructures in Polymers: Formation Synergetics, Requlations Methods and Influence on the Properties Materialovedenie, 9, 39–51.
34. Kozlov, G. V., Belousov, V. N., Mikitaev A. K. (1997). On Possibility of Shear Deformation Zone Realization in Central Part of Massive Polymer Samples. Fizika i Tekhnika Vysokikh Davlenii, 7(4), 107–113.
35. Ivanova, V. S., Balankin, A. S., Bunin, I. Zh., Oksogoev, A. A. (1994). Synergetics and Fractals in Material Science. Moscow, Nauka, 383 p.
36. Mandelbrot, B. B., Passoja, D. F., Pullay, A. J. (1984). Fractal Character of Fracture Surfaces of Metals. Nature, 308(5961), 721–722.
37. Huang, Z. H., Tian, J. F., Wang, Z. G. (1990). A Study of the Slit Island Analysis as a Method for Measuring Fractal Dimension of Fractured Surface. Scripta Metal., 27(7), 967–972.
38. Mikitaev, A. K., Kozlov, G. V. (2008). The Fractal Mechanics of Polymeric Materials. Nal chik, Publishers KBSU, 312 p.
39. Kozlov, G. V., Misra, R. D. K., Aphashagova, Z. Kh. (2009). A Particulate-Filled Polymer Nanocomposites Impact Toughness Structural Model. Nanotekhnika, 2, 71–74.
40. Williford, R. E. (1988). Multifractal Fracture. Scripta Metal, 22(11), 1749–1754.
41. Dolbin, I. V., Kozlov, G. V. (2004). The Structure Fractality and Fracture Energy of High Density Polyethylene. Proceedings of Intern. Sci. Conf. "Young People and Chemistry". Krasnoyarsk, KSU, 241–243.
42. Williford, R. E. (1988). Scaling Similarities between Fracture Surfaces, Energies and Sructure Parameters. Scripta Metal, 22(2), 197–200.
43. Bhattacharya, S. K., Brown, M. (1984). Micromechanisms of Crack Initiation in Thin Films and Thick Sections of Polyethylene. J. Mater. Sci., 19(10), 2767–2775.
44. Plati, E., Williams, J. G. (1975). The determination of the Fracture Parameters for Polymers in Impact. Polymer Engng. Sci., 15, 470–477.
45. Burya, A. I., Kozlov, G. V., Sviridenok, A. I., Malamatov, A. Kh. (1999). The Fractal Characteristics of High Density Polyethylene in High-Speed Fracture Conditions. Doklady NAN Belarusi, 43(1), 117–119.
46. Long, Q. Y., Sugin, L., Lung, C. W. (1991). Studies of the Fractal Dimension of a Fracture Surface Formed by Slow Stable Crack Propagation. J. Phys. D: Appl. Phys., 24(4), 602–607.
47. Malamatov, A. Kh., Kozlov, G. V., Mikitaev, A. K. (2006). To Question about Correlation of Impact Toughness and Fracture Surface Fractal Dimension for Polymers. Deformatsiya i Razrushenie, 10, 46–48.

48. Balankin, A. S., Izotov, A. D., Lazarev, V. B. (1993). Synergetics and Fractal Thermo-dynamics of Inorganic Materials. I. Thermomechanics of Multifractals. Neorganicheskie Metarialy, *29(4),* 451–457.

49. Bagryanskii, V. A., Malinovskii, V. K., Novikov, V. N., Pushchaeva, L. M., Sokolov, A. P. (1988). Inelastic Light Diffusion on Fractal Vibrational Modes in Polymers. Fizika Tverdogo Tela, *30(8),* 2360–2366.

50. Kozlov, G. V., Zaikov, G. E. (2004). Structure of the Polymer Amorphous State. Utrecht, Boston, Brill Academic Publishers, 465 p.

51. Kozlov, G. V., Sanditov, D. S. (1994). Anharmonic Effects and Physical-Mechanical Properties of Polymers. Novosibirsk, Nauka, 261 p.

52. Meakin, (1987). Stress Distribution for a Rigid Fractal Embedded in a Two-Dimensional Elastic Medium. Phys. Rev. A, *36(1),* 325–331.

53. Kozlov, G. V., Sanditov, D. S., Lipatov, Yu. S. (2004). Structural Analysis of Fluctuation Free Volume in Polymer Amorphous State. In: Achievements in Physics-Chemisry Field. Ed. Zaikov, G. E. a.a.Moscow, Khimiya, 412–474.

54. Malkin, A. Ya., Askadskii, A. A., Kovriga, V. V. (1978). The Measurement Methods of Polymers Mechanical Properties. Moscow, Khimiya, 336 p.

55. Kozlov, G. V., Shetov, R. A., Mikitaev, A. K. (1987). The Yield Stress Determination at Polymers Impact Loading by Sharpy. Vysokomolek. Soed. A, *29(9),* 2012–2013.

56. Kozlov, G. V., Mikitaev, A. K. (1988). Brittle and Ductile Fracture of Polyarylate and Polyarylatesulfone in Impact Tests Conditions. In: Polycondensation Processes and Polymers. Ed. Mikitaev, A. K. Nal chik, KBSU, 3–8.

57. Malamatov, A. Kh., Kozlov, G. V., Ligidov, M. Kh. (2006). On Polymers Superplasticity at Impact Loading. Plast. Massy, 6, 18–21.

58. Mikitaev, A. K., Kozlov, G. V., Zaikov, G. E. (2008). Polymer Nanocomposites: Variety of Structural Forms and Applications. New York, Nova Science Publishers Inc., 319 p.

59. Kozlov, G. V. The Dependence of Yield Stress on Strain Rate for Semicristalline Poly-mers. Manuscript deposited in VINITI RAN, Moscow, November 01 2002, 1885-B2002.

60. Matsuoka, S., Bair, H. E. (1977). The Temperature Drop in Glassy Polymers during De-formation. J. Appl. Phys., *48(10),* 4058–4062.

61. Petrie, S. E. B. (1976). The Effect of Excess Thermodynamic Properties versus Struc-ture Formation on the Physical Properties of Glassy Polymers. J. Macromol. Sci. Phys., *B12(2),* 225–247.

62. Wyzgoski, M. G. (1980). Physical Aging of Poly(acrylonitrile-butadiene-styrene). II. Defferential Scanning Calorimetry Measurements. J. Appl. Polymer Sci., *25(7),* 1455–1467.

63. Kozlov, G. V., Beloshenko, V. A., Gazaev, M. A., Lipatov, Yu. S. (1996). The Struc-tural Changes at Heat Aging of Cross-Linked Polymers. Vysokomolek. Soed. B, *38(8),* 1423–1426.

64. Kozlov, G. V., Beloshenko, V. A., Lipatov, Yu. S. (1998). The Fractal Treatment of Gross-Linked Polymers Physical Aging Process. Ukrainskii Khimicheskii Zhurnal, *64(3),* 56–59.

65. Blumen, A., Klafter, J., Zumofen, G. A. (1986). Reactions in Fractal Models of Disor-dered Systems In: Fractals in Physics. Ed. Pietronero L., Tosatti E. Amsterdam, Oxford, New York, Tokyo, North-Holland, 561–574.

66. Kozlov, G. V., Beloshenko, V. A., Varyukhin, V. N., Lipatov, Yu. S. (1999). Application of Cluster Model for the Description of Epoxy Polymer Structure and Properties. Polymer, *40(4)*, 1045–1051.
67. Kozlov, G. V., Dolbin, I. V., Zaikov, G. E. (2004). The theoretical Description of Amorphous Polymers Physical Aging. Zhurnal Prikladnoi Khimii, *77(2)*, 271–274.
68. Kozlov, G. V., Zaikov, G. E. (2001). The Generalized Description of Local Order in Polymers. In: Fractals and Local Order in Polymeric Materials. Kozlov, G. V., Zaikov, G. E., Ed., New York, Nova Science Publishers Inc., 55–63.
69. Family, F. (1984). Fractal Dimension and Grand Universality of Critical Phenomena. J. Stat. Phys., *36(5/6)*, 881–896.
70. Novikov, V. U., Kozlov, G. V. (2000). Structure and Properties of Polymers within the Frameworks of Fractal Approach. Uspekhi Khimii, *69(6)*, 572–599.
71. Bartenev, G. M., Zelenev, Yu. U. (1983). Physics and Mechanics of Polymers. Moscow, Vysshaya Shkola, 391 p.

CREEP

CONTENTS

A solid-phase polymers deformation description is always in the center of attention. Models, considering the indicated polymers as homogeneous (uniform) viscoelastic body or taking into consideration structure heterogeneity on the basis of "effective homogeneity" hypothesis are known. This hypothesis assumes the possibility of heterogeneity consideration with the help of one effective elasticity modulus introduction [1]. Both approaches result to multiparameter integral models, the application of which in practice is difficult enough, particularly in the range of finite strains. Such models complexity is predetermined to a considerable extent by nonlinear hereditary elasticity calculus formal application, which is assumed as an obligatory one at polymer materials deformation analysis [2]. The first from the described approaches was used for polyethylenes and nanocomposites on its basis, filled with ogranoclay, creep processes description [3, 4].

Unlike the approach, averaging mechanically heterogeneous body properties, it has been offered to conduct analysis of polymers deformation by another mode of their structural heterogeneity consideration [5, 6]. Structure heterogeneity or dynamical state in polymer materials bulk results to sharply differing deformation ability of these materials various structural components. Then it can be supposed that exactly deformation heterogeneity of polymer material structure results mainly to its viscoelastic properties nonlinearity [6]. The authors of Ref. [7] conducted the description of polyethylene and nanocomposites on its basis creep process with their structural heterogeneity appreciation within the frameworks of cluster model of polymers amorphous state structure [8].

Within the frameworks of the indicated structural model [8] it is supposed that semicrystalline polymer at all and polyethylene in particular consists of crystalline and amorphous phases, in addition the latter includes local order domains (clusters) and loosely packed matrix, which for polyethylenes in the case of testing at room temperature is in devitrificated state. It is obvious, that the last circumstance defines small elasticity modulus and high compliance of loosely packed matrix [6]. At introduction into polymer matrix organoclay to the mentioned above structural components two another ones are added: actually nanofiller and interfacial regions. Both semicrystalline polymer and nanocomposites on its basis have heterogeneous structure, in which devitrificated loosely packed matrix compliance will be essentially higher than the remaining structural components. This means that at loading firstly loosely packed matrix will be deformed and semicrystalline polymer or nanocomposite on its basis compliance will be defined by this very component content $\varphi_{l.m.}$. In this case the strain ε in creep testing at constant stress can be described according to the following relationship [7]:

$$\varepsilon \sim t^{\varphi_{l.m.}}, \quad (11.1)$$

where t is tests duration.

TABLE 11.1   The Structural and Mechanical Characteristics of HDPE and Nanocomposites on Its Basis [7]

Sample conventional sign	MMT contents, mas. %	HDPE – MA contents, mas. %	E, GPa
A	–	–	0.58
B	1.0	–	0.50
C	5.0	5.0	0.89

MMT is Na$^+$ – montmorillonite of mark cloisite 15A; HDPE – MA is HDPE, grafted by maleic anhydride (compatibilizer); E is elasticity modulus.

Proceeding from the said above, for loosely packed matrix relative fraction $\varphi_{l.m.}$ determination the following equation can be written [9]:

$$\varphi_{l.m.} = 1 - K - \varphi_{cl} - (\varphi_{n} + \varphi_{if}), \quad (11.2)$$

where K is crystallinity degree, φ_{cl}, φ_{n} and φ_{if} are the relative fractions of clusters, nanofiller (organoclay) and interfacial regions, respectively. It is obvious, that the values φ_{n} and φ_{if} are taken into account in the case of nanocomposites only [7].

$$K = 0,32C_{\infty}^{1/3}, \quad (11.3)$$

where C_{∞} is the characteristic ratio, which is polymer chain statistical flexibility indicator [11] and determined according to the Eq. (4.40).

For semicrystalline polymers the clusters relative fraction φ_{cl} is determined according to the following percolation relationship (analog of the Eq. (4.66)) [8]:

$$\varphi_{cl} = 0.03(1 - K)(T_{m} - T)^{0.55}, \quad (11.4)$$

where T_{m} and T are melting and testing temperatures, accordingly. For the considered polymer materials $T_{m} = 408$ K [8].

The nanofiller and interfacial regions total fraction ($\varphi_n + \varphi_{if}$) was estimated with the help of the formula [9]:

$$\frac{E_n}{E_m} = 1 + 11\left(\phi_n + \phi_{if}\right)^{1,7},$$

(11.5)

where E_n and E_m are elasticity moduli of nanocomposite and matrix polymer, respectively.

The conditional sign and composition of the considered polymeric materials are adduced in Table 11.1. Then, having determining structure pliable regions fraction or $\varphi_{l.m.}$ (the devitrificated loosely packed matrix elasticity modulus is approximately by three orders of magnitude smaller than the corresponding characteristic for the remaining structural components [12]), the curves $\varepsilon(t)$ can be described in the case of creep testing with the help of the Eq. (11.1). The value ε is given in relative units in such a way as that proportionality coefficient in the indicated relationship will be equal to one. The theory and experiment comparison is adduced in Fig. 11.1. As it follows from the plots of Fig. 11.1, the Eq. (11.1) describes well the creep experimental curves $\varepsilon(t)$ for both initial HDPE and nanocomposites on its basis. Let us note the principal distinction of approaches, applied by the authors of Ref. [3, 4] in the present work as well. In the last case very simple Eq. (11.1) was used, in which ε is a function of the structure state only. At the same time in Refs. [3, 4] for ε description within the frameworks of Maxwell, Kelvin, Burgers models and so on a large number of parameters was used, which are polymeric material property, but not its structural characteristics. In other words, in Ref. [7] the relation property-structure is obtained, that is the main distinction of the offered method from those used ones earlier [3, 4].

Thus, the structural model for polyethylene and nanocomposites on its basis creep process description is offered, taking into account the indicated polymeric materials structure heterogeneity. This treatment is given within the frameworks of the cluster model of polymers amorphous state structure and shown the good correspondence wit experiment [7].

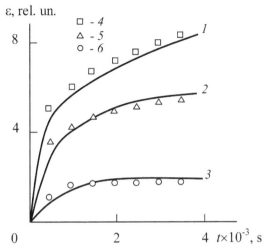

FIGURE 11.1   The dependences of strain ε on creep testing duration *t* for HDPE and nanocomposites on its basis. 1 ÷ 3 – the experimental data for samples A (1), B (2) and C (3); 4 ÷ 6 – calculation according to the Eq. (11.1) for samples A (4), B (5) and C (6) [7].

KEYWORDS

- **creep**
- **interfacial region**
- **nanofiller**
- **polymer**
- **strain**
- **test duration**

REFERENCES

1. Cristensen, R. (1982). The Introduction in Composites Mechanics. Moscow, Mir, 334 p.
2. Goldman, A. Ya. (1979). A Engineering Plastics Strength. Leningrad, Khimiya, 320 p.
3. Ranade, A., Nayak, K., Fairbrother, D., & D'Souza, N. (2005). Maleated and Non-Maleated Polyethylene-Montmorillonite Layered Silicate Blown Films: Greep, Dispersion and Crystallinity. Polymer, *46(21)*, 7323–7333.
4. Pegoretti, A., Dorigato, A., & Penati, A. (2007). Tensile Mechanical Response of Polyethylene-Clay Nanocomposites. EXPRESS Polymer Letters, *1(3)*, 123–131.

5. Averkin, B. A., Volodin, V. P., Kosterina, G. N., & Stepanov, A. B. (1984). Structure Heterogeneity Effect on Polymers Deformation Behavior. Vysokomolek. Soed. B, *26(5)*, 361–364.

6. Averkin, B. A., Egorov, E. A., Zhizhenkov, V. U., Slutsker, A. I., Stepanov, A. B., & Timofeev, V. S. (1989). A Semicrystalline Polymer Dynamical Heterogeneity Effect on their Deformation Mechanics. Vysokomolek. Soed. A, *31(10)*, 2173–2177.

7. Dzhangurazov, B. Zh., Kozlov, G. V., Ovcharenko, E. N., & Militaev, A. K. (2009). The Creep Process Simulation for Nanocomposites Polyethylene/Organoclay. Doklady Adygsk. *(Cherkessk.)* Intern. AN, *11(2)*, 86–89.

8. Kozlov, G. V., & Novikov, V. U. (2001). The Cluster Model of Polymers Amorphous State. Uspekhi Fizicheskikh Nauk, *171(7)*, 717–764.

9. Mikitaev, A. K., Kozlov, G. V., & Zaikov, G. E. (2008). Polymer Nanocomposites: Variety of Structural Forms and Applications. New York, Nova Science Publishers Inc., 319 p.

10. Aloev, V. Z., & Kozlov, G. V. (2002). The Physics of Orientational Phenomena in Polymeric Materials. Nal'chik, Poligrafservis IT, 288 p.

11. Budtov, V. P. (1992). Physical Chemistry of Polymer Solutions. Sankt-Peterburg, Khimija, 384 p.

12. Bartenev, G. M., & Frenkel, S. Ya. (1990). Polymer Physics. Leningrad, Khimiya, 432 p.

CHAPTER 12

MICROHARDNESS

CONTENTS

At present it is known [1–3] that microhardness H_v is the property, susceptible to morphological and structural changes in polymeric materials. For composite materials the availability of the filler, microhardness of which much exceeds corresponding characteristic for polymer matrix, is an additional powerful factor [4]. At pressing into polymer indentors sharpened in the shape of a cone or a pyramid the stressed state is localized in a small enough microvolume and it is supposed, that in such tests polymeric materials real structure is found [5]. In connection with the fact, that polymers structure (including cross-linked ones) is complex enough [6], the question arises, what structure component reacts on the indentor pressing.

The connection of microhardness, determined according to the results of the tests in a very localized microvolume, with such macroscopic characteristics of polymeric materials as elasticity modulus E and yield stress σ_Y, is another aspect of the problem. At present a large enough number of derived theoretically and received empirically relationships between H_v, E and σ_Y exists [7, 8].

The authors of Ref. [9] conducted cross-linked polymers microhardness description within the frameworks of the fractal (structural) models and the indicated parameter intercommunication with structure and mechanical characteristics clarification. The epoxy polymers structure description is given within the frameworks of the cluster model of polymers amorphous state structure [10], which allows to consider polymer as natural nanocomposites, in which nanoclusters play nanofiller role (this question will be considered in detail in chapter fifteen).

The authors of Ref. [9] used the cross-linked epoxy polymers based on diglycidyl ether of bisphenol A (ED-22) of anhydride curing (EP-2) [11]. Two series of EP-2 are used – one of them was cured at atmospheric pressure (EP-2–200) and another at hydrostatic pressure that allowed to obtain 10 samples of EP-2, differing by cross-linked networks topology [11].

Let us consider the intercommunication between microhardness H_v and other mechanical characteristics, in particular yield stress σ_Y, for the studied epoxy polymers. Tabor [12] found for metals, which were considered as perfectly plastic solid bodies, the following relationship between H_v and σ_Y:

$$\frac{H_v}{\sigma_Y} \approx c, \tag{12.1}$$

where c is the constant, which is approximately equal to 3.

The Eq. (12.1) assumes that the applied in microhardness tests pressure under the indentor is higher than the yield stress in quasistatic tests owing to restraints imposed by nondeformed polymer, surrounding the indentor. However, a number of authors of Ref. [3, 7, 8] showed, that the constant c value could differ essentially from 3 and be varied within the wide enough limits of ~1.5÷ 30.

The elasticity role in the indentation process was offered to account for the purpose of analysis spreading on wider range of solids. Hill obtained the following formula for solid body having elasticity modulus E and Poisson's ratio v [7]:

$$H_{\mathrm{v}} = \frac{2}{3}\left[1 + \ln\frac{E}{3(1-\nu)\sigma_Y}\right],\qquad(12.2)$$

and the empirical Marsch equation has the form [7]:

$$H_{\mathrm{v}} = \left(0,07 + 0,6\ln\frac{E}{\sigma_Y}\right)\sigma_Y\qquad(12.3)$$

The Eqs. (12.2) and (12.3) allow to give the ration H_{v}/σ_Y for epoxy polymers at the condition of the known E and σ_Y and the value v can be calculated with the help of the Eq. (2.20) by mechanical tests results.

Let us consider now the physical nature of the ratio H_{v}/σ_Y deviation from the constant $c \approx 3$ in the Eq. (12.1). The structure fractal dimension d_f can be calculated according to the Eq. (1.9) at the natural condition $d = 3$.

The Eqs. (1.9), (2.20) and (12.3) combination allows to obtain fractal variants of Hill and Marsch equations, accordingly [9]:

$$\frac{H_{\mathrm{v}}}{\sigma_Y} = \frac{2}{3}\left[1 + \ln\frac{2d_f}{(4-d_f)(3-d_f)}\right]\qquad(12.4)$$

and

$$H_{\mathrm{v}} = \left(0,07 + 0,6\ln\frac{3d_f}{3-d_f}\right)\sigma_Y\qquad(12.5)$$

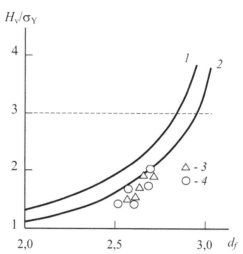

FIGURE 12.1   The dependences of the ratio H_v/σ_Y on structure fractal dimension d_f. 1, 2 – the theoretical calculation according to the Eqs. (12.4) (1) and (12.5) (2); 3, 4 – experimental data for epoxy polymers EP-2 (3) and EP-2–200 (4). The horizontal shaded line indicates Tabor's criterion $c = 3$ [9].

From the Eqs. (12.4) and (12.5) it follows, that the ratio H_v/σ_Y is defined only by solid body structure state, which is characterized by its main characteristic – fractal (Hausdorff) dimension d_f [13]. In Fig. 12.1, the theoretical dependences $(H_v/\sigma_Y)(d_f)$, calculated according to the Eqs. (12.4) and (12.5), are adduced. One can see, that they reveal full similarity, but absolute values H_v/σ_Y, calculated according to the Eq. (12.4), turn out to be on approximately 20% higher than analogous values, obtained according to the Eq. (12.5). The Tabor's condition (the Eq. (12.1) with $c \approx 3$) in the Eq. (12.5) case is reached at $d_f \approx 2.95$, that is, at the greatest fractal dimension value for real solids [14]. This circumstance assumes two consequences. Firstly, the indicated above Tabor's condition is correct for Euclidean solids only. For fractal objects $c < 3$ and in the typical for solids range of $d_f = 2.0 \div 2.95$ the value H_v/σ_Y varies within the limits of $1.5 \div 3.0$. Secondly, the value $c = 3$ reaching at $d_f = 2.95$ assumes the Eq. (12.5) higher precision in comparison with Eq. (12.4). The data of Fig. 12.1 are confirmed by this assumption, where the dependence of the experimental values H_v/σ_Y (points) on d_f value is shown. They correspond well to the curve, calculated according to the Eq. (12.5), but the correspondence with calculation according to the Eq. (12.4) is much worse [9].

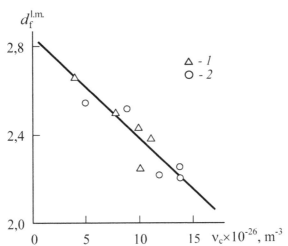

FIGURE 12.2   The dependence of loosely packed matrix structure fractal dimension $d_f^{\text{l.m.}}$ on cross-linking density n_c for epoxy polymers EP-2 (1) and EP-2–200 (2) [9].

It can be assumed that epoxy polymers structure loosely packed matrix sustains the main load in indentor pressing process as the least stiff its component. Its dimension $d_f^{\text{l.m.}}$ can be calculated according to the mixtures rule [9]:

$$d_f = d_f^{\text{cl}}\phi_{\text{cl}} + d_f^{\text{l.m.}}(1 - \phi_{\text{cl}}),\qquad(12.6)$$

where ϕ_{cl} is nanoclusters fractal dimension, ϕ_{cl} is their relative fraction.

The value d_f^{cl} is accepted equal to the greatest for real solids dimension (~2.95 [14]) in virtue of their dense package [10] and the value ϕ_{cl} can be calculated with the help of the Eq. (1.12).

In Fig. 12.2, the dependence of $d_f^{\text{l.m.}}$ on cross-linking density v_c increase results to $d_f^{\text{l.m.}}$ reduction or loosely packed matrix structure loosening. This effect should be reflected on the value H_v, which is confirmed by the data of Fig. 12.3, on which the dependence of H_v on $d_f^{\text{l.m.}}$ for the considered epoxy polymers is adduced. As one can see, the linear correlation $H_v\left(d_f^{\text{l.m.}}\right)$ is obtained, showing H_v growth at $d_f^{\text{l.m.}}$ increase., that is, at loosely packed matrix structure densification. The indicated correlation can be described analytically by the following empirical equation [9]:

$$H_v = 150\left(d_f^{\text{l.m.}} - 1\right),\text{ MPa.}\qquad(12.7)$$

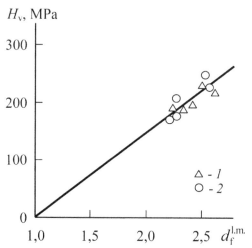

FIGURE 12.3 The dependence of microhardness H_v on loosely packed matrix structure fractal dimension $d_f^{l.m.}$ for epoxy polymers EP-2 (1) and EP-2–200 (2) [9].

The Eq. (12.7) shows, that the condition $H_v = 0$ can be realized for linear object with dimension $d_f = d = 1$ only (in fact for separate stretched linear macromolecule [15]), that is, for real solids, having dimension within the range of 2.0 ÷ 2.95, zero value H_v is unattainable. At minimum value $d_f^{l.m.} = 2.0$ the magnitude $H_v = 150$ MPa, for loosely packed matrix structure with the greatest density, that is, in the case, when nanoclusters and loosely packed matrix structures are undistinguishable, the value $H_v = 293$ MPa.

The loosely packed matrix microhardness $H_v^{l.m.}$ can also be calculated according to the mixtures rule [1, 8]:

$$H_v = H_v^{cl}\phi_{cl} + H_v^{l.m.}\left(1 - \phi_{cl}\right), \tag{12.8}$$

where H_v^{cl} is nanoclusters microhardness, by the indicated above reasons accepted equal to 293 MPa.

In Fig. 12.4, the dependences of H_v and $H_v^{l.m.}$ on K_{st} are adduced. As one can see, at $K_{st} = 0.50$ and 1.50 (the least values $v_c = (4 \div 9) \times 10^{26}$ m^{-3}) the values H_v and $H_v^{l.m.}$ are close and at higher values $v_c = (10 \div 14) \times 10^{26}$ m^{-3} in the case of $K_{st} = 0.75 \div 1.25$ the distinction between H_v and $H_v^{l.m.}$ can reach 80 MPa, that indicates high degree of loosely packed matrix loosening.

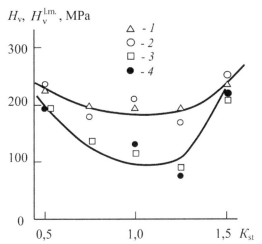

FIGURE 12.4   The dependences of microhardness of epoxy polymers H_v (1, 2) and loosely packed matrix $H_v^{l.m.}$ (3, 4) on the value of curing agent − oligomer ratio K_{st} for EP-2 (1, 3) and EP-2–200 (2, 4).

Hence, in the present chapter the Hill and Marsch equations fractal analogs are obtained, which has shown, that cross-linked epoxy polymers microhardness is defined by their structure only, characterized by its fractal dimension. Tabor's criterion is only fulfilled for Euclidean (or close to them) solids. The cross-linking degree enhancement results to loosely packed matrix loosening and corresponding reduction of cross-linked epoxy polymers microhardness. The similar results obtained for linear polyethylene and nanocomposites on its basis, filled with organoclay [16].

KEYWORDS

- elasticity
- epoxy polymer
- Hill equation
- Marsch equation
- microhardness
- Tabor's criterion

REFERENCES

1. Balta-Calleja, F. J., & Kilian, H. G. (1988). New Aspects of Yielding in Semicrystalline Polymers Related to Microstructure: Branched Polyethylene. Colloid Polymer Sci., *266(1)*, 29–34.
2. Balta-Calleja, F. J., Santa Cruz, C., Bayer, R. K., & Kilian, H. G. (1990). Microhardness and Surface Free Energy in Linear Polyethylene: the Role of Entanglements. Colloid Polymer Sci., *268(5)*, 440–446.
3. Aloev, V. Z., & Kozlov, G. V. (2002). Physics of Orientational Phenomena in Polymeric Materials. Nal chik, Poligrafservise IT, 288 p.
4. Perry, A. J., & Roweliffe, D. J. (1973). The Microhardness of Composite Materials. J. Mater. Sci. Lett, *8(6)*, 904–907.
5. Sanditov, D. S., & Bartenev, G. M. (1982). The Physical Properties of Disordered Structures. Novosibirsk, Nauka, 256 p.
6. Belousov, V. N., Beloshenko, V. A., Kozlov, G. V., & Lipatov, Yu. S. (1996). A Fluctuation Free Volume and Structure of Polymers. Ukrainskii Khimicheskii Zhurnal, *62(1)*, 62–65.
7. Kohlstedt, D. L. (1973). The Temperature Dependence of Microhardness of the Transition Metal Carbides. J. Mater. Sci., *8(6)*, 777–786.
8. Kozlov, G. V., Beloshenko, V. A., Aloev, V. Z., & Varyukhin, V. N. (2000). Microhardness of Ultra-High-Molecular Polyethylene and Componor on its Basis Produces by Solid-Phase Extrusion Method. Fiziko-Khimicheskaya Mekhanika Materialov, *36(3)*, 98–101.
9. Amirshikhova, Z. M., Kozlov, G. V., & Magomedov, G. M. (2010). The Fractal Model of Cross-Linked Polymers Microhardness. Inzhenernaya Fizika, *2*, 33–36.
10. Kozlov, G. V., Ovcharenko, E. N., & Mikitaev, A. K. (2009). Structure of the Polymers Amorphous State. Moscow, Publishers of the D.I. Mendeleev RKhTU, 392 p.
11. Kozlov, G. V., Beloshenko, V. A., Varyukhin, V. N., & Lipatov, Yu. S. (1999). Application of Cluster Model for the Description of Epoxy Polymer Structure and Properties. Polymer, *40(4)*, 1045–1051.
12. Tabor, D. (1951). The Hardness of Metals. New York, Oxford University Press, 329 p.
13. Kuzeev, I. R., Samigullin, G. Kh., Kulikov, D. V., & Zakirnichnaya, M. M. (1997). A Complex System in Nature and Engineering. Ufa, Publishers of USNTU, 225 p.
14. Balankin, A. S. (1991). Synergetics of Deformable Body. Moscow, Publishers of Ministry Defence SSSR, 404 p.
15. Alexander, S., & Orbach, R. (1982). Density of States on Fractals: "Fractons". J. Phys. Lett. (Paris), *43(17)*, L625–L631.
16. Dzhangurazov, B. Zh., Kozlov, G. V., Malamatov, A. Kh., & Mikitaev, A. K. (2010). The Structural Analysis of Nanocomposites Polymer/Organoclay Microhardness. Voprosy Materialovedeniya, *2*, 40–44.
17. Balta-Calleja, F. J., & Kilian, H. G. (1988). New Aspects of Yielding in Semicrystalline Polymers Related to Microstructure: Branched Polyethylene. Colloid Polymer Sci., *266(1)*, 29–34.
18. Balta-Calleja, F. J., Santa Cruz, C., Bayer, R. K., & Kilian, H. G. Microhardness and Surface Free Energy in Linear Polyethylene: the Role of Entanglements. Colloid Polymer Sci., *268(5)*, 440–446.

19. Aloev, V. Z., & Kozlov, G. V. (2002). Physics of Orientational Phenomena in Polymeric Materials. Nal chik, Poligrafservise IT, 288 p.

20. Perry, A. J., & Roweliffe, D. J. (1973). The Microhardness of Composite Materials. J. Mater. Sci. Lett, *8(6)*, 904–907.

21. Sanditov, D. S., & Bartenev, G. M. (1982). The Physical Properties of Disordered Structures. Novosibirsk, Nauka, 256 p.

22. Belousov, V. N., Beloshenko, V. A., Kozlov, G. V., & Lipatov, Yu. S. (1996). A Fluctuation Free Volume and Structure of Polymers. Ukrainskii Khimicheskii Zhurnal, *62(1)*, 62–65.

23. Kohlstedt, D. L. The Temperature Dependence of Microhardness of the Transition Metal Carbides. J. Mater. Sci., 1973, *8(6)*, 777–786.

24. Kozlov, G. V., Beloshenko, V. A., Aloev, V. Z., & Varyukhin, V. N. (2000). Microhardness of Ultra-High-Molecular Polyethylene and Componor on its Basis, Produces by Solid-Phase Extrusion Method. Fiziko-Khimicheskaya Mekhanika Materialov, *36(3)*, 98–101.

25. Amirshikhova, Z. M., Kozlov, G. V., & Magomedov, G. M. (2010). The Fractal Model of Cross-Linked Polymers Microhardness. Inzhenernaya Fizika, *2*, 33–36.

26. Kozlov, G. V., Ovcharenko, E. N., & Mikitaev, A. K. (2009). Structure of the Polymers Amorphous State. Moscow, Publishers of the D.I. Mendeleev RKhTU, 392 p.

27. Kozlov, G. V., Beloshenko, V. A., Varyukhin, V. N., & Lipatov, Yu. S. (1999). Application of Cluster Model for the Description of Epoxy Polymer Structure and Properties. Polymer, *40(4)*, 1045–1051.

28. Tabor, D. (1951). The Hardness of Metals. New York, Oxford University Press, 329 p.

29. Kuzeev, I. R., Samigullin, G. Kh., Kulikov, D. V., & Zakirnichnaya, M. M. (1997). A Complex System in Nature and Engineering. Ufa, Publishers of USNTU, 225 p.

30. Balankin, A. S. (1991). Synergetics of Deformable Body. Moscow, Publishers of Ministry Defence SSSR, 404 p.

31. Alexander, S., & Orbach, R. (1982). Density of States on Fractals: "Fractons". J. Phys. Lett. (Paris), *43(17)*, L625–L631.

32. Dzhangurazov, B. Zh., Kozlov, G. V., Malamatov, A. Kh., & Mikitaev, A. K. (2010). The Structural Analysis of Nanocomposites Polymer/Organoclay Microhardness. Voprosy Materialovedeniya, *2*, 40–44.

CHAPTER 13

THE POLYMERS STRUCTURE AND MECHANICAL PROPERTIES PREDICTION

CONTENTS

The polymeric materials structure and properties prediction is the most important polymer physics goal and the fact, that at present this goal is solved fragmentarily only, speaks on its complexity. Both polymers structure complexity and external factors (testing temperature, strain rate, defects availability, sample sizes and so on) large number, influencing on sample behavior in testing process, are those difficulties, with which scientists are clashed at this problem solution. Strictly speaking, the theoretically valid relationships are necessary for all these factors appreciation, but at present the majority of them are still not obtained. Therefore, further the empirical relationships, having clear physical grounds, will be used. The authors of Ref. [1] offered the methodology of mechanical behavior prediction for amorphous glassy polyarylatesulfone (PASF) film samples, received from different solvents, that allows to vary effectively their structure [2, 3]. For the mentioned goal solution the authors of Ref. [1] used the fractal analysis and cluster model of polymers amorphous state structure.

The polycondensation multiblock copolymer PASF film samples were studied, which were prepared by the method of 5% copolymer solutions in 9 different solvents (N, N-dimethylacetamide, chlorobenzene, dichloroethane, chloroform, methylene chloride, N, N-dimethylformamide, 1,4-dioxane, tetrachloroethane and tetrahydrofuran) pouring of glass substratum. As the data of Ref. [2] have shown, samples preparation by the indicated mode allows to vary their structure, estimated within the frameworks of the cluster model of polymers amorphous state structure [4, 5]. The samples were tensile-stressed with discontinuous strain (the range 0.75%) on the special installation, assembled on optical polarization microscope "Biolam" slide. After each discontinuous strain change the sample region, including a notch and local deformation zone at its tip, was photographed, that allows to determine characteristics of stable crack propagation process (see Fig. 5.2).

Polymers are often enough used as films, which were prepared from polymer solutions. As it is known [6], a solution change results to the essential variations of film samples of the same polymer. Therefore, a film sample structure prediction as a function of solvent characteristics, from which it was prepared, is the goal solution first stage. It is obvious, that the solubility parameter of solvent δ is its characteristic the best choice [7, 8]. The fractal dimension structure d_f was chosen as its characteristic [9], which can be determined according to the Eqs. (1.9) and (2.20).

In Fig. 13.1, the dependence of d_f on δ for 9 used for films preparation solvents is adduced, where the value δ was accepted according to the literary data [10].

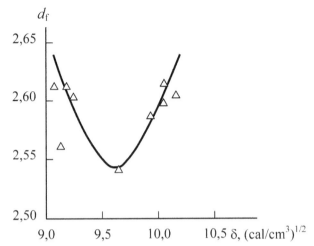

FIGURE 13.1   The dependence of PASF films structure fractal dimension d_f on solubility parameters d of solvent, from which a film was prepared [1].

The polymer solubility parameter δ_p value can be determined by its cohesion energy density W_c as follows [10]:

$$\delta_p = W_c^{1/2}$$

$$(13.1)$$

For PASF $W_c = 385$ MJ/m³ [10], from which δ_p for this polymer is equal to 9.61 (cal/cm³)$^{1/2}$. As it follows from the plot $d_f(\delta)$, adduced in Fig. 13.1, the d_f sharp increase occurs at difference absolute value $|\delta_p - \delta| = |\Delta\delta|$ growth and the least d_f value is reached at the condition $\delta_p = \delta$ ($|\Delta\delta| = 0$) [1]. This allows to obtain the following relationship for considered films value d_f estimation [1]:

$$d_f = 2.542 + 0.128|\Delta\delta|. \qquad (13.2)$$

In Fig. 13.2, the comparison of experimental (i.e., calculated according to the Eqs. (1.9) and (2.20)) and calculated according to the Eq. (13.2) values of polymer films structure fractal dimension is adduced.

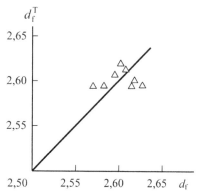

FIGURE 13.2   The relation between experimental d_f and theoretical d_f^T structure fractal dimensions values for PASF film samples. The straight line shows the relation 1:1 [1]

In its turn, the value d_f knowledge allows to determine the value v according to the Eq. (1.9) and proceeding from the last entanglements cluster network v_{cl} can be calculated and excess energy localization regions dimension D_f in polymer, which are for further calculations, can be estimated. The value v_{cl} was estimated according to the Eq. (1.10) and D_f – according to the Eq. (8.7).

As it is known [2, 12], within the frameworks of cluster model the elasticity modulus E value is defined by stiffness of amorphous polymers structure both components: local order domains (clusters) and loosely packed matrix. In Fig. 13.3, the dependences $E(v_{cl})$ are adduced, obtained for tensile tests three types: with constant strain rate, with strain discontinuous change and on stress relaxation. As one can see, the dependences $E(v_{cl})$ are approximated by three parallel straight lines, cutting on the axis E loosely packed matrix elasticity modulus $E_{l.m.}$ different values. The greatest value $E_{l.m.}$ is obtained in tensile tests with constant strain rate, the least one – at strain discontinuous change and in tests on stress relaxation $E_{l.m.} = 0$ [1].

The different relaxation processes proceeding degree for three types of tests, which is due to different temporal scale, is the cause of elasticity modulus such behavior. These processes, as it was to be expected, are realized in loosely packed matrix (more in detail see chapter two). The higher the indicated relaxation degree is (relaxation processes completion) the smaller D_f value is. This allows to approximate the obtained empirically linear correlations $E(D_f)$ by the common for all three tests types relationship [1]:

$$E = E_{l.m.} + 0.94(D_f - 3),\tag{13.3}$$

In Fig. 13.4, the comparison of experimental and calculated according to the Eq. (13.3) elasticity modulus values for PASF is adduced, from which their good conformity follows.

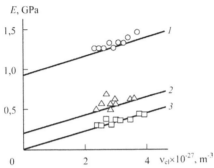

FIGURE 13.3   The dependences of elasticity modulus E on entanglements cluster network density n_{cl} in tests with constant strain rate (1), strain discontinuous change (2) and on stress relaxation (3) for PASF [1].

The strain energy release critical rate G_{Ic}, characterizing polymer local plasticity level (see the Eq. (5.13)), is coupled with material structure. The higher entanglements cluster network density v_{cl} is the more intensively shear local deformation mechanism is realized [13], and the larger G_{Ic} is. This means, that d_f growth should result to G_{Ic} reduction. Actually, for PASF considered samples such correlation was obtained and it can be approximated by the following relationship [1]:

$$G_{Ic} = 38 - 207(d_f - 2.5), \text{ kJ/m}^2. \tag{13.4}$$

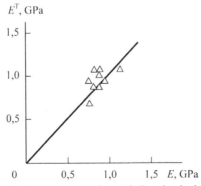

FIGURE 13.4   The relation between experimental E and calculated according to the Eq. (13.3) E^T elasticity modulus values for PASF film samples [1].

In Fig. 13.5, the comparison of experimental and calculated according to the Eq. (13.4) critical strain energy release rate values is adduced, which finds again their good conformity.

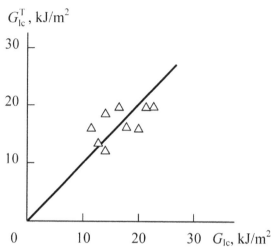

FIGURE 13.5   The relation between experimental G_{Ic} and calculated according to the Eq. (13.4) G_{Ic}^T critical strain energy release rate values for PASF film samples [1].

As it was shown in chapter five, the PASF film samples there found at tension the so-called shear deformation zone (ZD) [3] (see Fig. 5.2), through which the stable triangular crack propagates. This crack in propagation process maintains constant ratio of its opening δ_{cr} to length l_{cr} and in virtue of this is a self-similar object with dimension D_{cr}, determined according to the Eq. (5.9).

The Eqs. (8.1), (8.3) and (8.4) have been obtained above, which allow to estimate samples fracture stress σ_f as a function of D_{cr} and sharp notch length a. As it follows from the Eq. (5.9), the estimation of parameters δ_{cr} and l_{cr} or, more precisely, their critical values δ_c and l_{st} at the moment of instable crack propagation start, that is, sample disastrous fracture is necessary for the value D_{cr} calculation. Within the frameworks of dislocation analogies application for polymers inelastic deformation description for (more details see chapter four) the polymers yield stress value is determined as follows [14]:

$$\sigma_Y = \frac{Gb}{2\pi}\left(l_0 C_\infty v_{cl}\right)^{1/2},\tag{13.5}$$

where G is shear modulus, which can be calculated according to the Eq. (4.58), b is Burgers vector, estimated according to the Eq. (4.7), l_0 is polymer main chain skeletal bond length (for PASF $l_0 = 1.25$ Å [15]), C is characteristic ratio, which can be calculated according to the Eq. (4.40).

The value σ_Y in the case of ZD defines polymer drawing beginning, that is, local plasticity zone formation beginning. Therefore, in practice this stress can be assumed as polymer capacity for work upper boundary [1].

In Fig. 13.6, the comparison of experimental and calculated according to the Eq. (13.5) PASF yield stress values is adduced, which shows good enough these parameters conformity.

In its turn, within the frameworks of elastic fracture linear mechanics the value δ_c is determined as follows [16]:

$$\delta_c = \frac{G_{Ic}}{\sigma_Y}.$$
(13.6)

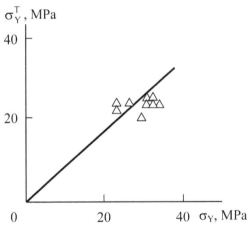

FIGURE 13.6   The relation between experimental s_Y and calculated according to the Eq. (13.5) σ_Y^T draw stress values for PASF film samples [1].

The Eqs. (1.9), (2.20) and (13.6) allow to obtain the following relationship for the value δ_c estimation [10]:

$$\delta_c = \frac{2\pi G_{Ic} d_f}{Eb\left(l_0 C_\infty \nu_{cl}\right)^{3/2}},$$
(13.7)

which is the Eq. (5.1) analog.

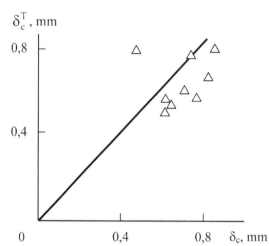

FIGURE 13.7 The relation between experimental d_c and calculated according to the Eq. (13.7) δ_c^T crack critical opening values for PASF film samples [1].

In Fig. 13.7, the comparison of experimental and calculated according to the Eq. (13.7) crack critical opening values is adduced, which shows their satisfactory conformity.

As it is known [16], within the frameworks of elastic fracture linear mechanics the value δ_c can be determined alternatively according to the Dugdale model:

$$\delta_c = \frac{K_{Ic}^2}{\sigma_Y E},\tag{13.8}$$

where K_{Ic} is critical stress intensity factor (stability to crack propagation) and local plasticity zone (ZD) length r_p can be estimated according to the Eq. (5.3).

The Eqs. (5.3) and (13.8) combination gives the relation of parameters δ_c and r_p [1]:

$$r_p = \frac{\pi}{8}\frac{E}{\sigma_Y}\delta_c,\tag{13.9}$$

and the Eqs. (1.9), (2.20) and (13.9) combination allows to obtain the following formula [1]:

$$r_p = 0{,}86\left(\frac{d_f}{3-d_f}\right)\delta_c, \qquad (13.10)$$

which is further used for value r_p calculation.

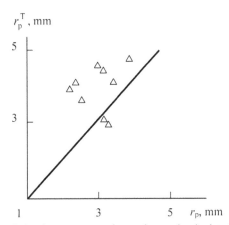

FIGURE 13.8   The relation between experimental r_p and calculated according to the Eq. (13.10) r_p^{T} ZD length values for PASF film samples [1].

In Fig. 13.8, the comparison of experimental and calculated theoretically local plasticity zone length values for PASF film samples is adduced. The data of this plot large enough scattering is due to the difficulty of experimental value r_p precise fixation, which is propagated practically continuously.

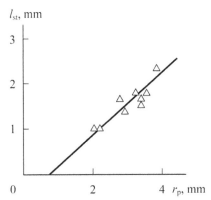

FIGURE 13.9   The dependence of stable crack limiting length l_{st} on ZD length r_p for PASF film samples [1].

In Fig. 13.9, the experimental relation of l_{st} and r_p is adduced. As one can see, the linear growth l_{st} at r_p increase is observed, that is, local plasticity intensification increases sample stability to crack propagation. At $l_{st} = 0$ the value $r_p \approx 0.6$ mm, in other words, at the values $r_p < 0.6$ mm the fracture will be realized by instable crack. The plot of Fig. 13.9 allows to estimate the value l_{st} according to the following empirical relationship [1]:

$$l_{st} = \frac{r_p - 0,4}{1,7}, \text{ mm.} \tag{13.11}$$

In Fig. 13.10, the comparison of experimental and calculated according to the Eq. (13.11) stable crack critical length values is adduced, which shows satisfactory correspondence of theory and experiment.

In Refs. [7, 17] it has been shown, that elasticity modulus E is the linear function of strain ε and at $\varepsilon = 0$ extrapolates to the theoretical limit E_o (see Fig. 3.2). In its turn, the theoretical (maximum) fracture stress σ_{th} of solid, having no defects, is estimated as follows [18]:

$$\sigma_{th} = 0.1 E_o. \tag{13.12}$$

For the considered polymer films two sources of σ_{th} decrease can be distinguished: mechanical stress concentration at stable crack tip and anharmonicity local "splash" owing to material structure modification in prefracture zone – ZD. The first factor can be taken into account by stress concentration coefficient K_s introduction (see the Eq. (5.10)) [7]:

$$K_s = 1 + 2\left(\frac{a + l_{st}}{\delta_c}\right)^{1/2}. \tag{13.13}$$

The another factor is taken into account by the ratio of Grüneisen parameters of nondeformed polymer γ_L and oriented material in ZD γ_{or} [1].

The Grüneisen parameter γ_L, characterizing intermolecular bonds anharmonicity level in polymer, is determined according to the following relationship [19]:

$$\gamma_L \approx 0,74\left(\frac{1 + \nu}{1 - 2\nu}\right). \tag{13.14}$$

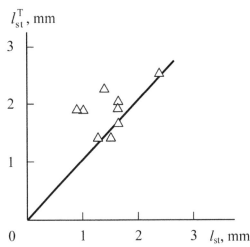

FIGURE 13.10 The relation between experimental l_{st} and calculated according to the Eq. (13.11) l_{st}^T values of stable crack critical length for PASF film samples [1].

The value γ_L with the Eqs. (1.9) and (13.14) appreciation can be determined as follows [4]:

$$\gamma_L \approx 0,37\left(\frac{d_f}{3-d_f}\right). \tag{13.15}$$

As oriented polymers studies showed [20], for them the value v at drawing ratio growth reached the magnitude ~0.425 very fast and further remains practically constant. Therefore, calculation with the Eq. (13.14) using gives the value $\gamma_{or} \approx 7.03$. Thus, the theoretical estimation of film samples fracture stress σ_f can be fulfilled as follows [21]:

$$\sigma_f = \frac{0,1E_0\gamma_L}{K_s\gamma_{or}} \tag{13.16}$$

The value E_0 for PASF was accepted equal to 8.2 GPa [17].

In Fig. 13.11, the comparison of experimental and calculated according to the Eq. (13.16) polymer films fracture stress values is adduced, which shows a good correspondence of theory and experiment.

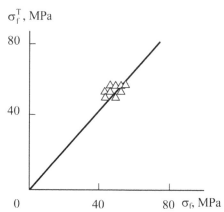

FIGURE 13.11   The relation between experimental s_f and calculated according to the Eq. (13.16) σ_f^T fracture stress values for PASF film samples [1].

Since PASF thin films are deformed in plane stress conditions, then the value K_{Ic} can be estimated according to the equation [16]:

$$K_{Ic}^2 = \frac{1,12\pi\left(a+l_{st}\right)\sigma_f^2}{\left(1-\nu^2\right)}.$$ (13.17)

In Fig. 13.12, the comparison of experimental and calculated according to the Eq. (13.17) critical stress intensity factor values is shown, which demonstrates the indicated parameters good conformity.

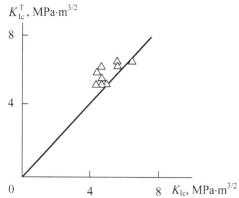

FIGURE 13.12   The relation between experimental K_{Ic} and calculated according to the Eq. (13.17) K_{Ic}^T critical stress intensity factor (stability to crack propagation) values for PASF film samples [1].

Let us pay attention in conclusion to two important, in our opinion, circumstances: firstly, it has been found out (Fig. 13.13), that the strain of stable crack advancement beginning ε_b grows at v_{cl} increase according to the following empirical approximation [1]:

$$\varepsilon \approx 1.62 \times 10^{-29} v_{cl}. \tag{13.18}$$

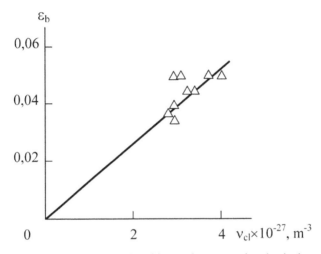

FIGURE 13.13 The dependence of stable crack propagation beginning strain e_b on entanglements cluster network density n_{cl} for PASF film samples [1].

And at last, for fractal cracks the following empirical relationship is valid [22]:

$$G_{Ic} \sim l_{st}^{D_c - 1} \tag{13.19}$$

As it follows from the plot of Fig. 13.14, the Eq. (13.19) is valid for PASF considered film samples that confirms the correctness of the fractal analysis application for their fracture description [1].

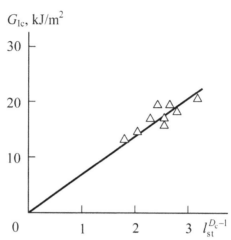

FIGURE 13.14 The fractal relation between critical strain energy release rate G_{Ic} and critical stable crack length l_{st}, corresponding to the Eq. (13.19), for PASF film samples [1].

TABLE 13.1 The Experimental D_{cr} and Theoretical D_{cr}^{T} Values of Crack Fractal Dimension for PASF Films, Prepared For Polymer Solutions in Different Solvents [1]

Solvent	D_{cr}	D_{cr}^{T}
N, N-dimethylacetamide	1.75	1.71
Chlorobenzene	1.69	1.72
Dichloroethane	1.64	1.72
Chloroform	1.71	1.72
N, N-dimethylformamide	1.61	1.70
1,4-dioxane	1.72	1.71
Tetrachloroethane	1.81	1.69
Methylene chloride	1.82	1.75
Tetrahydrofuran	1.73	1.71

The values D_{cr}, adduced in Table 13.1, demonstrate that all considered PASF films should be possessed by the stationary crack self-stopping property (see the Eq. (8.4)).

The experimental confirmation of this postulate is adduced in Fig. 13.15 for PASF film sample, prepared from polymer solution in methylene chloride (the experimental value D_{cr} = 1.82, the theoretical one D_{cr}^T = 1.75). As one can see, the fracture stress σ_f growth at sharp notch length a enhancement is observed.

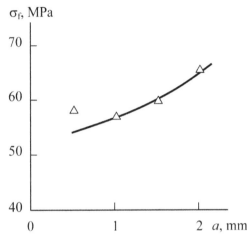

FIGURE 13.15   The dependence of fracture stress s_f on sharp notch length a for PASF film sample, prepared from solution in methylene chloride [1].

In Table 13.2, the greatest relative errors Δ for each parameter are adduced, in addition all the indicated parameters were divided into two groups – geometrical ones (δ_c, r_p and l_{st}) and also structural and mechanical ones. One can see that the first group parameters error in general is larger than the second one. This circumstance is not to many theoretical treatment mistakes, as to the difficulty of geometrical parameters precise measurement directly before the film fracture owing to both ZD and stable crack fast propagation. For the remaining parameters errors are overstated owing to the usage of the data for only one sample. This circumstance is due to experiment large duration.

TABLE 13.2   The Greatest Relative Errors D of Determining Parameters [1]

The group of parameters	Parameter	Δ, %	Δ$_{av}$, %
Geometrical	l_{st}	33	
parameters	r_p	31	29
	δ_c	23	
	d_f	0.8	
Structural and	E	22	
mechanical parameters	G_{Ic}	18	16
	σ_Y	23	
	σ_f	13	
	K_{Ic}	20	

Hence, the adduced in the present chapter techniques can be considered as the physical base for computer prediction and simulation of film polymer samples at quasistatic tension. It is important to emphasize that in theoretical calculation not a single experimentally determined parameter is not used, but the initial characteristics (δ, δ_p or W_c, E_0, l_0) are taken from literary sources [1].

KEYWORDS

- mechanical properties
- polymer
- prediction
- stable crack
- stress concentration
- structure

REFERENCES

1. Shogenov, V. N., Kozlov, G. V., & Mikitaev, A. K. (2007). The Prediction of Mechanical Behavior, Structure and Properties of Film Polymer Samples at Quasistatic Tension. In: Polycondensation Reactions and Polymers. Nal'chik, Publishers KBSU, 252–270.
2. Shogenov, V. N., Belousov, V. N., Potapov, V. V., Kozlov, G. V., & Prut, E. V. (1991). The Glassy Polyarylatesulfone Curves Stress-Strain Description within the Frameworks of High-Elasticity Concepts. Vysokomolek. Soed. A, *33(1)*, 155–160.
3. Kozlov, G. V., Beloshenko, V. A., Shogenov, V. N., & Lipatov, Yu. S. (1995). Local Deformation of Polyarylatesulfone Films. Doklady NAN Ukraine, *5*, 100–102.
4. Kozlov, G. V., & Novikov, V. U. (2001). The Cluster Model of Polymers Amorphous State. Uspekhi Fizicheskikh Nauk, *171(7)*, 717–764.
5. Kozlov, G. V., & Zaikov, G. E. (2004). Structure of the Polymer Amorphous State. Utrecht, Boston, Brill Academic Publishers, 465 p.
6. Budtov, V. P. (1992). Physical Chemistry of Polymer Solutions. Sankt-Peterburg, Khimiya, 384 p.
7. Kozlov, G. V., & Sanditov, D. S. (1994). Anharmonic Effects and Physical-Mechanical Properties of Polymers. Novosibirsk, Nauka, 261 p.
8. Kozlov, G. V., Temiraev, K. B., & Kaloev, N. I. (1998). The Solvent Nature Influence on Polyarylate Structure and Formation Mechanism in the Low-Temperature Polycondensation Conditions. Doklady AN, *362(4)*, 489–492.
9. Kuzeev, I. R., Samigullin, G. Kh., Kulikov, D. V., & Zakirnichnaya, M. M. (1997). A Complex Systems in Nature and Engineering Ufa, Publishers USTTU, 225 p.
10. Wiehe, I. (1995). A. Polygon Mapping with Two-Dimensional Solubility Parameters. Ind. Engng. Chem. Res. *34(2)*, 661–673.
11. Kozlov, G. V., Dolbin, I. V., Mashukov, N. I., Burmistr, M. V., & Korenyako, V. A. (2001). A Macromolecular Coils in Diluted Solutions Fractal Dimension Prediction on the Basis of Two-Dimensional Solubility Parameter Model. Voprosy Khimii i Khimicheskoi Tekhnologii, *6*, 71–77.
12. Beloshenko, V. A., Kozlov, G. V., Stroganov, I. V., & Stroganov, V. F. (1994). Properties of Loosely-Packed Matrix of Epoxidiane Polymers, Modified by Adamantane Carbone Acids. Fizika i Tekhnika Vysokikh Davlenii, *4(3–4)*, 113–117.
13. Henkee, C. S., & Kramer, E. J. (1984). Crazing and Shear Deformation in Cross-Linked Polystyrene. J. Polymer Sci.: Polymer Phys. Ed., *22(4)*, 721–737.
14. Belousov, V. N., Kozlov, G. V., Machukov, N. I., & Lipatov, Yu. S. (1993). The Application of Dislocation Analogues for Yielding Process Description in Crystallizable Polymers. Doklady AN, *328(6)*, 706–708.
15. Aharoni, S. M. (1983). On Entanglements of Flexible and Rodlike Polymers. Macromolecules, *16(9)*, 1722–1728.
16. Bucknall, C. B. (1977). Toughened Plastics. London, Applied Science, 318 p.
17. Kozlov, G. V., Shetov, R. A., & Mikitaev, A. K. (1987). Methods of Elasticity Modulus Measurement in Polymers Impact Tests. Vysokomolek. Soed. *29(5)*, 1109–1110.
18. Bartenev, G. M., & Razumovskaya, I. V. (1960). The Theoretical Strength and Critical Fracture Stress of Solids. Doklady AN SSSR, *133(2)*, 341–344.
19. Sanditov, D. S., & Bartenev, G. M. (1982). The Physical Properties of Disordered Structures. Novosibirsk, Nauka, 256 p.

20. Beloshenko, V. A., Kozlov, G. V., Slobodina, V. G., Prut, E. V., & Grinev, V. G. (1995). Thermal Shrinkage of Ultra-High-Molecular Polyethylene and Polymerization filled Compositions on its Basis. Vysokomolek. Soed. B, *37(6)*, 1089–1992.
21. Shogenov, V. N., Kozlov, G. V., & Mikitaev, A. K. (1989). Prediction of Rigid-Chain Polymers Fracture Process Parameters. Vysokomolek. Soed. B, *31(11)*, 809–811.
22. Goldstein, R. V., & Mosolov, A. B. (1991). A Crack with Fractal Surfaces. Doklady AN SSSR, *319(4)*, 840–844.

CHAPTER 14

FRACTAL MECHANICS OF ORIENTED POLYMERS

CONTENTS

A polymers orientation by uniaxial tension can be accompanied by essential volume changes. These changes are realized by different mechanisms, defined by polymeric material structure. So, at uniaxial drawing of semi-crystalline high density polyethylene (HDPE) volume change up to 30% was observed, which is due to cracks formation, oriented perpendicularly to tension direction [1]. At solid-phase extrusion of polymerization-filled compositions (componors) on the basis of ultra-high-molecular polyethylene (UHMPE) volume increase is due to interfacial boundaries polymer matrix-filler breakdown and it can be reached of ~10% [2]. However, in the case of melt uniaxial tension volume changes are nit observed. It has been assumed [1] that in this case at melt orientation macromolecules high mobility and structure ordered elements absence are ensured viscous medium tension without microvoids formation and such oriented melt crystallization results to formation of the system, not consisting of pores and other continuum interruptions.

The authors of Ref. [3] substantiated in very general terms and with fractal analysis methods application the described above behavior of polymer materials at uniaxial tension.

As Balankin shown [4], the relative change of excitation region volume in deformed body can be presented in the form:

$$\delta V_e = \left(1 - v_e\right)\frac{\sigma_{or}}{E_{or}} = \left(1 - 2v\right)\frac{\sigma}{E} \pm \delta V_{rel} + \delta V_d, \qquad (14.1)$$

where δV_e is excitation region volume, v and v_e are Poisson's ratio for initial and oriented polymers, respectively; σ and σ_{or} are fracture stress, E and E_{or} are elasticity modulus for initial and oriented polymers, accordingly. The first member in right part of the Eq. (14.1) is connected with elastic strains, the second one – with stress relaxation by plastic deformation, the third one – with micro, meso- and macrodefects formation. If the defects accumulation results always to volume increase, then volume change, coupled with plastic deformation, has sigh, opposite one to elastic component: "minus" at $\sigma > 0$ and "plus" at $\sigma < 0$ (compression stress).

The dependence of oriented polymer Poisson's ratio v_e on the parameter

$$\Delta = \frac{\delta V_{rel} \mp V_d}{\delta V_e} \qquad (14.2)$$

has the following form [4]:

$$V_e = \frac{v + 0,5\Delta}{1 + \Delta}. \tag{14.3}$$

The value of v for polymer melt can be estimated as equal to 0.5. It is obvious, that in this case $v_e = v$, since Poisson's ratio value cannot exceed of 0.5 (Le Chatelier-Brau principle [4]). Then from the Eq. (14.3) $\Delta = 0$ follows and from the Eqs. (14.1) and (14.2) the condition follows [3]:

$$-\delta V_{rel} + \delta V_d = 0. \tag{14.4}$$

In other words, the fractal analysis predicts volume changes absence for any body with $v = 0.5$ (including polymer melt). Owing to compensation of volume change, induces micro, meso- and macrodefects formation, by volume change, induced by plastic deformation [3].

The situation changes in the case of solid-phase (semicrystalline) polymer uniaxial drawing. As the experimental estimations shown [5], the Poisson's ratio value for initial polymeric materials (componors UHMPE-Al and UHMPE-bauxite) $v \approx 0.36$ and for these materials extrudates with draw ratio $\lambda \geq 3-v_e \approx 0.43$. From the Eq. (14.3) it follows that $\Delta \approx 0.857$. This means componors volume obligatory increase, expressed in cracks formation on interfacial boundaries polymer matrix-filler [3]:

$$\delta V_{rel}\Delta = -\delta V_{rel} + \delta V_d. \tag{14.5}$$

Since $\Delta \approx 0.857$ and $\delta V_e \approx 0.1$, then from the Eq. (14.5) it follows that in this case plastic deformation cannot be compensated volume increase, which is due to formation of defects – cracks on interfacial boundaries.

One more important aspect is necessary to note. From the Eq. (14.3) it follows, that in any case $\delta V_e \approx \delta V$. The opposite sigh of the inequality ($v_e < v$) means volume decreasing or polymeric material densification that is impossible [6].

The fractal dimension of initial d_f and oriented d_f^{or} polymeric materials are connected one another by the following relationship [4]:

$$d_f^{or} = \frac{d_f + 3\Delta}{1 + 3\Delta}. \tag{14.6}$$

As it was noted above, the value Δ is always larger or equal to zero. This means, according to the Eq. (14.6), solid-phase polymeric material structure fractal dimension increase at uniaxial drawing, which is confirmed experimentally [5, 7].

Thus, the stated above results demonstrated, that volume changes availability or absence in uniaxial tension process is due to structure type. If the structure is Euclidean object (dimension $d = 3$, $v = 0.5$ [8]), then volume changes are absent, if it is fractal object ($2 \leq d_f < 3.0 \leq v < 0.5$ [7]), then volume changes are obligatory [3].

In Ref. [5] it has been shown, that in solid-phase extrusion process of polymerization-filled compositions on the basis of UHMPE structure sharp change occurs, characterized by its fractal dimension d_f, within the range of extrusion draw ratio $\lambda = 1 \div 3$, after that up to $\lambda = 9$, the value d_f remains practically invariable. The same compositions structural changes analysis shows, that sharp drop of local order regions (clusters) relative fraction $\varphi_{l.m.}$ change in the range of $1 \div 9$ [7] and crystallinity degree K linear growth in the same range of λ [9] are observed. Therefore, the question arises on mechanisms, defining such cardinal structural changes of polymerization-filled compositions extrudates and, as consequence, their properties changes [2, 10]. In Ref. [11] the indicated mechanisms analysis was fulfilled within the frameworks of deformable body synergetics and fractal analysis on the example of two polymerization-filled compositions – UHMPE-Al and UHMPE-bauxite.

As it has been noted above, in the indicated compositions solid-phase extrusion process the sharp drop of φ_{cl} (from ~0.26 up to ~0.05) within the range of $\lambda = 1 \div 3$ is observed, that results to corresponding d_f increase from 2.63 up to ~2.87 [5]. As it is known [12], between the parameters d_f and φ_{cl} the intercommunication exists, expressed by the Eq. (1.12). In Fig. 14.1 the dependences, calculated according to the Eqs. (1.9) and (1.12) comparison is shown. As one can see, the good conformity between them is observed, that confirms the made above conclusion – local order regions decay (φ_{cl} decrease) in solid-phase extrusion process is the d_f growth cause [11].

As it was shown earlier, both clusters [13] and crystallites [14] are dissipative structures (DS). In Fig. 14.2, the dependences of crystallinity degree K on λ for both indicated compositions are adduced. The essential K increase at λ growth is observed. So, for compositions UHMPE-Al the value K increases almost twice at λ change from 1 up to 9. The plots of Figs. 14.1 and 14.2 show that the DS type change from local order regions to crystalline regions occurs in solid-phase extrusion process, in addition this process is expressed most strongly within the range of $\lambda = 1 \div 3$. Therefore, it follows to suppose, that shown in Fig. 14.1 sharp d_f increase within the indicated λ range is due to DS type spontaneous change at polymers structure critical state achievement in excess energy cumulation, when shape change cannot be compensated by volume change [15]. This conclusion is confirmed by the data of Ref. [7], where it has been shown, that loosely packed matrix relative fraction $\varphi_{1.m.}$ reaches it's the greatest value (~0.39) exactly at $\lambda = 3$. The further growth $\varphi_{1.m.}$ impossibility defines DS type change in solid-phase extrusion process. In its turn, DS type change defines deformation mechanism change from brittle to ductile one. As it is known [15], deformation character within the frameworks of multifractal analysis is defined by the correlation dimension d_c value, controlling energy dissipation level: at brittle and quasibrittle deformation type $1 < d_c \leq 2$ and at ductile one $- 2 \leq d_c \leq 3$. The boundary value λ between indicated deformation types can be estimated from the following considerations. The d_c value at brittle deformation type is connected with relative transverse strain ψ by the relationship [15]:

$$d_c = \frac{1 - \psi}{\psi}. \tag{14.7}$$

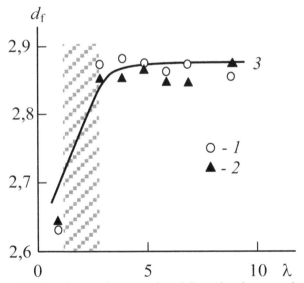

FIGURE 14.1 The dependences of structure fractal dimension d_f on extrusion draw ratio λ, calculated according to the Eq. (1.9) (1, 2) and the Eq. (1.12) (3) for UHMPE-Al (1, 3) and UHMPE-bauxite (2, 3). The shaded region shows the λ range, corresponding DS spontaneous change [11].

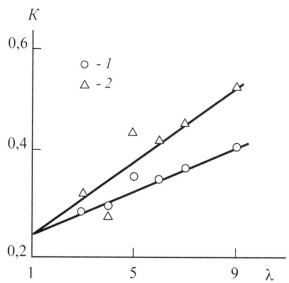

FIGURE 14.2   The dependences of crystallinity degree K on extrusion draw ratio l for UHMPE-Al (1) and UHMPE-bauxite (2) [11].

For the boundary value $d_c = 2$ (brittle deformation) let us obtain from the Eq. (14.7) the greatest value $\psi_m = 0.333$. In its turn, the longitudinal boundary strain ε_m can be estimated as follows [15]:

$$\varepsilon_m = \frac{1}{d_c} = \frac{\psi_m}{1 - \psi_m}. \tag{14.8}$$

At $d_c = 2$ ($\psi_m = 0.333$) the ε_m value is equal to 0.5 or, with the Eq. (4.39) appreciation, $\lambda_m = 1.5$. The greatest value ψ_m, corresponding to the DS type change, is equal to 0.65 [15]. Then let us obtain from the Eq. (14.8) $\varepsilon_m = 1.86$ or $\lambda_m = 2.86$. Thus, the DS type change range was estimated as $\lambda_m \approx 1.50 \div 2.86$ and it is shown in Fig. 14.1 by the shaded region. It is easy to see, that it corresponds well to the range of compositions structure fractal dimension d_f sharp change.

As it has been noted earlier [10], the indicated compositions extrudates structure at $\lambda \geq 3$ is characterized by high values of Poisson's ratio ν ($\nu \approx 0.43$), which are equal to values ν_Y at polymers yielding [16], and small stability to shear strain. Such structure state was defined as "freezed yielding" [10]. This circumstance allows to suppose, that a typical for plastic deformation state characterized by the criterion (4.44), is reached in solid-phase extrusion process.

At plastic or ductile deformation mechanism the value d_c can be determined as follows [15]:

$$d_c = \frac{2}{3\psi(1-\psi)} . \tag{14.9}$$

At the criterion (4.44) fulfillment $d_c = d_f = 2.87$ and at $\lambda \geq 3$ and $\psi = 0.78$ according to the criterion (14.9). This corresponds well to the ductile deformation achievement condition at $\psi > 0.65$ [15].

And at last, let us estimate the value of limiting extrusion draw ratio λ_{lim}, which can be reached in solid-phase extrusion process, on the basis of considered above model. The extrusion draw ratio λ there changed at expense of different diameter dies usage and calculated according to the formula [18]:

$$\lambda = \frac{d_b^2}{d_d^2} , \tag{14.10}$$

where d_b, d_d are billet and calibrating die section diameters, respectively. In its turn, the value ψ can be written as follows [11]:

$$\psi = \frac{d_b - d_d}{d_b} . \tag{14.11}$$

Assuming the limiting value $d_c = 3$, from the Eq. (14.9) $\psi = 0.79$ and from the Eq. (4.12) $d_b = 4.76 d_d$ can be obtained. Let us obtain the limiting value $\lambda_{lim} = 22.7$ according to the Eq. (14.10). This value is close to limiting values of λ for considered extrusion mode, cited in Ref. [19], if a billets preparation special methods are not applied.

Hence, the stated above results shown, that sharp structure change in solid-phase extrusion process of polymerization-filled compositions UHMPE-Al and UHMPE-bauxite is due to deformation mechanism change from brittle to ductile one. In its turn, the indicated mechanism change is induced to dissipative structures type spontaneous change at achievement of criterion, when shape change cannot be compensated by volume change. The usage of deformable solid body synergetics and fractal analysis methods allows to estimate limiting draw ratio in solid-phase extrusion process [11].

The reduction of elasticity modulus E of oriented (extruded) amorphous polymers in comparison with initial polymer is there main feature [20].

The authors of Ref. [21] supposed, that in orientational drawing process of poly(metyl methacrylate) (PMMA) the following structure changes occur: the transition to more equilibrium structure owing to molecular package improvement and internal stresses relaxation. The quantitative structural model absence not allows the authors of Ref. [21] to give direct proofs of their suppositions. In Ref. [22] such treatment was fulfilled on the example of extruded amorphous polyarylates DV and DF-10 with the cluster model of polymers amorphous state structure using [12, 23].

In Fig. 14.3, the dependence of macromolecular entanglements cluster network density v_{cl}, which is as a first approximation measure of segments number in local order regions (clusters) and, hence, measure of such order degree, on extrusion draw ratio λ for polyarylates DV and DF-10 is adduced. The corresponding value v_{cl} of sample, prepared by pressing under pressure, was accepted as the value v_{cl} for nonoriented DV sample [24]. As one can see, polyarylates extrusion results to essential (more than in 1.5 times) sample ordering degree increase, in addition this effect is reached even at $\lambda \approx 2$ and a further λ growth the value v_{cl} is practically not changed.

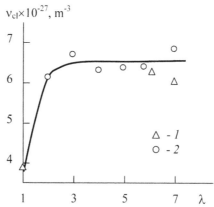

FIGURE 14.3 The dependence of entanglements cluster network density v_{cl} on extrusion draw ratio λ for polyarylates DV (1) and DF-10 (2) [22].

The increase of v_{cl} according to the Eq. (1.11) means φ_{cl} growth and φ_{cl} enhancement according to the Eq. (1.12) assumes d_f reduction. Thus, the comparison of the plots of Fig. 14.1 and 14.3 assumes, that for semicrystalline and amorphous glassy polymer λ growth results to structure diametrically opposite changes – for the first this is d_f enhancement, for the second – its reduction.

In Ref. [25], it has been shown that the value E is defined by contributions by both clusters and loosely packed matrix. This fact is reflected in Fig. 14.4 by the dependence $E(v_{cl})$ for DV film samples, plotted according o the data of Ref. [25]. However, the dependence of relaxation modulus $E = 0$ at $v_{cl} = 0$. This shows, that stress relaxation is realized completely in loosely packed matrix of amorphous glassy polymer (see also Figs. 2.5 and 2.6). If the data $E(v_{cl})$ for extrudates DV and DF-10 traced on the plots of Fig. 14.4, then it turns out that they lie on the straight line $E(v_{cl})$, but not on $E(v_{cl})$. The last circumstance assumes stress relaxation in loosely packed matrix of extrudates DV and DF-10. Let us note, that the adduced above results explained causes of that fact, that extrudates elasticity modulus is smaller and yield stress σ_Y is larger than for the same polymer film samples [24]. Reduction E for extrudates is due to mentioned above stress relaxation in loosely packed matrix and, hence, its contribution in value E disappearance (Fig. 14.4). The σ_Y growth is due to v_{cl} higher values for extrudates in comparison with film samples (Fig. 14.3), since the value σ_Y is defined by values v_{cl} only and is independent on loosely packed matrix properties (the Eq. (13.5)). The indirect confirmation of this conclusion follows from the data of Ref. [26], where polycarbonate samples tests, extrudated at different extrusion temperatures T_e, shown, that the smallest E values were obtained at $T_e > T_g$, in addition in this case the value E of extruded polycarbonate proves to be lower than initial sample elasticity modulus.

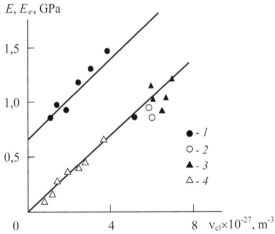

FIGURE 14.4   The dependences of elasticity modulus E (1, 2, 3) and relaxation modulus E_∞ (4) on entanglements cluster network density v_{cl} for film samples DV (1, 4) and extrudates DV (2) and DF-10 (3) [22].

Thus, the stated above results assumed that structure ordering degree increase and stresses relaxation in loosely packed matrix are main structural changes at polyarylates solid-phase extrusion (at $T_e > T_g$).

In connection with fractal geometry ideas [27] in modern physics active and numerous attempts are undertaken to explain the dependences of type:

$$\Phi(Z) = A_{v_{fr}} Z^{-v_{fr}} ,$$ (14.12)

meeting in different fields of natural sciences. In the Eq. (14.12), v_{fr} is some fractional power exponent, $A_{v_{fr}}$ is the constant, depending on v_{fr}, Z is some intensive variable (time, temperature, stress and so on). Although the mathematics of fractional calculus at present is well elaborated [28, 29], wide application of fractional integrals and derivatives is restrained owing to one simple cause-they have no clear physical interpretation. If their clear physical interpretation were found approximately the same, as the usual integrals and derivatives have, then it would undoubtedly enlarge their application field in physical [30]. Therefore, one of goals of the Refs. [31, 32] is physical interpretation of fractional exponent v_{fr} in the polymers case.

At present the tendency of polymers mechanics main principles revision is marked. One of the intensively developing trends is connected with fractal conception using [34]. The wide field of this conception in physics different branches is due to two features. The first is connected with using of notions of Hausdorff-Bezikovich fractional dimension geometry. This helped to describe adequately systems with complex spatial structure, what cannot be done within the frameworks of Euclidean geometry. The second feature is connected with fractional integration and differentiation calculus using [35].

Within the frameworks of this formalism to account for consistently nonlinear phenomenon complex nature is a success, such as memory effects and spatial correlations. In addition the earlier known solutions are not only reproduced, but their nontrivial generalization is given. Another important feature is connected with fractal structures self-similarity using. Unlike the traditional methods of system description on the basis of averaging different procedures, when microscopic level "erasing" occurs, in fractal conception medium self-affine structure and thus within the frameworks of this conception system micro and macroscopic description levels are united. Exactly such method is important for complex multicomponent systems, discovered far from thermodynamic equilibrium state [35], which are polymers [12]. The authors of Refs. [31, 32] are attempted two indicated trends combination.

The authors of Ref. [36] offered to use for polymers stress-strain (σ–ε) curves description the following general model of viscoelastic body, based on the fractional order derivatives, which has the appearance:

$$\sigma + \sum_{j=1}^{m} b_j D_{l_i}^{\alpha_j} \sigma = E_0 \varepsilon \sum_{j=1}^{m} E_j D_{l_i}^{\beta_j} \varepsilon , \qquad (14.13)$$

where $\sigma = \sigma$ (*t*), $\varepsilon = \varepsilon$ (t) are stress and strain in time t moment, b_j, E_j, α_1, β_1 are assigned values and D_{l_i} is an operator of fractional differentiation of order v_{fr} [36].

After a number of assumptions the authors of Ref. [36] have come to the conclusion, that the yield part of the curve $\sigma - \varepsilon$ can be described with the help of fractional differentiation calculus. For this purpose the following equation was used [36]:

$$\sigma = D_{l_i}^{v_{fr}} \varepsilon . \qquad (14.14)$$

However, in Ref. [36] although the good conformity of theory and experiment was obtained, but parameters D_{l_i} and v_{fr} were not identified within the frameworks of polymers structure or properties. Therefore, the goals propounded above were solved on the example of yield process description of polymerization-filled compositions (componors) on the basis UHMPE, prepared by solid-phase extrusion method [2].

At solid body deformation the heat flow is formed, which is due to deformation. The thermodynamics first law establishes that the internal energy change in sample dU is equal to the sum of work dW, carried out on a sample, and the heat flow dQ into sample (see the Eq. (4.31)). This relation is valid for any deformation, reversible or irreversible. There are two thermodynamically irreversible cases, for which $dQ = -dW$; uniaxial deformation of Newtonian liquid and ideal elastoplastic deformation. For solid-phase polymers deformation has an essentially different character: the ratio Q/W is not equal to one and varies within the limits of 0.35 ÷ 0.75, depending on testing conditions [37]. In other words, for these materials thermodynamically ideal plasticity is not realized. The cause of such effect is thermodynamically nonequilibrium nature or fractality of solid-phase polymers structure. Within the frameworks of fractal analysis it has been shown that this results to polymers yielding process realization not in the entire sample volume, but in its part only.

It is easy to see, that at $D_{l_i}^{v_{fr}}$ replacement on Young's modulus E the Eq. (14.14) transforms into a classical Hooke law. The appearance in this equation of operator $D_{l_i}^{v_{fr}}$ is due the indicated above polymers structure fractality. Proceeding from the said above, it can be supposed that the operator $D_{l_i}^{v_{fr}}$ should be written as $E^{v_{fr}}$.

Hence, a solid-phase polymers deformation process is realized in fractal space with the dimension, which is equal to structure dimension d_f. In such space the deformation process can be presented schematically as the "devil's staircase" [39]. Its horizontal sections correspond to temporal intervals, where deformation is absent. In this case deformation process is described with using of fractal time t, which belongs to the points of Cantor's set [30]. If Euclidean object deformation is considered then time belongs to real numbers set.

For processes evolution with fractal time the mathematics of fractional integration and differentiation is used [39]. As it has been shown in Ref. [30], in this case the fractional exponent v_{fr} coincides with Cantor's set fractal dimension and indicates system states fraction, preserved during all evolution time t. Let us remind, that Cantor's set is considered in one-dimensional Euclidean space ($d = 1$) and therefore, its fractal dimension $d_f < 1$ in virtue of fractal definition [39]. For fractal objects in Euclidean spaces with higher dimensions ($d > 1$) the fractional part of d_f should be accepted as v_{fr} according to the Eq. (4.32) [40]. Then the v_{fr} value characterizes the fractal (polymer structure) fraction, which is invariable in deformation process [31].

For $d = 3$ and $2 \le d_f < 3$ (the fractal object in three-dimensional Euclidean space) the following modification of the Eq. (14.14) can be written:

$$\sigma_Y = \varepsilon_Y E^{d_f - 2}, \qquad (14.15)$$

where ε_Y is yield strain.

In Fig. 14.5, the comparison of experimental σ_Y and calculated according to the Eq. (14.15) σ_Y^T yield stress values for extrudates of UHMPE, componors UHMPE-Al and UHMPE-bauxite, prepared with various extrusion draw ratio, is accepted, from which the good conformity of theory and experiment follows.

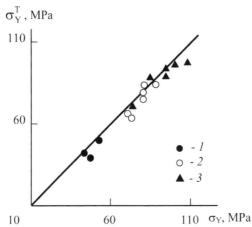

FIGURE 14.5   The comparison of experimental σ_Y and calculated according to the Eq. (14.15) σ_Y^T yield stress values for UHMPE (1), UHMPE-Al (2) and UHMPE-bauxite (3) [31].

From the Eq. (14.15) it follows, that polymers structure fractality ($d_f < d$) results to yield stress essential reduction. From the point of view of thermodynamics σ_Y indicated reduction is due to accumulation in sample of internal (latent) energy, the relative fraction of which is equal to about v_{fr} [41]. For Euclidean solids ($d_f = d$) the Eq. (14.15) gives Hooke law. At the same time for the indicated materials extrudates σ_Y strong increase in comparison with initial samples is due to E and d_f simultaneous growth. It is follows to note also, that the Eq. (14.15) can be used for description of polymers deformation on the elasticity part (at $d_f = d$) and on cold flow plateau (at $d_f = d$ and elasticity modulus replacement on strain hardening modulus) [32].

And in conclusion we will dwell upon the units measurement agreement in both parts of the Eq. (14.15). Since for the convenience of considerations σ_Y value is determined in MPa, then the constant coefficient $C = (1 \text{MPa})^{3-d_f}$ according to the generally accepted technique should be introduced into the right-hand part of the Eq. (14.15) [42]. Units measurement change results to C change. So, at using GPa $C = \left(10^{-3}\right)^{3-d_f}$ and so on.

Hence, the stated above results shown the conformity of fractional derivatives method and traditional fractal analysis, using Hausdorff dimension d^f notion. The physical significance of fractional exponent and its determination method were elucidated. The theoretical analysis is given the good quantitative correspondence to experiment [31].

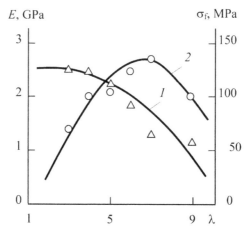

FIGURE 14.6 The dependences of elasticity modulus E (1) and fracture stress σ_f (2) on extrusion draw ratio λ for componors UHMPE – bauxite [44].

The authors number was offered earlier [43] the new method of polymerization-filled compositions on the basis of UHMPE processing – powder billet ram extrusion, allowing to prepare rod-like products with mechanical properties high level. The studies, carried out on systems UHMPE-kaolin, UHMPE-Al, shown [2], that at large draw ratios λ a componors strengthening reduction occurs. The obtained result comparison with the data by extrudates UHMPE and its compositions with smaller kaolin contents fracture [43] allows to suppose, that fracture stress extreme change can be connected with the filler availability. The Ref. [44] is devoted to this question clarification.

As and in the case [45], the studied componors fracture stress σ_f depending on λ changes extremely and similarly o elasticity modulus E, reaching the greatest values in the region of $\lambda \approx 5$ (Fig. 14.6) at bauxite content 40 mas. %. The maximum for componors UHMPE-Al with Al content 70 mas. % is disposed at smaller λ values that at the same filler content 54 mas. % [45]. Hence, at filler contents increase strengthening reduction process is displayed earlier (at smaller λ). The λ increasing induces also extrudates density ρ monotonous reduction also (Fig. 14.7).

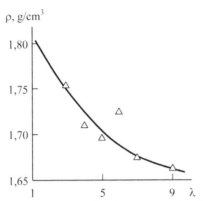

FIGURE 14.7   The dependence of samples density ρ on extrusion draw ratio λ for componor UHMPE-Al [44].

In Fig. 14.8, the change of samples limiting draw ratio λ_{lim}, corresponding to their fracture in mechanical tests, as a function extrusion draw ratio λ is shown. As one can seem the λ_{lim} increase at λ growth is observed. Such behavior λ_{lim} is differed from traditionally observed one for oriented polymers, when λ increasing decreases λ_{lim} owing to molecular chains mobility exhausting [19]. Personally, it has been established, that for matrix UHMPE λ enhancement from 3 up to 5 is accompanied by λ_{lim} reduction from 1.18 up to 1.12 [44].

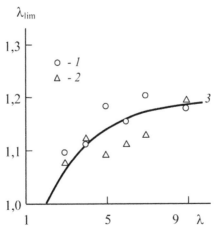

FIGURE 14.8   The dependences of limiting draw ratio λ_{lim} on extrusion draw ratio λ for componors UHMPE-Al (1, 3) and UHMPE-bauxite (2, 3). 3 – the theoretical dependence $\lambda_{lim}(\lambda)$, calculated according to the Eq. (14.16) [44].

The cited facts in the aggregate with established ones earlier [43, 45] allow to confirm, that extruded componors fracture process is connected directly with filler presence, namely, with samples continuity violation owing to interfacial regions polymer-filler loosening and multiple microcracks formation. In favor of this mechanism E decreasing at large extrusion draw ratios is testified, that can be due to polymer-filler adhesional strength reduction [46], ρ decrease and λ_{lim} increase, observed at λ enhancement (Figs. 14.6 ÷ 14.8).

The data of fracture surface studies (Fig. 14.9) corresponds also to made conclusion. At small λ fracture surface relief is homogeneous one that indicated good adhesion between UHMPE and filler. In the case of $\lambda \geq 7$ can be see relief heterogeneity, which is due to adhesion loss in interfacial boundary polymer-filler and, as consequence, material fracture on this boundary. A microcracks aggregation, formed during extrusion, and their fracture occurs at componors samples deformation in mechanical tests process [2].

The dependence $\rho(\lambda)$ and mechanism of fracture knowledge allow to use the fractal analysis methods [47] for theoretical calculation of limiting draw ratio λ_{lim}, which is connected with the value ρ according to the following relationship [47]:

$$\rho = \rho_0 \lambda_{lim}^{-\alpha}, \qquad (14.16)$$

where ρ_0 is the density of nondeformed material and

$$\alpha = d - d_{fr}, \qquad (14.17)$$

where d is dimension of Euclidean space, in which a fractal is considered (it is obvious, that in our case $d = 3$), d_{fr} is fracture surface fractal dimension, which for quasibrittle fracture is determined according to the Eq. (4.50).

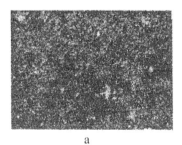 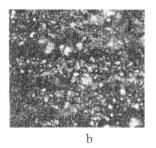

a   b

FIGURE 14.9 The microphotographs of extruded componors UNMPE-Al surfaces of fracture: $\lambda = 3$ (a) and $\lambda = 7$ (b). Enlargement 200× [2].

In Fig. 14.8 the dependences $\lambda_{lim}(\lambda)$, calculated according to the Eq. (14.16). As one can see, the good conformity of theory and experiment is obtained, that is, the fractal analysis notions usage allows the considered componors fracture process adequate description. Let us note, that the calculation λ_{lim} is carried out by extruded samples density, which are not subjected to further deformation in mechanical tests. This means, that microcracks formation occurs mainly at extrusion. At deformation in mechanical tests their merging is realized, resulting in the end to sample fracture as a whole [44].

Lately a numerous and successful attempts of the multifractal formalism application for different materials structure and properties description [48–50], including polymer ones [51, 52], are undertaken. As a matter of fact, this means the indicated formalism usage for applied goals, for example, in engineering science. At present the correlations of multifractal spectrum specific characteristics (Renyi dimensions, "latent ordering" parameter and so on) and traditional parameters, described polymeric material properties (elasticity modulus, Grüneisen parameter and so on) [53, 54] are obtained. At present for polymer particulate-filled composites there exists the simple technique of multifractal spectra calculation, based on these materials two-phase nature [34, 51, 54, 55]. In Ref. [56] the indicated technique was used for estimation of extrudated componors structure characteristic Renyi dimensions with the purpose of study of these materials structure changes at extrusion draw ratio variation.

As it is known [16], the any material structure, including polymer composites, can be described either as multifractal, or as regular fractal. In the first case for materials structure description the dimensions spectrum was used, for example, generalized Renyi dimensions spectrum $D_q(q)$, where q is index ($q = -\infty, \ldots, 0, \ldots, +\infty$), which has specific sigmoid shape [39]. In the second case all generalized Renyi dimensions are equal, that is, $D_{-\infty} = \ldots = D_0 = \ldots = D_{+\infty}$ and is this case the indicated spectrum is degenerated in straight line, parallel to q axis [57]. Thus, for a formalism type definition, describing material structure, it is enough to determine characteristic Renyi dimensions $D_{-\infty}$ and $D_{+\infty}$ (in practice there used usually dimensions D_{-40} at $q = -40$ and D_{40} at q = 40, accordingly) and in the case $D_{-40} \neq D_{40}$ the structure represents itself multifractal and in the case $D_{-40} = D_{40}$ – regular fractal [16, 39, 57].

In Ref. [51], the Halsey multifractal formalism [39], modified by Williford [58], application for particulate-filled polymer composites structure and properties description was considered. The Cantor set ("dust") was used as mathematical model. It is assumed, that section of length l_1 and probabilistic measure p_1 characterizes a filler and l_2 and p_2 – a polymer matrix, respective-ly. Such attribution is due to that fact that the section, characterized by index 1 is attributed to brittle fracture branch according to [58] and filler contents increase raises composites brittleness. If in Cantor construction from initial section of length 1 its middle part is removed, then two remaining thirds length will be equal to ~0.667 [39]. Then this value follows to divide propor-tionally to filler particles (aggregates of particles) size and distance between filler neighboring particles surface, using their averaged values, and this will be corresponded to scales l_1 and l_2.

In the considered case this procedure can be concretized as follows [56]. Since for polymerization-filled compositions some appreciable filler par-ticles aggregation is not observed in virtue of their preparation method fea-tures, then for initial componors the distance between filler particles b_p is determined as follows [51]:

$$b_p = \left[\left(\frac{4\pi}{3\varphi_n} \right)^{1/2} - 2 \right] d_p, \qquad (14.18)$$

where φ_n is filler volume contents, d_p its particles diameter.

For prepared by solid-phase extrusion samples it is assumed, that the value b_p changes proportionally to extrusion draw ratio λ.

The total componors fracture probability p in mechanical tests is obvi-ously equal to one. Further this value is divided at the condition $p = p_1 + p_2$ as follows. In Fig. 14.10, the dependence of fracture stress σ_f on extrusion draw ratio λ for componor UHMPE-Al is shown. As one can see, the σ_f growth is ceased and changes to drop(compare with the plot of Fig. 14.6). Such the dependence $\sigma_f(\lambda)$ type is due to interfacial boundaries polymer-filler fracture at $\lambda > 5$ [44]. Hence, at $\lambda \le 5$ the fracture of polymer matrix and interfacial boundary is equally probable ($p_1 = p_2 = 0.5$) and at $\lambda > 5$ the second fracture probability is higher ($p_1 > p_2$). In addition it is assumed, that σ_f at $\lambda > 5$ is equal to interfacial boundary strength σ_{if} and for $\sigma_{if} = \sigma_f = 0$ the condition $p_1 = p$ will b obvious one. An intermediate values p_2 are determined according to the equation [55]:

$$p_2 = \frac{0,5\sigma_{if}}{\tau_m} \qquad (14.19)$$

where τ_m is polymer matrix shear strength, equal to shear yield stress τ_Y [34]:

$$\tau_m = \tau_Y = \frac{\sigma_Y}{\sqrt{3}} \tag{14.20}$$

The number of characteristic points are existed for a multifractal diagram. So, at $q = \infty$ $(q = 40)$ [34]:

$$D_{40} = \frac{\ln p_1}{\ln l_1} \tag{14.21}$$

and at $q = -\infty$ $(q = -40)$ [34]:

$$D_{-40} = \frac{\ln p_2}{\ln l_2}. \tag{14.22}$$

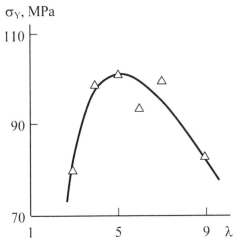

FIGURE 14.10   The dependence of fracture stress s_f on extrusion draw ratio l for componor UHMPE-Al [56].

The Eqs. (14.21) and (14.22) together with estimated by considered above method parameters p_1, l_1, p_2 and l_2 allow to calculate the dimensions D_{40} and D_{-40} [55]. In Fig. 14.11 the dependences of Renyi characteristic dimensions D_{-40} and D_{40} on extrusion draw ratio λ for componors UHMPE-Al and UHMPE-bauxite. As one can see, at the definite values λ (λ_{cr}) the componors structure transition from multifractal (canonical spectrum, D_q grows at q increase [59]) to regular fractal $(D_{-40} = D_{40})$ occurs and then at $\lambda > \lambda_{cr}$ – again to multifractal (pseudospectrum, D_q decreases at q growth).

The data of Figs. 4.10 and 4.11 comparison shows, that the value λ_{cr}, obtained according to the dependences $D_{-40}(q)$ and $D_{40}(q)$, corresponds to the value λ, at which fracture stress σ_f drop (Fig. 14.10) or interfacial boundaries polymer-filler fracture begins [2]. Thus, within the frameworks of multifractal formalism interfacial boundaries fracture of componors in solid-phase extrusion process is realized by polymer matrix structure regular fractal state achievement [56].

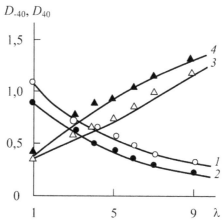

FIGURE 14.11 The dependences of critical Renyi dimensions D_{40} (1, 2) and D_{-40} (3, 4) on extrusion draw ratio λ for componors UHMPE-Al (1, 3) and UHMPE-bauxite (2, 4) [56].

As it follows from the data of Fig. 14.11, the values λ_{cr} for componor UHMPE-bauxite ($\varphi_n = 0.167$) is smaller than corresponding parameter for UHMPE-Al ($\varphi_n = 0.260$). This assumes λ_{cr} decrease at φ_n reduction. As it is known [44], for considered componors extrudates fracture strain ε_f (or λ_f, see Fig. 14.8) enhancement at λ growth is observed, that is due to interfacial boundaries fracture effect [2]. Therefore, the value ε_f is the most sensitive indicator of this structural effect. In Fig. 14.12 the dependences $\varepsilon_f(\lambda)$ for UHMPE-Al and UHMPE-bauxite are adduced, from which it follows that for the second from indicated componors ε_f growth (and, hence, interfacial boundaries fracture [44]) at λ increasing begins earlier than for the first (at $\lambda > 5$ and $\lambda > 3$, respectively). The cited threshold values λ correspond well to the values λ_{cr}, estimated from the data of Fig. 14.11. Thus, the theoretical estimations results (the data of Fig. 14.11) correspond well to experiment (the data of Fig. 14.12).

As it has been noted above, at λ_{cr} a multifractal spectrum type change occurs, which supposes, that the considered componors structure can be characterized as quasiperiodic system and calculated for it adaptability recourse R_a according to the equation [59]:

$$R_a = \frac{D_{40}q^{max} - D_{-40}q^{min}}{D_{40}q^{max} - D_1},$$

(14.23)

where D_1 is the structure information dimension, determining according to [55].

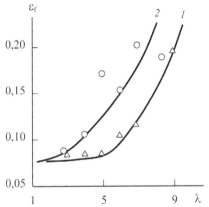

FIGURE 14.12   The dependences of fracture strain ε_f on extrusion draw ratio λ for componors UHMPE-Al (1) and UHMPE – bauxite (2) [56].

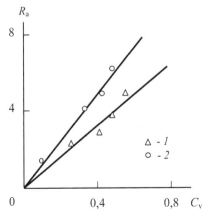

FIGURE 14.13   The dependences of adaptability resource R_a on voids relative fraction C_v for componors UHMPE – Al (1) and UHMPE – bauxite (2) [56].

It can be supposed, that adaptability resource R_a is connected with voids on componors interfacial boundaries formation during these boundaries fracture process and this allows extrusion up to higher draw ratios. Let us remind, that the similar mechanism absence for UHMPE results to the fact, that this polymer breaks down in extrusion process at $\lambda > 5$ [2]. The interfacial boundaries fracture degree can be characterized with the help of the voids relative fraction C_v, the values of which were accepted according to the data of Ref. [2]. In Fig. 14.13 the dependences $R_a(C_v)$ shown, which prove to be linear ones and passing through coordinates origin. These dependences correspond well to made above suppositions. At void absence ($C_v = 0$) the adaptability resource is equal to zero also, that is, componors extrusion higher than λ_{cr} is impossible. The componors samples fracture criterion during extrusion process is obviously such: $C_v = 1.0$. From the data of Fig. 14.13, it follows that the limiting values of adaptability resource R_a^{lim} correspond to this criterion: ~7.8 for UHMPE – Al and ~11.3 for UHMPE – bauxite. The R_a^{lim} estimation at $\lambda = 11$, that is, at the conditions, where componors samples fracture during solid-phase extrusion process is confirmed experimentally, was given the following values: 8.95 for UHMPE – Al and 11.45 – for UHMPE – bauxite, that correspond well to corresponding estimations according to the plots of Fig. 14.13.

Let us consider the physical picture of dissipative structures type change at $\lambda = \lambda_{cr}$, which was postulated above (see Fig. 14.1) [11]. As it has been shown in Ref. [2], at $\lambda = \lambda_{cr}$ cluster relative fraction φ_{cl} decreasing from 0.35 up to the smallest value ~0.05 occurs and at further λ enhancement the value φ_{cl} is not changed. At the same time crystallinity degree K growth at λ enhancement owing to orientational crystallization is observed (see Fig. 14.2). Thus, the dissipative structures type change at λ_{cr} means the transition from clusters in amorphous phase to crystallites, which are formed during orientational crystallization process. Since voids formation on interfacial boundaries polymer-filler begins at λ_{cr} also, then it follows to expect the correlation between parameters C_v and K. Actually, as it follows from the data of Fig. 14.14, such correlation exists, although it has large enough data scattering. This correlation can be written analytically as follows [56]:

$$C_v = 1.60(K - K_{in}),\qquad\qquad(14.24)$$

where K_{in} is crystallinity degree of nonoriented componor.

It is significant, that the condition $C_v = 0$ is realized at $K = K_{in}$, in other words, the voids on interfacial boundaries are connected not with total crystallinity degree of samples, but with that its part, which is due to orientational crystallization.

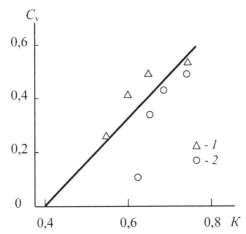

FIGURE 14.14   The dependence of voids relative fraction C_v on crystallinity degree K for componors UHMPE-Al(1) and UHMPE-bauxite (2) [56].

As it has been noted in Ref. [60], the generalized Renyi dimension $D_{-\infty}$ (or D_{-40}) characterizes the most concentrated system set. In Ref. [55] it has been shown, that for amorphous polymer matrix the cluster are such set and the linear correlation $D_{-40}(\varphi_{cl})$ was obtained. It is obvious that for the considered componors with polymer semicrystalline matrix the crystalline regions will be the most concentrated set that assumes a definite correlation between D_{-40} and K. Actually, such correlation exists, as it follows from the data of Fig. 14.15 and it is described by the simple analytical relationship [56]:

$$D_{-40} = 1.52 \ K. \tag{14.25}$$

Hence, the cited above results shown correctness and expediency of multifractal formalism in it's the simplest variant for analysis of structure changes at polymerization-filled compositions on the basis of UHMPE during solid-phase extrusion. The observed experimentally during extrusion process effects were received the quantitative description within the frameworks of this formalism. Let us note purely geometrical character of main multifractal characteristics calculation, independent on polymer matrix and filler properties [56].

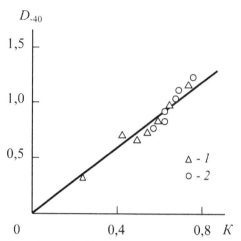

FIGURE 14.15   The dependence of critical generalized Renyi dimension D_{-40} on crystallinity degree K for componors UHMPE-Al (1) and UHMPE-bauxite (2) [56].

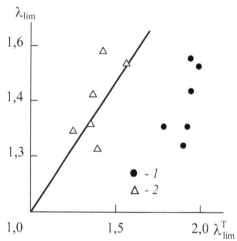

FIGURE 14.16   The relation between draw ratios at fracture for extrudates DF-10, determined experimentally λ_{lim} and calculated theoretically λ_{lim}^{T} according to the Eqs. (5.7) (1) and (7.8) (2) [24].

The limiting draw ratio at fracture λ_{lim} increase at extrusion draw ratio λ growth is one from the most interesting features of the amorphous polymers extrudates [26, 61]. Such effect was observed and in the case of DF-10 extrudates [2]. Let us consider the value λ_{lim} (λ_{lim}^{T}) theoretical estimation possibility with the fractal analysis methods using. Within the frameworks of high-elasticity theory the value λ^T_{lim} is determined according to the Eq. (5.7). According to the data of Ref. [62] the Eq. (5.7) describes well the results for amorphous glassy polymers. The fractal variant of λ^T_{lim} estimation was given by the Eq. (7.8). In Fig. 14.16 the comparison of experimental λ_{lim} and calculated according to the Eqs. (5.7) and (7.8) λ^T_{lim} values of draw ration at fracture for extrudates DF-10. The good conformity between λ_{lim} and λ^T_{lim} obtained according to the fractal model (the Eq. (7.8)), is observed, whereas the Eq. (5.7) is given the overstated λ^T_{lim}. The indicated discrepancy cause is that fact that high-elasticity theory is not taken into account chains mobility "freezing" in glassy state (i.e., it assumes $D_{ch} = 2.0$). In the more general terms it can be said, that the high-elasticity theory equations are correct for a structures with Euclidean geometry, which (or good approximation to which) are rubbers. At the same time the glassy polymers structure is fractal object and their properties correct description is possible within the frameworks of fractal analysis only [8].

KEYWORDS

- **adaptability resource**
- **dimensions spectrum**
- **draw ratio**
- **extrusion**
- **fractional derivative**
- **oriented polymer**

REFERENCES

1. El'yashevich, G .K., Karpov, E. A., Lavrent'ev, V. K., Poddybnyi, V. I., Genina, M. A., & Zabashta, Yu. F. (1993). The Noncrystalline Regions Formation in Polyethylene at Stretching High Degress. Vysokomolek. Soed, A, *35(6)*, 681–685.
2. Aloev, V. Z., & Kozlov, G. V. (2002). Physics of Orientational Phenomena in Polymeric Materials. Nal'chik, Poligrafservise IT. 288 p.
3. Aloev, V. Z., Kozlov, G. V., & Dolbin, I. V. (2000). The Fractal Ayalysis of Vilume Changes at Polymers Uniaxial Tension. Vestnik KBSU, series fizicheskie nauki *(5)*, 41–42.
4. Balankin, A. S. (1991). Synergetics of Deformable Body. Moscow, Publishers of Ministry Defence SSSR, 404 p.

5. Kozlov, G. V., Beloshenko, V. A., Varyukhin, V. N., & Gazaev, M. A. (1998). The Fractal Model of Oriented Polymers Thermal Shrinkage. Prikladnaya Mekhanika i Tekhnicheskaya Fizika, *39(1)*, 160–163.

6. Kozlov, G. V., & Sanditov, D. S. (1994). Anharmonic Effects and Physical-Mechanical Properties of Polymers. Novosibirsk, Nauka, 261 p.

7. Kozlov, G. V., Beloshenko, V. A., Varyukhin, V. N., & Novikov, V. U. (1997). The Order and Fractality of Semicrystalline Polymers. Zhurnal Fizicheskikh Issledovanii, *1(2)*, 204–207.

8. Kozlov, G. V., & Novikov, V. U. (1998). Synergetics and Fractal Analysis of Cross-Linked Polymers. Moscow, Klassika, 112 p.

9. Beloshenko, V. A., Kozlov, G. V., Slobodina, V. G., Prut, E. V., & Grinev, V. G. (1995). Thermal Shrinkage of Extrudates of Ultra-High-Molecular Polyethylene and Polymerization-Filled Compositions on its Basis. Vysokomolek. Soed. B, *37(6)*, 1089–1092.

10. Beloshenko, V. A., Kozlov, G. V., Varyukhin, V. N., & Slobodina, V. G. (1997). Properties of Ultra-High-Molecular Polyethylene and Related Polymerization-Filled Composites Produces by Solid-State Extrusion. Acta Polymerica, *48(5–6)*, 181–192.

11. Kozlov, G. V., Aloev, V. Z., Novikov, V. U., Beloshenko, V. A., & Zaikov, G. E. (2001). The Change of Deformation Mechanism and Structure in Solid-Phase Extrusion Process of Polymerization-Filled Compositions. Plast. Massy, *3*, 21–23.

12. Kozlov, G. V., Ovcharenko, E. N., & Mikitaev, A. K. (2009). Structure of the Polymers Amorphous State. Moscow, Publishers of the D.I. Mendeleev RKhTU, 392 p.

13. Kozlov, G. V., Shustov, G. B., & Temiraev, K. B. (1997). The Nature of Dissipative Structures in Polymers Amorphous State. Vestnik KBSU, series khimicheskie nauki, *2*, 50–52.

14. Lindenmeyer, H. (1979). Polymer Morphology as Dissipative Structure. Polymer J., *11(8)*, 677–679.

15. Balankin, A. S., Ivanova, V. S., & Breusov, V. P. (1992). Collective Effects in Metals Fracture Kinetics and Dissipative Structures Fractal Dimension Spontaneous Change at Ductile-Brittle Transition. Doklady AN, *322(6)*, 1080–1086.

16. Balankin, A. S., Bugrimov, A. L., Kozlov, G. V., Mikitaev, A. K., & Sanditov, D. S. (1992). The Fractal Structure and Physical-Mechanical Properties of Amorphous Glassy Polymers. Doklady AN, *326(3)*, 463–466.

17. Kozlov, G. V., Yanovskii, Yu. G., & Karnet, Yu. N. (2008). Generalized Fractal Model of Yielding Process of Amorphous Glassy Polymers. Mekhanika Kompozitsionnykh Materialov i Konsrutsii, *14(2)*, 174–187.

18. Beloshenko, V. A., Grinev, V. G., Kuznetsov, E. N., Novokshonova, L. A., Slobodina, V. G., Kudinova, O. I., Rudakov, V. M., & Tarasova, G. M. (1994). The Solid-Phase Extrusion of Compositions on the Basis of Polyethylene. Fizika i Tekhnika Vysokikh Davlenii, *4(1)*, 91–95.

19. The Supermodulus Polymers. Ed. Ciferri, A., Ward, I., Leningrad, & Khimiya, (1983), 272 p.

20. De. Rudder, J. L., & Filisko, F. (1997). Mechanical Property and Physical Structure Changes in Highly Hot-Drawn Polycarbonate. J. Appl. Phys., *48(10)*, 4026–4031.

21. Shishkin, N. I., Milagin, M. F., & Gabaraeva, A. D. (1963). A Molecular Network and Orientational Processes in Amorphous Polystyrene. Fizika Tverdogo Tela, *5(12)*, 3453–3462.

22. Kozlov, G. V., Temiraev, K. B., & Shustov, G. B. (1998). The Structure Change in Extrusion Process of Amorphous Polyarylates. Plast. Massy, 9, 27–29.
23. Kozlov, G. V., & Novikov, V. U. (2001). The Cluster Model of Polymers Amorphous State. Uspekhi Fizicheskikh Nauk, 171(7), 717–764.
24. Aloev, V. Z., Kozlov, G. V., Dolbin, I. V., & Beloshenko, V. A. (2001). Structure and Properties of Extruded Polyarylate. Izvestiya KBNC RAN, 1, 70–78.
25. Shogenov, V. N., Belousov, V. N., Potapov, V. V., Kozlov, G. V., & Prut, E. V. (1991). The Glassy Polyarylatesulfone Curves Stress-Strain Description within the Frameworks of High-Elasticity Concepts. Vysokomolek. Soed. A., 33(1), 155–160.
26. Inoue, N., Nakayama, N., & Ariyama T. (1981). Hydrostatic Extrusion of Amorphous Polymers and Properties of Extrudates. J. Macromol. Sci.-Phys., B 19(2), 543–563.
27. Mandelbrot, B. B. (1982). The Fractal Geometry of Nature. San-Francisco, W.H.Freeman and Comp., 459 p.
28. Oldham, K., & Spanier, Y. (1973). Fractional Calculus. London, New York, Academic Press, 329 p.
29. Samko, S. G., Kilbas, A. A., & Marichev, O. I. (1987). Integrals and Derivatives of Fractional Order and their Some Applications. Minsk, Nauka i Tekhnika, 688 p.
30. Nigmatullin, R. R. (1992). Fractional Integral and its Physical Interpretation. Teoretishaya i Matematicheskaya Fizika, 90(3), 354–367.
31. Kozlov, G. V., Aloev, V. Z., & Yanovskii, Yu. G. (2003). The Simulation of Polymerization-Filled Compositions Extrudates Yielding on the Basis of Derivatives of Fractional Order. Inzhenernaya Fizika, 3, 31–33.
32. Kozlov, G. V., Aloev, V. Z., Yanovskii, Yu. G., & Zaikov, G. E. (2005). Yielding Process of Polymerization-Filled Composites: a Description within the Frameworks of the Fractional order Derivatives Theory. J. Balkan Tribologic. Assoc., 11(5), 221–226.
33. Mikitaev, A. K., & Kozlov, G. V. (2008). Fractal Mechanics of Polymeric Materials. Nal'chik, Publishers of KBSU, 312 p.
34. Kozlov, G. V., Yanovskii, Yu. G., & Zaikov, G. E. (2010). Structure and Properties of Particulate-Filled Polymer Composites: the Fractal Analysis New York, Nova Science Publishers Inc. 282 p.
35. Meilanov, R. R., Sveshnikova, D. A., & Shabanov, O. M. (2001). Sorption Kinetics in Systems with Fractal Structure. Izvestiya VUZov, Severo-Kavkazsk. Region, estestv. Nauki, 1, 63–66.
36. Kekharsaeva, E. R., Mikitaev, A. K., & Aleroev, I. S. (2001). Model of Stress-Strain Characteristics of Chlor-Containing Polyesters on the Basis of Derivatives of Fractional Order. Plast. Massy, 3, 35.
37. Adams, G. W., & Farris, J. (1989). Latent Energy of Deformation of Amorpous Polymers. 1. Deformation Calorimetry. Polymer, 30(9), 1824–1828.
38. Balankin, A. S., & Bugrimov, A. L. (1992). The Fractal Theory of Polymers Plasticity. Vysokomolek. Soed. A, 34(10), 135–139.
39. Halsey, I. C., Jensen, M. H., Kadanov, L. P., Procaccia, I., & Shraiman, B. I. (1986). Fractal Measures and their Singularities: the Characterization of Strange Sets. Phys. Rev. A, 33(2), 1141–1151.
40. Kozlov, G. V., Shystov, G. B., & Zaikov, G. E. (2003). Fractal Analysis of Themooxidative Degradation Process of Polymeric Melts. J. Appl. Polymer Sci, 89(9), 2378–2381.

41. Kozlov, G. V., Sanditov, D. S., & Ovcharenko, N. (2001). The Plastic Deformation Energy and Amorphous Glassy Polymers Structure. Mater of II-th Interdiscriplinary Symposium "Fractals and Applied Synergetics FaAs-01". Moscow, Publishers of MSOU, 81–83.

42. Yuang, Z. H., Tian, I. F., & Wang, Z. G. (1990). Comments on some of the Fractal Equations. Mater. Sci. Engng., A *128(1–3)*, L13–L14.

43. Beloshenko, V. A., Slobodina, V. G., Grinev, V. G., & Prut, E. V. (1994). The Solid-Phase Extrusion of Polymerization-Filled Polyethylene. Vysokomolek. Soed. B, *36(6)*, 1–21–1024.

44. Kozlov, G. V., Beloshenko, V. A., & Slobodina, V. G. (1996). The Fracture Mechanism of Extrudated Componors. Plast. Massy *(3)*, 14–16.

45. Aloev, V. Z., Beloshenko, V. A., & Abazekhov, M. M. (2000). The Study of Extrudated Componors Yielding Process. Vestnik KBSU, Series Fizicheskie Nauki, *4*, 50–51.

46. Knunyants, N. N., Lyapunova, M. A., Manevich, L. I., Oshmyan, V. G., & Shaulov, A. Yu. (1986). The Simulation of Nonideal Adhesional Coupling Influence on Elastic Properties of Particulate-Filled Composite. Mekhanika Kompozitnykh Materialov, *21(2)*, 231–234.

47. Balankin, A. S. (1991). A Fractals Elastic Properties, Transverse Strains Effect and Solids Free Fracture Dynamics. Doklady AN SSSR, *319(5)*, 1098–1101.

48. Vstovskii, G. V., Kolmakov, L. G., & Terent ev, V. F. (1993). The Multifractal Analysis of Molybdenum Near-Surface Layers Fracture Features. Izvestiya AN, Metally *4*, 164–178.

49. Kolmakov, A. G. (1996). The Intercommunication of Multifractal Characteristics of Molybdenum Surface Structures with its Mechanical Properties. Izvestiya AN, Metally *6*, 37–43.

50. Semenov, B. I., Agibalov, S. N., & Kolmakov, A. G. (1999). The Description of Casting Aluminium-Matrix Composite Structure with Multifractal Formalism Method Using. Materialovedenie, *5*, 25–33.

51. Novikov, V. U., Kozlov, G. V., & Bilibin, A. V. (1998). Polymer Composites Fracture Analysis within the Frameworks of Multifractal Formalism. Materialovedenie, *10*, 14–19.

52. Kozlov, G. V. (2002). The Multifractal Analysis of Diffusion Processes in Semisrystalline Polyethylene and it's Melt. In: New Perspectives in Chemistry and Biochemistry. Ed. Zaikov, G. E. New York, Nova Science Publishers Inc., 57–65.

53. Novikov, V. U., Kozitskii, D. V., Deev, I. S., & Kobets, L. P. (2001) The Multifractal Parametrization of Epoxy Polymers Deformed Structure. Materialovedenie, *11*, 2–10.

54. Kozlov, G. V., Dolbin, I, B., & Lipatov, Yu. S. (2001). The Physical Significance of Multifractal Formalism Ordering Parameter for Particulate-Filled Polymer Composites. Proceedings of Internat Symposium "Order, Disorder and Oxides Properties", ODPO-2001, Sochi, September 27–29, 174–180.

55. Kozlov, G. V., & Ovcharenko, E. N. (2001). The Intercommunication of Multifractal Characteristics and Structure Parameters for Particulate-Filled Polymer Composites. Izvestiya KBNC RAN, *2*, 81–85.

56. Kozlov, G. V., Aloev, V. Z., & Yanovskii, Yu. G. (2004). The Multifractal Analysis of Polymerization-Filled Compositions Solid-Phase Extrusion Processes. Mekhanika Kompozitsionnykh Materialov i Konstruktsii, *2*, 267–275.

57. Hayakawa, Y., Sato, S., & Matsushita, M. (1986). Scaling Structure of the Growth-Probability Distribution in Diffusion-Limited Aggregation Processes. Phys. Rev. A., *33(2)*, 1141–1151.

58. Williford, R. E. (1988). Multifractal Fracture. Scripta Metal, *22(11)*, 1749–1754.

59. Ivanova, V. S., Kuzeev, I. R., & Zakirnichnaya, M. M. (1998). Synergetics and Fractals. Universality of Metals Mechanical Behavior. Ufa, Publishers of UGNTU, 366 p.

60. Novikov, V. U., Kozitskii. D. V., & Ivanova, D. V. (1999). The Materials Structure Analysis Computer Technique Development with Multifractal Formalism Using. Materialovedenie, *8*, 12–16.

61. Berestnev, B. I., Enikolopov, N. S., Tsygankov, S. A., & Shishkova, N. K. (1985). A Polymer Hydrostatic Extrusion. Doklady AN USSR, *4*, 47–49.

62. Haward, R. N., & Thackray, G. (1968). The Use of a Mathematical Model to Describe: Isothermal Stress-Strain Curves in Glassy Thermoplastics Proc. Roy. Soc. London, A *302(1471)*, 453–472.

CHAPTER 15

POLYMERS AS NATURAL COMPOSITES: STRUCTURE AND PROPERTIES

CONTENTS

15.1 INTRODUCTION

The idea of different classes polymers representation as composites is not new. Even 35 years ago Kardos and Raisoni [1] offered to use composite models for the description of semicrystalline polymers properties number and obtained prediction of the indicated polymers stiffness and thermal strains to a precision of ±20%. They considered semicrystalline polymer as composite, in which matrix is the amorphous and the crystallites are a filler. The authors of Ref. [1] also supposed that other polymers, for example, hybrid polymer systems, in which two components with different mechanical properties were present obviously, can be simulated by a similar method.

In Ref. [2] it has been pointed out, that the most important consequence from works by supramolecular formation study is the conclusion that physical-mechanical properties depend in the first place on molecular structure, but are realized through supramolecular formations. At scales interval and studies methods resolving ability of polymers structure the nanoparticle size can be changed within the limits of 1÷100 and more nanometers. The polymer crystallites size makes up 10÷20 nm. The macromolecule can be included in several crystallites, since at molecular weight of order of 6×10^4 its length makes up more than 400 nm. These reasoning's point out, that macromolecular formations and polymer systems in virtue of their structure features are always nanostructural systems.

However, in the cited above works the amorphous glassy polymers consideration as natural composites (nanocomposites) is absent, although they are one of the most important classes of polymeric materials. This gap reason is quite enough, that is, polymers amorphous state quantitative model absence. However, such model appearance lately [3–5] allows to consider the amorphous glassy polymers (both linear and cross-linked ones) as natural nanocomposites, in which local order regions (clusters) are nanofiller and surrounded them loosely packed matrix of amorphous polymers structure is matrix of nanocomposite. Proceeding from the said above, in the present chapter description of amorphous glassy polymers as natural nanocomposites, their limiting characteristics determination and practical recommendation by the indicated polymers properties improvement will be given.

15.2 NATURAL NANOCOMPOSITES STRUCTURE

The synergetics principles revealed structure adaptation mechanism to external influence and are universal ones for self-organization laws of spatial structures in dynamical systems of different nature. The structure adaptation is the reformation process of structure, which loses stability, with the new more stable structure self-organization. The fractal (multifractal) structure, which is impossible to describe within the framework of Euclidean geometry, are formed in reformation process. A wide spectrum of natural and artificial topological forms, the feature of which is self-similar hierarchically organized structure, which amorphous glassy polymers possessed [6], belongs to fractal structures.

The authors of Refs. [7, 8] considered the typical amorphous glassy polymer (polycarbonate) structure change within the frameworks of solid body synergetics.

The local order region, consisting of several densely packed collinear segments of various polymer chains (for more details see chapter one) according to a signs number should be attributed to the nanoparticles (nanoclusters) [9]:

1) their size makes up 2÷5 nm;
2) they are formed by self-assemble method and adapted to the external influence (e.g., temperature change results to segments number per one nanocluster change);
3) the each statistical segment represents an atoms group and boundaries between these groups are coherent owing to collinear arrangement of one segment relative to another.

The main structure parameter of cluster model-nanoclusters relative fraction φ_{cl}, which is polymers structure order parameter in strict physical sense of this tern, can be calculated according to the Eq. (1.11). In its turn, the polymer structure fractal dimension d_f value is determined according to the Eqs. (1.9) and (2.20).

In Fig 15.1, the dependence of φ_{cl} on testing temperature T for PC is shown, which can be approximated by the broken line, where points of folding (bifurcation points) correspond to energy dissipation mechanism change, coupling with the threshold values φ_{cl} reaching. So, in Fig, 15.1 T_1 corresponds to structure "freezing" temperature T_0 [4], T_2 to loosely packed matrix glass transition temperature T'_g [11] and T_3 to polymer glass transition temperature T_g.

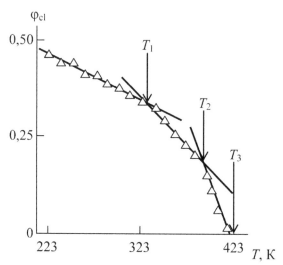

FIGURE 15.1   The dependence of nanoclusters relative fraction φ_{cl} on testing temperature T for PC. The critical temperatures of bifurcation points are indicated by arrows (explanations are given in the text) [18].

Within the frameworks of solid body synergetics it has been shown [12], that at structures self-organization the adaptation universal algorithm, described by the Eq. (9.3), is realized at transition from previous point of structure instability to subsequent one. The value $m = 1$ corresponds to structure minimum adaptivity and $m = m^*$ to maximum one. In Ref. [12] the table is adduced, in which values A_m, m and Δ_i are given, determined by the gold proportion rule and corresponding to spectrum of structure stability measure invariant magnitudes for the alive and lifeness nature systems. The indicated table usage facilitates determination of the interconnected by the power law stability and adaptivity of structure to external influence [12].

Using as the critical magnitudes of governing parameter the values φ_{cl} in the indicated bifurcation points T_0, T_g' and T_g (ϕ_{cl}' and T_{cl}^*, accordingly) together with the mentioned in table data [12], values A_m, Δ_i and for PC can be obtained, which are adduced in Table 15.1. As it follows from the data of Table 15.1, systematic reduction of parameters A_m and Δ_i at the condition $m = 1 = \text{const}$ is observed. Hence, within the frameworks of solid body synergetics temperature T_g' can be characterized as bifurcation point ordering-degradation of nanostructure and T_g – as nanostructure degradation-chaos [12].

It is easy to see, that Δ_i decrease corresponds to bifurcation point critical temperature increase.

TABLE 15.1   The Critical Parameters of Nanocluster Structure State For PC [8]

The temperature range	ϕ'_{cl}	ϕ^*_{cl}	A_m	Δ_i	m	m^*
213÷333 K	0.528	0.330	0.623	0.618	1	1
333÷390 K	0.330	0.153	0.465	0.465	1	2
390÷425 K	0.153	0.049	0.324	0.324	1	8

Therefore, critical temperatures T_{cr} (T_0, T_g' and T_g) values increase should be expected at nanocluster structure stability measure Δ_i reduction. In Fig 15.2 the dependence of T_{cr} in Δ_i reciprocal value for PC is adduced, on which corresponding values for polyarylate (PAr) are also plotted. This correlation proved to be linear one and has two characteristic points. At Δ_i = 1 the linear dependence $T_{cr}(\Delta_i^{-1})$ extrapolates to T_{cr} = 293K, that is, this means, that at the indicated Δ_i value glassy polymer turns into rubber-like state at the used testing temperature T = 293K. From the data of the determined by gold proportion law Δ_i = 0.213 at m = 1 follows [12]. In the plot of Fig. 15.2 the greatest for polymers critical temperature T_{cr} = T_{ll}. (T_{ll} is the temperature of "liquid 1 to liquid 2" transition), defining the transition to "structureless liquid" [13], corresponds to this minimum magnitude. For polymers this means the absence of even dynamical short-lived local order [13].

Hence, the stated above results allow to give the following interpretation of critical temperatures T_g' and T_g of amorphous glassy polymers structure within the frameworks of solid body synergetics. These temperatures correspond to governing parameter (nanocluster contents) ϕ_{cl} critical values, at which reaching one of the main principles of synergetics is realized-subordination principle, when a variables set is controlled by one (or several) variable, which is an order parameter. Let us also note reformations number m = 1 corresponds to structure formation mechanism particle-cluster [4, 5].

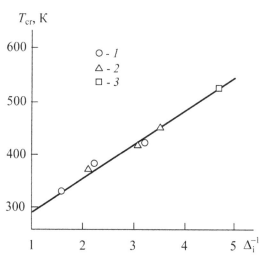

FIGURE 15.2   The dependence of critical temperatures T_{cr} on reciprocal value of nanocluster structure stability measure Δ_i for PC (1) and PAr(2), $3 - T_{ll}$ value for PC [19].

The authors of Ref. [14, 15] considered synergetics principles application for the description of behavior of separate nanocluster structure, characterized by the integral parameter φ_{cl} nanoclusters in the system for the same amorphous glassy polymers. This aspect is very important, since, as it will be shown is subsequent sections, just separate nanoclusters characteristics define natural nanocomposites properties by critical mode. One from the criterions of nanoparticle definition has been obtained in Ref. [16]: atoms number N_{at} in it should not exceed $10^3 \div 10^4$. In Ref. [15] this criterion was applied to PC local order regions, having the greatest number of statistical segments $n_{cl} = 20$. Since nanocluster is amorphous analog of crystallite with the stretched chains and at its functionality F a number of chains emerging from it is accepted, then the value n_{cl} is determined as follows [4]:

$$n_{cl} = \frac{F}{2},\qquad(15.1)$$

where the value F was calculated according to the Eq. (1.7).

The statistical segment volume simulated as a cylinder, is equal to $l_{st}S$ and further the volume per one atom of substance (PC) a^3 can be calculated according to the equation [17]:

$$a^3 = \frac{M}{\rho N_A p}, \qquad (15.2)$$

where M is repeated link molar mass, ρ is polymer density, N_A is Avogadro number, p is atoms number in a repeated link.

For PC $M = 264$ g/mole, $\rho = 1200$ kg/m^3 and $p = 37$. Then $a^3 = 9.54$ Å^3 and the value N_{at} can be estimated according to the following simple equation [17]:

$$N_{at} = \frac{l_{st} \cdot S \cdot n_{cl}}{a^3}. \qquad (15.3)$$

For PC $N_{at} = 193$ atoms per one nanocluster (for $n_{cl} = 20$) is obtained. It is obvious that the indicated value N_{at} corresponds well to the adduced above nanoparticle definition criterion ($N_{at} = 10^3 \div 10^4$) [9, 17].

Let us consider synergetics of nanoclusters formation in PC and PAr. Using in the Eq. (9.3) as governing parameter critical magnitudes n_{cl} values at testing temperature T consecutive change and the indicated above the table of the determined by gold proportion law values A_m, m and Δ_i, the dependence $\Delta(T)$ can be obtained, which is adduced in Fig 15.3. As it follows from this figure data, the nanoclusters stability within the temperature range of $313 \div 393$K is approximately constant and small ($\Delta_i \approx 0.232$ at minimum value $\Delta_i \approx 0.213$) and at $T > 393$K fast growth Δ_i (nanoclusters stability enhancement) begins for both considered polymers.

This plot can be explained within the frameworks of a cluster model [3–5]. In Fig 15.3 glass transition temperatures of loosely packed matrix T_g', which are approximately 50 K lower than polymer macroscopic glass transition temperature T_g, are indicated by vertical shaded lines. At T_g' instable nanoclusters, that is, having small n_{cl} decay occurs. At the same time stable and, hence, more steady nanoclusters remain as a structural element, that results to Δ_i growth [14].

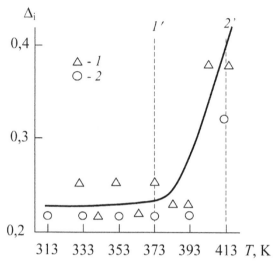

FIGURE 15.3   The dependence of nanoclusters stability measure Δ_i on testing temperature T for PC(1) and PAR(2). The vertical shaded lines indicate temperature T'_g for PC (1') and PAR (2') [14].

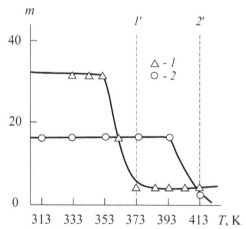

FIGURE 15.4   The dependences of reformations number m for nanoclusters on testing temperature T. The designations are the same as in Fig. 15.3 [14].

In Fig. 15.4, the dependences of reformations number m on testing temperature T for PC and PAr are adduced. At relatively low temperatures ($T < T'_g$) segments number in nanoclusters is large and segment joining (separation) to nanoclusters occurs easily enough, that explains large values m. At T

→ T'_g reformations number reduces sharply and at $T > T'_g$ $m \approx 4$. Since at $T >$ T'_g in the system only stable clusters remain, then it is necessary to assume, that large m at $T < T'_g$ are due to reformation of just instable nanoclusters [15].

In Fig. 15.5, the dependence of n_{cl} on m is adduced. As one can see, even small m enhancement within the range of 2÷16 results to sharp increasing in segments number per one nanocluster. At $m \approx 32$, the dependence $n_{cl}(m)$ attains asymptotic branch for both studied polymers. This supposes that n_{cl} ≈ 16 is the greatest magnitude for nanoclusters and for $m \geq 32$ this term belongs equally to both joining and separation of such segment from nanocluster.

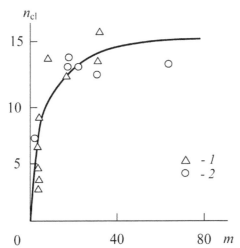

FIGURE 15.5   The dependence of segments number per one nanocluster n_{cl} on reformations number m for PC (1) and PAR (2) [14].

In Fig 15.6, the relationship of stability measure Δ_i and reformations number m for nanoclusters in PC and PAr is adduced. As it follows from the data of this figure, at $m \geq 16$ (or, according to the data of Fig 15.5, $n_{cl} \geq$ 12) Δ_i value attains its minimum asymptotic magnitude $\Delta_i = 0.213$ [12]. This means, that for the indicated n_{cl} values nanoclusters in PC and PAr structure are adopted well to the external influence change ($A_m \geq 0.91$).

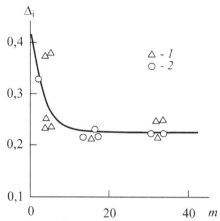

FIGURE 15.6   The dependence of stability measure Δ_i on reformation number m for PC (1) and PAR (2) [14].

Nanoclusters formation synergetics is directly connected with the studied polymers structure macroscopic characteristics. As it has been noted above, the fractal structure, characterized by the dimension d_f, is formed as a result of nanoclusters reformations. In Fig. 15.7 the dependence $d_f(\Delta_i)$ for the considered polymers is adduced, from which d_f increase at Δ_i growth follows. This means, that the increasing of possible reformations number m, resulting to Δ_i reduction (Fig. 15.6), defines the growth of segments number in nanoclusters, the latter relative fraction φ_{cl} enhancement and, as consequence, d_f reduction [3–5].

And let us note in conclusion the following aspect, obtaining from the plot $\Delta_i(T)$ (Fig. 15.3) extrapolation to maximum magnitude $\Delta_i \approx 1.0$. The indicated Δ_i value is reached approximately at $T \approx 458$ K that corresponds to mean glass transition temperature for PC and Par. Within the frameworks of the cluster model T_g reaching means polymer nanocluster structure decay [3–5] and, in its turn, realization at T_g of the condition $\Delta_i \approx 1.0$ means, that the "degenerated" nanocluster, consisting of one statistical segment or simply statistical segment, possesses the greatest stability measure. Several such segments joining up in nanocluster mains its stability reduction (see Figs. 15.5 and 15.6), that is the cause of glassy polymers structure thermodynamical nonequilibrium [14].

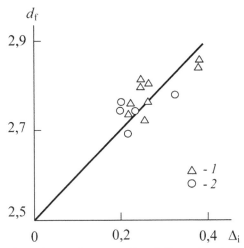

FIGURE 15.7   The dependence of structure fractal dimension d_f on stability measure of nanoclusters Δ_i for PC (1) and PAR (2) [14].

Therefore, the stated above results showed synergetics principles applicability for the description of association (dissociation) processes of polymer segments in local order domains (nanoclusters) in case of amorphous glassy polymers. Such conclusion can be a priori, since a nanoclusters are dissipative structures [6]. Testing temperature increase rises nanoclusters stability measure at the expense of possible reformations number reduction [14, 15].

As it has been shown lately, the notion "nanoparticle" (nanocluster) gets well over the limits of purely dimensional definition and means substance state specific character in sizes nanoscale. The nanoparticles, sizes of which are within the range of order of $1 \div 100$ nm, are already not classical macroscopic objects. They represent themselves the boundary state between macro and microworld and in virtue of this they have specific features number, to which the following ones are attributed:

1.   nanoparticles are self-organizing nonequilibrium structures, which submit to synergetics laws;
2.   they possess very mature surface;
3.   nanoparticles possess quantum (wave) properties.

For the nanoworld structures in the form of nanoparticles (nanoclusters) their size, defining the surface energy critical level, is the information parameter of feedback [19].

The first from the indicated points was considered in detail above. The authors of Refs. [20, 21] showed that nanoclusters surface fractal dimension changes within the range of $2.15 \div 2.85$ that is their well developed surface sign. And at last, let us consider quantum (wave) aspect of nanoclusters nature on the example of PC [22]. Structural levels hierarchy formation and development "scenario" in this case can be presented with the aid of iterated process [23]:

$$l_k = \langle a \rangle B_\lambda^k; \quad \lambda_k = \langle a \rangle B_\lambda^{k+1}; \quad k = 0, 1, 2,..., \tag{15.4}$$

where l_k is specific spatial scale of structural changes, $_k$ is length of irradiation sequence, which is due to structure reformation, k is structural hierarchy sublevel number, $B_\lambda = \lambda_b/\langle a \rangle = 2.61$ is discretely wave criterion of microfracture, λ_b is the smallest length of acoustic irradiation sequence.

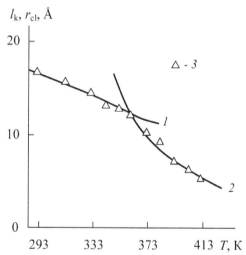

FIGURE 15.8   The dependences of structural changes specific spatial scale l_k at $B_\lambda = 1.06$ (1) and 1.19 (2) and nanoclusters radius r_{cl} (3) on testing temperature T for PC [22].

In Fig. 15.8, the dependences of l_k and nanoclusters radius r_{cl} on T are adduced, where l_k was determined according to the Eq. (15.4) and the value r_{cl} was calculated according to the Eq. (4.64). As it follows from the data of Fig. 15.8, the values l_k and r_{cl} agree within the whole studied temperatures range. Let us note, that if in Ref. [23] the value $B_\lambda = 2.61$, then for PC the indicated above agreement was obtained at $B_\lambda = 1.19$ and 1.06. This distinction confirms the thesis about distinction of synergetics laws in reference to nano-microworld objects (let us remind, that the condition $B_\lambda = 2.61$ is valid even in the case of earthquakes [14]). It is interesting to note, that B_λ change occurs at glass transition temperature of loosely packed matrix, that is, approximately at $T_g - 50$ K [11].

Hence, the stated above results demonstrated that the nanocluster possessed all nanoparticles properties, that is, they belonged to substance intermediate state-nanoworld.

And in completion of the present section let us note one more important feature of natural nanocomposites structure. In Refs. [24, 25] the interfacial regions absence in amorphous glassy polymers, treated as natural nanocomposites, was shown. This means, that such nanocomposites structure represents a nanofiller (nanoclusters), immersed in matrix (loosely packed matrix of amorphous polymer structure), that is, unlike polymer nanocomposites with inorganic nanofiller (artificial nanocomposites) they have only two structural components.

15.3   THE NATURAL NANOCOMPOSITES REINFORCEMENT

As it is well-known [26], very often a filler introduction in polymer matrix is carried out for the last stiffness enhancement. Therefore, the reinforcement degree of polymer composites, defined as a composite and matrix polymer elasticity moduli ratio, is one of their most important characteristics.

At amorphous glassy polymers as natural nanocomposites treatment the estimation of filling degree or nanoclusters relative fraction φ_{cl} has an important significance. Therefore, the authors of Ref. [27] carried out the comparison of the indicated parameter estimation different methods, one of which is EPR-spectroscopy (the method of spin probes). The indicated method allows to study amorphous polymer structural heterogeneity, using radicals distribution character. As it is known [28], the method, based on the parameter d_1/d_c – the ratio of spectrum extreme components total intensity to central component intensity-measurement is the simplest and most suitable method of nitroxyl radicals local concentrations determination. The value of dipole-dipole interaction ΔH_{dd} is directly proportional to spin probes concentration C_w [29]:

$$\Delta H_{dd} = A \cdot C_w, \tag{15.5}$$

where $A = 5 \times 10^{-20}$ Ersted·cm^3 in the case of radicals chaotic distribution.

On the basis of the Eq. (15.5) the relationship was obtained, which allows to calculate the average distance r between two paramagnetic probes [29]:

$$r = 38\left(\Delta H_{dd}\right)^{-1/3}, \text{Å} \qquad (15.6)$$

where ΔH_{dd} is given in Ersteds.

In Fig. 15.9, the dependence of d_1/d_c on mean distance r between chaotically distributed in amorphous PC radicals-probes is adduced. For PC at $T = 77K$ the values of $d_1/d_c = 0.38 \div 0.40$ were obtained. One can make an assumption about volume fractions relation for the ordered domains (nanoclusters) and loosely packed matrix of amorphous PC. The indicated value d_1/d_c means, that in PC at probes statistical distribution 0.40 of its volume is accessible for radicals and approximately 0.60 of volume remains unoccupied by spin probes, that is, the nanoclusters relative fraction $_{cl}$ according to the EPR method makes up approximately $0.60 \div 0.62$.

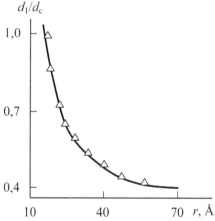

FIGURE 15.9   The dependence of parameter d_1/d_c of EPR spectrum on the value of mean distance r between radicals for PC [27].

This result corresponds well to the experimental data of Yech [30] and Perepechko [31], who obtained the values 0.60 and 0.63 for densely packed regions relative fraction in amorphous polymers.

The authors of Ref. [11] fulfilled $_{cl}$ estimation with the aid of reversed gas chromatography and obtained the following magnitudes of this parameter for PC, poly (methyl methacrylate) and polysulfone: 0.70, 0.60 and 0.65, accordingly (Table 15.2).

Within the frameworks of the cluster model φ_{cl} estimation can be fulfilled by the percolation relationship (the Eq. (4.66)) usage. Let us note, that in the given case the temperature of polymers structure quasiequilibrium state attainment, lower of which φ_{cl} value does not change, that is, T_0 [32], is accepted as testing temperature T. The calculation φ_{cl} results according to the Eq. (4.66) for the mentioned above polymers are adduced in Table 15.2, which correspond well to other authors estimations.

Proceeding from the circumstance, that radicals-probes are concentrated mainly in intercluster regions, the nanocluster size can be estimated, which in amorphous PC should be approximately equal to mean distance r between two paramagnetic probes, that is, ~50 Å (Fig. 15.9). This value corresponds well to the experimental data, obtained by dark-field electron microscopy method ($\approx 30 \div 100$ Å) [33].

Within the frameworks of the cluster model the distance between two neighboring nanoclusters can be estimated according to the Eq. (4.63) as $2R_{cl}$. The estimation $2R_{cl}$ by this mode gives the value 53.1 Å (at F = 41) that corresponds excellently to the method EPR data.

Thus, the Ref. [27] results showed, that the obtained by EPR method natural nanocomposites (amorphous glassy polymers) structure characteristics corresponded completely to both the cluster model theoretical calculations and other authors estimations. In other words, EPR data are experimental confirmation of the cluster model of polymers amorphous state structure.

The treatment of amorphous glassy polymers as natural nanocomposites allows to use for their elasticity modulus E_p (and, hence, the reinforcement degree $E_p/E_{l.m.}$, where $E_{l.m.}$ is loosely packed matrix elasticity modulus) description theories, developed for polymer composites reinforcement degree description [9, 17]. The authors of Ref. [34] showed correctness of particulate-filled polymer nanocomposites reinforcement of two concepts on the example of amorphous PC. For theoretical estimation of particulate-filled polymer nanocomposites reinforcement degree E_n/E_m two equations can be used. The first from them has the look [35]:

$$\frac{E_n}{E_m} = 1 + \varphi_n^{1,7}, \tag{15.7}$$

where E_n and E_m are elasticity moduli of nanocomposites and matrix polymer, accordingly, φ_n is nanofiller volume contents.

The second equation offered by the authors of Ref. [36] is:

$$\frac{E_n}{E_m} = 1 + \frac{0,19W_n l_{st}}{D_p^{1/2}},\tag{15.8}$$

where W_n is nanofiller mass contents in mas.%, D_p is nanofiller particles diameter in nm.

Let us consider included in the Eqs. (15.7) and (15.8) parameters estimation methods. It is obvious, that in the case of natural nanocomposites one should accept: $E_n = E_p$, $E_m = E_{1.m.}$ and $\varphi_n = \varphi_{cl}$, the value of the latter can be estimated according to the Eq. (4.66).

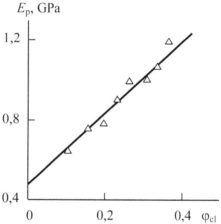

FIGURE 15.10   The dependence of elasticity modulus E_p on nanoclusters relative fraction φ_{cl} for PC [34].

The mass fraction of nanoclusters W_{cl} can be calculated as follows [37]:

$$W_{cl} = \rho \varphi_{cl},\tag{15.9}$$

where ρ is nanofiller (nanoclusters) density, which is equal to 1300 kg/m³ for PC.

The value $E_{1.m.}$ can be determined by the construction of $E_p(\varphi_{cl})$ plotting, which is adduced in Fig. 15.10. As one can see, this plot is approximately linear and its extrapolation to $\varphi_{cl} = 0$ gives the value $E_{1.m.}$ And at last, as it follows from the nanoclusters definition (see chapter one) one should accept $D_p \approx l_{st}$ for them and then the Eq. (15.8) accepts the following look [34]:

In Fig. 15.11 the comparison of theoretical calculation according to the Eqs. (15.7) and (15.10) with experimental values of reinforcement degree

$E_p/E_{1.m.}$ for PC is adduced. As one can see, both indicated equations give a good enough correspondence with the experiment: their average discrepancy makes up 5.6% in the Eq. (15.7) case and 9.6% for the Eq. (15.10). In other words, in both cases the average discrepancy does not exceed an experimental error for mechanical tests. This means, that both considered methods can be used for PC elasticity modulus prediction. Besides, it necessary to note, that the percolation relationship (the Eq. (15.7)) qualitatively describes the dependence $E_p/E_{1.m.}(\varphi_{cl})$ better, than the empirical relationship (the Eq. (15.10)).

$$\frac{E_n}{E_m} = 1 + 0,19\rho\phi_{cl}l_{st}^{1/2} . \qquad (15.10)$$

The obtained results allowed to make another important conclusion. As it is known, the percolation relationship (the Eq. (15.7)) assumes, that nanofiller is percolation system (polymer composite) solid-body component and in virtue of this circumstance defines this system elasticity modulus. However, for artificial polymer particulate-filled nanocomposites, consisting of polymer matrix and inorganic nanofiller, the Eq. (15.7) in the cited form gives the understated values of reinforcement degree. The authors of Refs. [9, 17] showed, that for such nanocomposites the sum ($\varphi_n+\varphi_{if}$), where φ_{if} was interfacial regions relative fraction, was a solid-body component (see the Eq. (11.5)). The correspondence of experimental data and calculation according to the Eq. (15.7) demonstrates, that amorphous polymer is the specific nanocomposite, in which interfacial regions are absent [24, 25]. This important circumstance is necessary to take into consideration at amorphous glassy polymers structure and properties description while simulating them as natural nanocomposites. Besides, one should note, that unlike micromechanical models the Eqs. (15.7) and (15.10) do not take into account nanofiller elasticity modulus, which is substantially differed for PC nanoclusters and inorganic nanofillers [34].

Another mode of natural nanocomposites reinforcement degree description is micromechanical models application, developed for polymer composites mechanical behavior description [1, 37–39]. So, Takayanagi and Kerner models are often used for the description of reinforcement degree on composition for the indicated materials [38, 39]. The authors of Ref. [40] used the mentioned models for theoretical treatment of natural nanocomposites reinforcement degree temperature dependence on the example of PC.

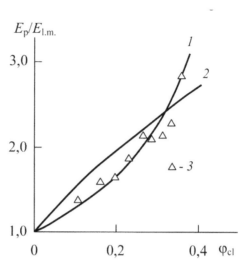

FIGURE 15.11 The dependences of reinforcement degree $E_p/E_{1.m}$ on nanoclusters relative fraction φ_{cl} for PC. 1 – calculation according to the Eq. (15.7); 2 – calculation according to the Eq. (15.10); 3 – the experimental data [34].

Takayanagi model belongs to a micromechanical composite models group, allowing empirical description of composite response upon mechanical influence on the basis of constituent it elements properties. One of the possible expressions within the frameworks of this model has the following look [38]:

$$\frac{G_c}{G_m} = \frac{\phi_m G_m + (\alpha + \phi_f) G_f}{(1 + \alpha \phi_f) G_m + \alpha \phi_m G_f},$$
(15.11)

where G_c, G_m and G_f are shear moduli of composite, polymer matrix and filler, accordingly, φ_m and φ_f are polymer matrix and filler relative fractions, respectively, α is a fitted parameter.

Kerner equation is identical to the Eq. (15.11), but for it the parameter α does not fit and has the following analytical expression [38]:

$$\alpha_m = \frac{2(4 - 5v_m)}{(7 - 5v_m)},$$
(15.12)

where α_m and v_m are parameter α and Poisson's ratio for polymer matrix.

Let us consider determination methods of the Eqs. (15.11) and (15.12) parameters, which are necessary for the indicated equations application in the case of natural nanocomposites, Firstly, it is obvious, that in the last case one should accept: $G_c = G_p$, $G_m = G_{l.m.}$, $G_f = G_{cl}$, where G_p, $G_{l.m.}$ and G_{cl} are shear moduli of polymer, loosely packed matrix and nanoclusters, accordingly, and also $\varphi_f = \varphi_{cl}$, where φ_{cl} is determined according to the percolation relationship (the Eq. (4.66)). Young's modulus for loosely packed matrix and nanoclusters can be received from the data of Fig. 15.10 by the dependence $E_p(\varphi_{cl})$ extrapolation to $\varphi_{cl} = 1.0$, respectively. The corresponding shear moduli were calculated according to the general Eq. (4.58). The value of nanoclusters fractal dimension d_f^{cl} in virtue of their dense package is accepted equal to the greatest dimension for real solids ($d_f^{l.m.} = 2.95$ [40]) and loosely packed matrix fractal dimension $d_f^{l.m.}$ can be estimated with the help of the Eq. (12.6).

However, the calculation according to the Eqs. (15.11) and (15.12) does not give a good correspondence to the experiment, especially for the temperature range of $T = 373 \div 413$ K in PC case. As it is known [38], in empirical modifications of Kerner equation it is usually supposed, that nominal concentration scale differs from mechanically effective filler fraction ϕ_f^{ef}, which can be written accounting for the designations used above for natural nanocomposites as follows [41].

$$\phi_f^{ef} = \frac{\left(G_p - G_{l.m.}\right)\left(G_{l.m.} + \alpha_{l.m.}G_{cl}\right)}{\left(G_{cl} - G_{l.m.}\right)\left(G_{l.m.} + \alpha_{l.m.}G_p\right)}, \tag{15.13}$$

where $\alpha_{l.m.} = \alpha_m$. The value $\alpha_{l.m.}$ can be determined according to the Eq. (15.12), estimating Poisson's ratio of loosely packed matrix $v_{l.m.}$ by the known values $d_f^{l.m.}$ according to the Eq. (1.9).

Besides, one more empirical modification ϕ_f^{ef} exists, which can be written as follows [41]:

$$\phi_{cl_2}^{ef} = \phi_{cl} + c\left(\frac{\phi_{cl}}{2r_{cl}}\right)^{2/3}, \tag{15.14}$$

where c is empirical coefficient of order one r_{cl} is nanocluster radius, determined according to the Eq. (4.64).

At the value ϕ_f^{ef} calculation according to the Eq. (15.14) magnitude c was accepted equal to 1.0 for the temperature range of $T = 293 \div 363$ K and equal to 1.2 – for the range of $T = 373 \div 413$ K and $2r_{cl}$ is given in nm. In Fig. 15.12 the comparison of values ϕ_f^{ef}, calculated according to the Eqs.

(15.13) and (15.14) (ϕ_{cl}^{ef} and $\phi_{cl_2}^{ef}$, accordingly) is adduced. As one can see, a good enough conformity of the values $\phi_{cl_1}^{ef}$, estimated by both methods, is obtained (the average discrepancy of $\phi_{cl_1}^{ef}$ and ϕ_{cl}^{ef} makes up slightly larger than 20%). Let us note, that the effective value φ_{cl} exceeds essentially the nominal one, determined according to the Eq. (4.66): within the range of $T = 293 \div 363K$ by about 70% and within the range of $T = 373 \div 413K$ – almost in three times.

In Fig. 15.13, the comparison of experimental and calculated according to Kerner equation (the Eq. (15.11)) with the Eqs. (15.13) and (15.14) using values of reinforcement degree by shear modulus $G_p/G_{1.m.}$ as a function of testing temperature T for PC is adduced. As one can see, in this case at the usage of nanoclusters effective concentration scale (ϕ_{cl}^{ef} instead of φ_{cl}) the good conformity of theory and experiment is obtained (their average discrepancy makes up 6%).

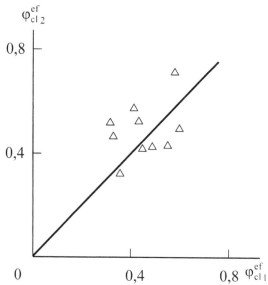

FIGURE 15.12   The comparison of nanoclusters effective concentration scale $\phi_{cl_1}^{ef}$ and $\phi_{cl_2}^{ef}$, calculated according to the Eqs. (15.13) and (15.14), respectively, for PC. A straight line shows the relation 1:1 [41].

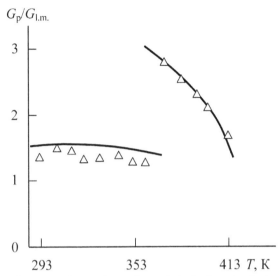

FIGURE 15.13 The comparison of experimental (points) and calculated according to the Eqs. (15.11), (15.13) and (15.14) (solid lines) values of reinforcement degree by shear modulus $G_p/G_{l.m.}$ as a function of testing temperature T for PC [41].

Hence, the stated above results have shown the modified Kerner equation application correctness for natural nanocomposites elastic response description. Really this fact by itself confirms the possibility of amorphous glassy polymers treatment as nanocomposites. Microcomposite models usage gives the clear notion about factors, influencing polymers stiffness.

15.4 INTERCOMPONENT ADHESION IN NATURAL NANOCOMPOSITES

Amorphous glassy polymers as natural nanocomposites puts forward to the foreground their study intercomponent interactions, that is, interactions nanoclusters – loosely packed matrix. This problem plays always one of the main roles at multiphase (multicomponent) systems consideration, since the indicated interactions or interfacial adhesion level defines to a great extent such systems properties [42]. Therefore, the authors of Ref. [43] studied the physical principles of intercomponent adhesion for natural nanocomposites on the example of PC.

The authors of Ref. [44] considered three main cases of the dependence of reinforcement degree E_c/E_m on φ_f. In this work the authors have shown, that there are the following main types of the dependences $E_c/E_m(\varphi_f)$ exist:

(1) the ideal adhesion between filler and polymer matrix, described by Kerner equation (perfect adhesion), which can be approximated by the following relationship:

$$\frac{E_c}{E_m} = 1 + 11,64\phi_f - 44,4\phi_f^2 + 96,3\phi_f^3; \qquad (15.15)$$

(2) zero adhesional strength at a large friction coefficient between filler and polymer matrix, which is described by the equation:

$$\frac{E_c}{E_m} = 1 + \phi_f; \qquad (15.16)$$

(3) the complete absence of interaction and ideal slippage between filler and polymer matrix, when composite elasticity modulus is defined practically by polymer cross-section and connected with the filling degree by the equation:

$$\frac{E_c}{E_m} = 1 - \phi_f^{2/3}. \qquad (15.17)$$

In Fig. 15.14, the theoretical dependences $E_p/E_{1.m.}(\varphi_{cl})$ plotted according to the Eqs. (15.15) $\div$ (15.17), as well as experimental data (points) for PC are shown. As it follows from the adduced in Fig. 15.14 comparison at $T = 293 \div 363$ K the experimental data correspond well to the Eq. (15.16), that is, in this case zero adhesional strength at a large friction coefficient is observed. At $T = 373 \div 413$ K the experimental data correspond to the Eq. (15.15), that is, the perfect adhesion between nanoclusters and loosely packed matrix is observed. Thus, the adduced in Fig. 15.14 data demonstrated, that depending on testing temperature two types of interactions nanoclusters – loosely packed matrix are observed: either perfect adhesion or large friction between them. For quantitative estimation of these interactions it is necessary to determine their level, which can be made with the help of the parameter b_m, which is determined according to the equation [45]:

$$\sigma_f^c = \sigma_f^m K_s - b_m \phi_f, \qquad (15.18)$$

where σ_f^c and σ_f^m are fracture stress of composite and polymer matrix, respectively, K_s is stress concentration coefficient. It is obvious, that since b_m increase results to σ_f^c reduction, then this means interfacial adhesion level decrease.

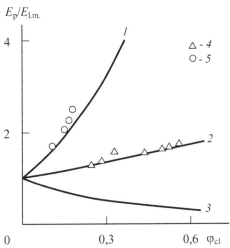

FIGURE 15.14 The dependences of reinforcement degree $E_p/E_{1.m}$ on nanoclusters relative fraction φ_{cl}. 1÷3 – the theoretical dependences, corresponding to the Eqs. (15.15) ÷ (15.17), accordingly; 4, 5 – the experimental data for PC within the temperature ranges: 293÷363K(4) and 373÷413K(5) [43].

The true fracture stress σ_f^{tr} for PC, taking into account sample cross-section change in a deformation process, was used as σ_f^c for natural nano-composites, which can be determined according to the known formula:

$$\sigma_f^{tr} = \sigma_f^n \left(1 + \varepsilon_f\right),\qquad\qquad 15.19$$

where σ_f^m is nominal (engineering) fracture stress, ε_f is strain at fracture. The value σ_f^m, which is accepted equal to loosely packed matrix strength $\sigma_f^{1.m.}$, was determined by graphic method, namely, by the dependence $\sigma_f^{1.m}$ (φ_{cl}) plotting, which proves to be linear, and by subsequent extrapolation of it to $\varphi_{cl} = 0$, that gives $\sigma_f^{1.m.}=40$ MPa [43].

And at last, the value Ks can be determined with the help of the following equation [39]:

$$\sigma_f^{tr} = \sigma_f^{l.m.}\left(1 - \phi_{cl}^{2/3}\right)K_s.$$

(15.20)

The parameter b_m calculation according to the stated above technique shows its decrease (intercomponent adhesion level enhancement) at testing temperature raising within the range of $b_m \approx 500 \div 130$.

For interactions nanoclusters – loosely packed matrix estimation within the range of $T = 293\div373$K the authors of Ref. [48] used the model of Witten-Sander clusters friction, stated in Ref. [46]. This model application is due to the circumstance, that amorphous glassy polymer structure can be presented as an indicated clusters large number set [47]. According to this model, Witten-Sander clusters generalized friction coefficient t can be written as follows [46]:

$$f = \ln c + \beta \cdot \ln n_{cl},$$

(15.21)

where c is constant, β is coefficient, n_{cl} is statistical segments number per one nanocluster.

The coefficient β value is determined as follows [46]:

$$\beta = \left(d_f^{cl}\right)^{-1},$$

(15.22)

where d_f^{cl} is nanocluster structure fractal dimension, which is equal, as before, to 2.95 [40].

In Fig. 15.15, the dependence $b_m(f)$ is adduced, which is broken down into two parts. On the first of them, corresponding to the range of $T = 293\div363$ K, the intercomponent interaction level is intensified at f decreasing (i.e., b_m reduction is observed and on the second one, corresponding to the range of $T = 373 \div 413$ K, b_m = const independent on value f. These results correspond completely to the data of Fig. 15.14, where in the first from the indicated temperature ranges the value $E_p/E_{l.m.}$ is defined by nanoclusters friction and in the second one by adhesion and, hence, it does not depend on friction coefficient.

As it has been shown in Ref. [48], the interfacial (or intercomponent) adhesion level depends on a number of accessible for the formation interfacial (intercomponent) bond sites (nodes) on the filler (nanocluster) particle surface N_u, which is determined as follows [49]:

$$N_u = L^{d_u},$$

(15.23)

where L is filler particle size, d_u is fractal dimension of accessible for contact ("nonscreened") indicated particle surface.

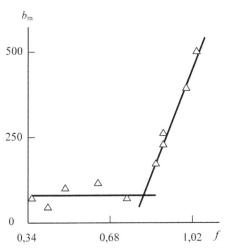

FIGURE 15.15   The dependence of parameter b_m on generalized friction coefficient f for PC [43].

One should choose the nanocluster characteristic size as L for the natural nanocomposite, which is equal to statistical segment l_{st}, determined according to the Eq. (2.15), and the dimension d_u is determined according to the following relationship [49]:

$$d_u = (d_{surf} - 1) + \left(\frac{d - d_{surf}}{d_w}\right), \tag{15.24}$$

where d_{surf} is nanocluster surface fractal dimension, d_w is dimension of random walk on this surface, estimated according to Aarony-Stauffer rule [49]:

$$d_w = d_{surf} + 1. \tag{15.25}$$

The following technique was used for the dimension d_{surf} calculation. First the nanocluster diameter $D_{cl} = 2r_{cl}$ was determined according to the Eq. (4.64) and then its specific surface S_u was estimated [35]:

$$S_u = \frac{6}{\rho_{cl} D_{cl}}, \tag{15.26}$$

where ρ_{cl} is the nanocluster density, equal to 1300 kg/m³ in the PC case.

And at last, the dimension d_{surf} was calculated with the help of the equation [20]:

$$S_u = 5,25 \times 10^3 \left(\frac{D_{cl}}{2} \right)^{d_{surf} - d}. \tag{15.27}$$

In Fig. 15.16, the dependence $b_m(N_u)$ for PC is adduced, which is broken down into two parts similarly to the dependence $b_m(f)$ (Fig15.15). At $T = 293 \div 363$ K the value b_m is independent on N_u, since nanocluster – loosely packed matrix interactions are defined by their friction coefficient. Within the range of $T = 373 \div 413$ K intercomponent adhesion level enhancement (b_m reduction) at active sites number N_u growth is observed, as was to be expected. Thus, the data of both Figs. 15.16 and 15.15 correspond to Fig. 15.14 results.

With regard to the data of Figs. 15.15 and 15.16 two remarks should be made. Firstly, the transition from one reinforcement mechanism to another corresponds to loosely packed matrix glass transition temperature, which is approximately equal to $T_g - 50$K [11]. Secondly, the extrapolation of Fig. 15.16 plot to $b_m = 0$ gives the value $N_u \approx 71$, that corresponds approximately to polymer structure dimension $d_f = 2.86$.

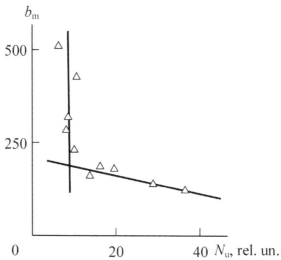

FIGURE 15.16 The dependence of parameter b_m on nanocluster surface active ("nonscreened") sites number N_u for PC [43].

In this theme completion an interesting structural aspect of intercomponent adhesion in natural nanocomposites (polymers) should be noted.
Despite the considered above different mechanisms of reinforcement and
nanoclusters-loosely packed matrix interaction realization the common
dependence $b_m(\varphi_{cl})$ is obtained for the entire studied temperature range of
$293 \div 413K$, which is shown in Fig. 15.17. This dependence is linear, that
allows to determine the limiting values $b_m \approx 970$ at $\varphi_{cl} = 1.0$ and $b_m = 0$ at
$\varphi_{cl} = 0$. Besides, let us note, that the shown in Figs. 15.14÷15.16 structural
transition is realized at $\varphi_{cl} \approx 0.26$ [43].

Hence, the stated above results have demonstrated, that intercomponent
adhesion level in natural nanocomposites (polymers) has structural origin
and is defined by nanoclusters relative fraction. In two temperature ranges
two different reinforcement mechanisms are realized, which are due to large
friction between nanoclusters and loosely packed matrix and also perfect (by
Kerner) adhesion between them. These mechanisms can be described successfully within the frameworks of fractal analysis.

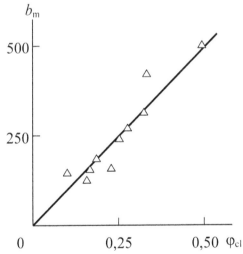

FIGURE 15.17   The dependence of parameter b_m on nanoclusters relative fraction φ_{cl} for
PC [43].

The further study of intercomponent adhesion in natural nanocomposites was fulfilled in Ref. [50]. In Fig. 15.18, the dependence $b_m(T)$ for PC is
shown, from which b_m reduction or intercomponent adhesion level enhancement at testing temperature growth follows. In the same figure the maximum

value b_m for nanocomposites polypropylene/Na$^+$-montmorillonite [9] was shown by a horizontal shaded line. As one can see, b_m values for PC within the temperature range of $T = 373 \div 413$ K by absolute value are close to the corresponding parameter for the indicated nanocomposite, that indicates high enough intercomponent adhesion level for PC within this temperature range.

Let us note an important structural aspect of the dependence $b_m(T)$, shown in Fig. 15.18. According to the cluster model [4], the decay of instable nano-clusters occurs at temperature $T'_g \approx T_g - 50$ K, holding back loosely packed matrix in glassy state, owing to which this structural component is devitrifi-cated within the temperature range of $d_f^t \div T_g$. Such effect results to rapid re-duction of polymer mechanical properties within the indicated temperature range [51]. As it follows from the data of Fig. 15.18, precisely in this tem-perature range the highest intercomponent adhesion level is observed and its value approaches to the corresponding characteristic for nanocomposites polypropylene/Na$^+$-montmorillonite.

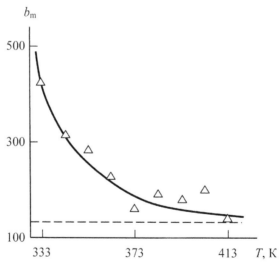

FIGURE 15.18   The dependence of parameter b_m on testing temperature T for PC. The horizontal shaded line shows the maximum value b_m for nanocomposites polypropylene/Na$^+$-montmorillonite [50].

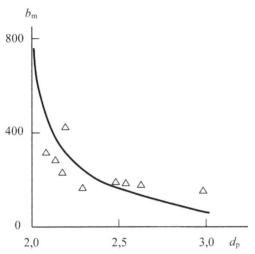

FIGURE 15.19   The dependence of parameter b_m on nanoclusters surface fractal dimension d_{surf} for PC [50].

It can be supposed with a high probability degree that adhesion level depends on the structure of nanoclusters surface, coming into contact with loosely packed matrix, which is characterized by the dimension d_{surf}. In Fig. 15.19, the dependence $b_m(d_{surf})$ for PC is adduced, from which rapid reduction b_m (or intercomponent adhesion level enhancement) follows at d_{surf} growth or, roughly speaking, at nanoclusters surface roughness enhancement.

The authors of Ref. [48] showed that the interfacial adhesion level for composites polyhydroxyether/graphite was raised at the decrease of polymer matrix and filler particles surface fractal dimensions difference. The similar approach was used by the authors of Ref. [50], who calculated nanoclusters d_f^{cl} and loosely packed matrix $d_f^{l.m.}$ fractal dimensions difference Δd_f:

$$\Delta d_f = d_f^{cl} - d_f^{l.m.} , \tag{15.28}$$

where d_f^{cl} is accepted equal to real solids maximum dimension ($d_f^{cl} = 2.95$[40]) in virtue of their dense packing and the value d_f^{cl} was calculated according to the mixtures rule (the Eq. (12.6)).

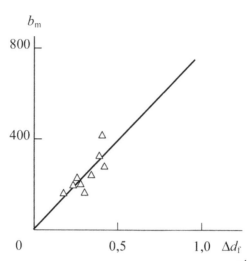

FIGURE 15.20   The dependence of parameter b_m on nanoclusters d_f^{cl} and loosely packed matrix d_f^{cl} structures fractal dimensions difference Δd_f for PC [50].

In Fig. 15.20, the dependence of b_m on the difference Δd_f is adduced, from which b_m decrease or intercomponent adhesion level enhancement at Δd_f reduction or values d_f^{cl} and $d_f^{l.m.}$ growing similarity follows. This dependence demonstrates, that the greatest intercomponent adhesion level, corresponding to $b_m = 0$, is reached at $\Delta d_f = 0.95$ and is equal to ~780.

The data of Figs. 15.14 and 15.18 combination shows, that the value $b_m \approx 200$ corresponds to perfect adhesion by Kerner. In its turn, the Figs. 15.16 and 15.17 plots data demonstrated, that the value $b_m \approx 200$ could be obtained either at $d_{surf} > 2.5$ or at $\Delta d_f < 0.3$, accordingly. The obtained earlier results showed [24], that the condition $d_{surf} > 2.5$ was reached at $r_{cl} < 7.5$Å or $T > 373$ K, that again corresponded well to the stated above results. And at last, the $\Delta d_f \approx 0.3$ or $T_g' \approx 2.65$ according to the Eq. (12.6) was also obtained at $T \approx 373$K.

Hence, at the indicated above conditions fulfillment within the temperature range of $T < T_g'$ for PC perfect intercomponent adhesion can be obtained, corresponding to Kerner equation, and then the value E_p estimation should be carried out according to the Eq. (15.15). At $T = 293$ K ($\varphi_{cl} = 0.56$, $E_m = 0.85$GPa) the value E_p will be equal to 8.9 GPa, that approximately in 6 times larger, than the value E_p for serial industrial PC brands at the indicated temperature.

Let us note the practically important feature of the obtained above results. As it was shown, the perfect intercomponent adhesion corresponds to $b_m \approx 200$, but not $b_m = 0$. This means, that the real adhesion in natural nanocomposites can be higher than the perfect one by Kerner, which was shown experimentally on the example of particulate-filled polymer nanocomposites [17, 52]. This effect was named as nanoadhesion and its realization gives large possibilities for elasticity modulus increase of both natural and artificial nanocomposites. So, the introduction in aromatic polyamide (phenylone) of 0.3 mas.% aerosil only at nanoadhesion availability gives the same nanocomposite elasticity modulus enhancement effect, as the introduction of 3 mas. % of organoclay, which at present is assumed as one of the most effective nanofillers [9]. This assumes, that the value $E_p = 8.9$ GPa for PC is not a limiting one, at any rate, theoretically. Let us note in addition, that the indicated E_p values can be obtained at the natural nanocomposites nanofiller (nanoclusters) elasticity modulus magnitude $E_{cl} = 2.0$ GPa, that is, at the condition $E_{cl} < E_p$. Such result possibility follows from the polymer composites structure fractal concept [53], namely, the model [44], in which the Eqs. (15.15) ÷ (15.17) do not contain nanofiller elasticity modulus, and reinforcement percolation model [35] (see the Eq. (11.5)).

The condition $d_{surf} < 2.5$, that is, $r_{cl} < 7.5$ Å or $N_{cl} < 5$, in practice can be realized by the nanosystems mechanosynthesis principles using, the grounds of which are stated in Ref. [54]. However, another more simple and, hence, more technological method of desirable structure attainment realization is possible, one from which will be considered in subsequent section.

Hence, the stated above results demonstrated, that the adhesion level between natural nanocomposite structural components depended on nanoclusters and loosely packed matrix structures closeness. This level change can result to polymer elasticity modulus significant increase. A number of this effect practical realization methods was considered [50].

The mentioned above dependence of intercomponent adhesion level on nanoclusters radius r_{cl} assumes more general dependence of parameter b_m on nanoclusters geometry. The authors of Ref. [55] carried out calculation of accessible for contact sites of nanoclusters surface and loosely packed matrix number N_u according to the Eq. (15.23) for two cases. the nanocluster is simulated as a cylinder with diameter D_{cl} and length l_{st}, where l_{st} is statistical segment length, therefore, in the first case its butt-end is contacting with loosely packed matrix nanocluster surface and then $L = D_{cl}$ and in the second case with its side (cylindrical) surface and then $L = l_{st}$. In Fig. 15.21 the dependences of parameter b_m on value N_u, corresponding to the two considered above cases, are adduced. As one can see, in both cases, for the range of $T = 293 \div 363$ K l_{st}, where interactions nanoclusters – loosely packed matrix are characterized by powerful friction between them, the value b_m does not depend on N_u, as it was expected. For the range of $T = 373 \div 413$ K, where between nanoclusters and loosely packed matrix perfect adhesion is observed, the linear dependences $b_m(N_u)$ are obtained. However, at using value D_{cl} as Lb_m reduction or intercomponent adhesion level enhancement at N_u decreasing is obtained and at $N_u = 0$ b_m value reaches its minimum magnitude $b_m = 0$. In other words, in this case the minimum level of intercomponent adhesion is reached at intercomponent bonds formation sites (nodes) absence that is physically incorrect [48]. And on the contrary at the condition $L = l_{st} b_m$ the reduction (intercomponent adhesion level enhancement) at the increase of contacts number N_u between nanoclusters and loosely packed matrix is observed, that is obvious from the physical point of view. Thus, the data of Fig. 15.21 indicate unequivocally, that the intercomponent adhesion is realized over side (cylindrical) nanoclusters surface and butt-end surfaces in this effect formation do not participate.

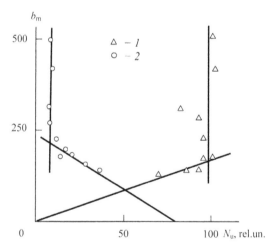

FIGURE 15.21 The dependences of parameter b_m on a number of accessible for intercomponent bonds formation sizes on nanocluster surface N_u at the condition $L = D_{cl}$ (1) and $L = l_{st}$ (2) for PC [55].

Let us consider geometrical aspects intercomponent interactions in natural nanocomposites. In Fig. 15.22 the dependence of nanoclusters butt-end S_b and side (cylindrical) S_c surfaces areas on testing temperature T for PC are adduced. As one can see, the following criterion corresponds to the transition from strong friction to perfect adhesion at $T = 373K$ [55]:

$$S_b \approx S_c. \qquad (15.29)$$

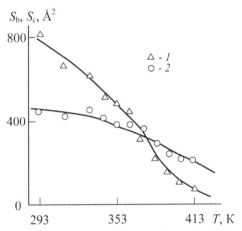

FIGURE 15.22 The dependences of nanoclusters butt-end $S_b(1)$ and cylindrical $S_c(2)$ surfaces areas on testing temperature T for PC [55].

Hence, the intercomponent interaction type transition from the large friction nanoclusters − loosely packed matrix to the perfect adhesion between them is defined by nanoclusters geometry: at $S_b > S_c$ the interactions of the first type is realized and at $S_b < S_c$ − the second one. Proceeding from this, it is expected that intercomponent interactions level is defined by the ratio S_b/S_c. Actually, the adduced in Fig. 15.23 data demonstrate b_m reduction at the indicated ratio decrease, but at the criterion (15.29) realization or $S_b/S_c \approx 1$ S_b/S_c decreasing does not result to b_m reduction and at $S_b/S_c < 1$ intercomponent adhesion level remains maximum high and constant. [55].

Hence, the stated above results have demonstrated, that interactions nanoclusters-loosely packed matrix type (large friction or perfect adhesion) is defined by nanoclusters butt-end and side (cylindrical) surfaces areas ratio or their geometry if the first from the mentioned areas is larger that the second one then a large friction nanoclusters-loosely packed matrix is realized; if the second one exceeds the first one, then between the indicated structural components perfect adhesion is realized. In the second from the indicated cases intercomponent adhesion level does not depend on the mentioned areas ratio and remains maximum high and constant. In other words, the adhesion nanoclusters-loosely packed matrix is realized by nanoclusters cylindrical surface.

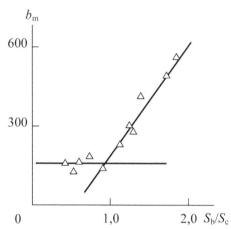

FIGURE 15.23   The dependence of parameter b_m on nanoclusters butt-end and cylindrical surfaces are ratio S_b/S_c value for PC [55].

The stated above results were experimentally confirmed by the EPR-spectroscopy method [56]. The Eqs. (4.63), (15.1) and (15.6) comparison shows, that dipole-dipole interaction energy ΔH_{dd} has structural origin, namely [56]:

$$\Delta H_{dd} \approx \left(\frac{V_{cl}}{n_{cl}} \right). \tag{15.30}$$

As estimations according to the Eq. (15.30) showed, within the temperature range of $T = 293 \div 413K$ for PC ΔH_{dd} increasing from 0.118 up to 0.328 Ersteds was observed.

Let us consider dipole-dipole interaction energy ΔH_{dd} intercommunication with nanoclusters geometry. In Fig. 15.24 the dependence of ΔH_{dd} on the ratio S_c/S_b for PC is adduced. As one can see, the linear growth ΔH_{dd} at ratio S_c/S_b increasing is observed, that is, either at S_c enhancement or at S_b reduction. Such character of the adduced in Fig. 15.24 dependence indicates unequivocally, that the contact nanoclusters-loosely packed matrix is realized on nanocluster cylindrical surface. Such effect was to be expected, since emerging from the butt-end surface statistically distributed polymer chains complicated the indicated contact realization unlike relatively smooth cylindrical surfaces. It is natural to suppose, that dipole-dipole interactions intensification or ΔH_{dd} increasing results to natural nanocomposites elasticity modulus E_p enhancement. The second as natural supposition at PC consideration as nanocomposite is the influence on the value E_p of nanoclusters (nanofiller) relative fraction φ_{cl}, which is determined according to the percolation relationship (the Eq. (4.66)).

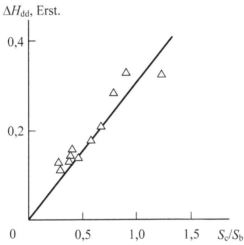

FIGURE 15.24   The dependence of dipole-dipole interaction energy ΔH_{dd} on nanoclusters cylindrical S_c and butt-end S_b surfaces areas ratio for PC [56].

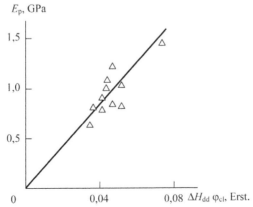

FIGURE 15.25   The dependence of elasticity modulus E_p on complex argument $(\Delta H_{dd}\varphi_{cl})$ for PC [56].

In Fig. 15.25, the dependence of elasticity modulus E_p on complex argument $(\Delta H_{dd}\varphi_{cl})$ for PC is presented. As one can see, this dependence is a linear one, passes through coordinates origin and is described analytically by the following empirical equation [56].

$$E_p = 21(\Delta H_{dd}\varphi_{cl}), \text{ GPa,} \qquad (15.31)$$

which with the appreciation of the Eq. (15.30) can be rewritten as follows [56]:

$$E_p = 21 \times 10^{-26}\left(\frac{f_{cl}V_{cl}}{n_{cl}}\right), \text{ GPa.} \qquad (15.32)$$

The Eq. (15.32) demonstrates clearly, that the value E_p and, hence polymer reinforcement degree is a function of its structural characteristics, described within the frameworks of the cluster model [3–5]. Let us note, that since parameters v_{cl} and φ_{cl} are a function of testing temperature, then the parameter n_{cl} is the most suitable factor for the value E_p regulation for practical purposes. In Fig. 15.26 the dependence $E_p(n_{cl})$ for PC at $T = 293$ K is adduced, calculated according to the Eq. (15.32), where the values v_{cl} and φ_{cl} were calculated according to the Eqs. (1.11) and (4.66), accordingly. As one can see, at small n_{cl} (<10) the sharp growth E_p is observed and at the smallest possible value $n_{cl} = 2$ the magnitude $E_p \approx 13.5$GPa. Since for PC $E_{l.m.} = 0.85$GPa, then it gives the greatest reinforcement degree $E_p/E_m \approx 15.9$. Let us note, that the greatest attainable reinforcement degree for artificial nanocomposites (polymers filled with inorganic nanofiller) cannot exceed 12 [9]. It is notable, that the shown in Fig. 15.26 dependence $E_p(n_{cl})$ for PC is identical completely by dependence shape to the dependence of elasticity modulus of nanofiller particles diameter for elastomeric nanocomposites [57].

Hence, the presented above results have shown that elasticity modulus of amorphous glassy polycarbonate, considered as natural nanocomposite, are defined completely by its suprasegmental structure state. This state can be described quantitatively within the frameworks of the cluster model of polymers amorphous state structure and characterized by local order level. Natural nanocomposites reinforcement degree can essentially exceed analogous parameter for artificial nanocomposites [56].

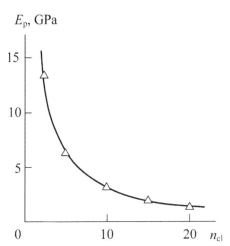

FIGURE 15.26   The dependence of elasticity modulus E_p on segments number n_{cl} per one nanocluster, calculated according to the Eq. (15.32) for PC at $T = 293$K [56].

As it has been shown above (see the Eqs. (15.7) and (15.15)), the nanocluster relative fraction increasing results to polymers elasticity modulus enhancement similarly to nanofiller contents enhancement in artificial nanocomposites. Therefore, the necessity of quantitative description and subsequent comparison of reinforcement degree for the two indicated above nanocomposites classes appears. The authors of Ref. [58, 59] fulfilled the comparative analysis of reinforcement degree by nanoclusters and by layered silicate (organoclay) for polyarylate and nanocomposite epoxy polymer/Na⁺—montmorillonite [60], accordingly.

In Fig. 15.27 theoretical dependences of reinforcement degree E_n/E_m on nanofiller contents φ_n, calculated according to the Eqs. (15.15) ÷ (15.17), are adduced. Besides, in the same figure the experimental values (E_n/E_m) for nanocomposites epoxy polymer Na⁺-montmorillonite (EP/MMT) at $T < T_g$ and $T > T_g$ (where T and T_g are testing and glass transition temperatures, respectively) are indicated by points. As one can see, for glassy epoxy matrix the experimental data correspond to the Eq. (15.16), that is, zero adhesional strength at a large friction coefficient and for devitrificated matrix – to the Eq. (15.15), that is, the perfect adhesion between nanofiller and polymer matrix, described by Kerner equation. Let us note that the authors of Ref. [17] explained the distinction indicated above by a much larger length of epoxy polymer segment in the second case.

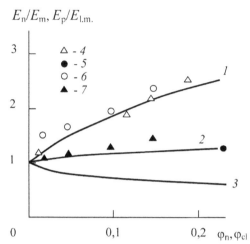

FIGURE 15.27 The dependences of reinforcement degree E_n/E_m and $E_p/E_{l.m.}$ on the contents of nanofiller φ_n and nanoclusters φ_{cl}, accordingly. 1÷3 – theoretical dependences (E_n/E_m) (φ_n), corresponding to the Eqs. (15.15)÷(15.17); 4.5 – the experimental data $(E_p/E_{l.m.})$ (φ_n) for Par at $T = T_g' \div T_g$ (4) and $T < T_g'$ (5); 6, 7 – the experimental data (E_n/E_m) (φ_n) for EP/MMT at $T > T_g$ (6) and $T < T_g$ (7) [59].

To obtain the similar comparison for natural nanocomposite (polymer) is impossible, since at $T \geq T_g$ nanoclusters are disintegrated and polymer ceases to be quasi-two-phase system [5]. However, within the frameworks of two-stage glass transition concept [11] it has been shown, that at temperature T_g', which is approximately equal to $T_g - 50$ K, instable (small) nanoclusters decay occurs, that results to loosely packed matrix devitrification at the indicated temperature [5]. Thus, within the range of temperature $T_g' \div T_g$ natural nanocomposite (polymer) is an analog of nanocomposite with glassy matrix [58]. As one can see, for the temperatures within the range of $T = T_g' \div T_g$ ($\varphi_{cl} = 0.06 \div 0.19$) the value $E_p/E_{l.m.}$ corresponds to the Eq. (15.15), that is, perfect adhesion nanoclusters-loosely packed matrix and at $T < T_g'$ ($\varphi_{cl} > 0.24$) – to the Eq. (15.16), that is, to zero adhesional strength at a large friction coefficient. Hence, the data of Fig. 15.27 demonstrated clearly the complete similarity, both qualitative and quantitative, of natural (Par) and artificial (EP/MMT) nanocomposites reinforcement degree behavior. Another microcomposite model (e.g., accounting for the layered silicate particles strong anisotropy) application can change the picture quantitatively only. The data of Fig. 15.27 qualitatively give the correspondence of reinforcement degree of nanocomposites indicated classes at the identical initial conditions.

Hence, the analogy in behavior of reinforcement degree of polyarylate by nanoclusters and nanocomposite epoxy polymer/Na^+-montmorillonite by layered silicate gives another reason for the consideration of polymer as natural nanocomposite. Again strong influence of interfacial (intercomponent) adhesion level on nanocomposites of any class reinforcement degree is confirmed [17].

15.5 THE METHODS OF NATURAL NANOCOMPOSITES NANOSTRUCTURE REGULATION

As it has been noted above, at present it is generally acknowledged [2], that macromolecular formations and polymer systems are always natural nanostructural systems in virtue of their structure features. In this connection the question of using this feature for polymeric materials properties and operating characteristics improvement arises. It is obvious enough that for structure-properties relationships receiving the quantitative nanostructural model of the indicated materials is necessary. It is also obvious that if the dependence of specific property on material structure state is unequivocal, then there will be quite sufficient modes to achieve this state. The cluster model of such state [3–5] is the most suitable for polymers amorphous state structure description. It has been shown, that this model basic structural element (cluster) is nanoparticles (nanocluster) (see Section 15.1). The cluster model was used successfully for cross-linked polymers structure and properties description [61]. Therefore, the authors of Ref. [62] fulfilled nanostructures regulation modes and of the latter influence on rarely cross-linked epoxy polymer properties study within the frameworks of the indicated model.

In Ref. [62], the studied object was an epoxy polymer on the basis of resin UP5–181, cured by iso-methyltetrahydrophthalic anhydride in the ratio by mass 1:0.56. Testing specimens were obtained by the hydrostatic extrusion method. The indicated method choice is due to the fact, that high hydrostatic pressure imposition in deformation process prevents the defects formation and growth, resulting to the material failure [64]. The extrusion strain ε_e was calculated according to the Eqs. (14.10) and (4.39) and makes up 0.14, 0.25, 0.36, 0.43 and 0.52. The obtained by hydrostatic extrusion specimens were annealed at maximum temperature 353 K during 15 min.

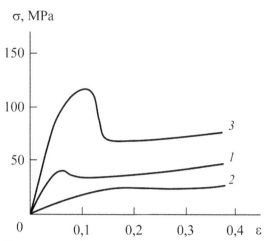

FIGURE 15.28   The stress – strain (σ – ε) diagrams for initial (1), extruded up to $\varepsilon_e = 0.52$ (2) and annealed (3) REP samples [62]

The hydrostatic extrusion and subsequent annealing of rarely cross-linked epoxy polymer (REP) result to very essential changes of its mechanical behavior and properties, in addition unexpected ones enough. The qualitative changes of REP mechanical behavior can be monitored according to the corresponding changes of the stress – strain (σ – ε) diagrams, shown in Fig. 15.28. The initial REP shows the expected enough behavior and both its elasticity modulus E and yield stress σ_Y are typical for such polymers at testing temperature T being distant from glass transition temperature T_g on about 40 K [51]. The small ($\approx$ 3 MPa) stress drop beyond yield stress is observed, that is also typical for amorphous polymers [61]. However, REP extrusion up to $_e$ = 0.52 results to stress drop $\Delta\sigma_Y$ ("yield tooth") disappearance and to the essential E and σ_Y reduction. Besides, the diagram σ – ε itself is now more like the similar diagram for rubber, than for glassy polymer. This specimen annealing at maximum temperature T_{an} = 353 K gives no less strong, but diametrically opposite effect – yield stress and elasticity modulus increase sharply (the latter in about twice in comparison with the initial REP and more than one order in comparison with the extruded specimen). Besides, the strongly pronounced "yield tooth" appears. Let us note, that specimen shrinkage at annealing is small ($\approx$10%), that makes up about 20% of ε_e [62].

The common picture of parameters E and σ_Y change as a function of ε_e is presented in Figs. 15.29 and 15.30, accordingly. As one can see, both indicated parameters showed common tendencies at ε_e change: up to $\varepsilon_e \approx 0.36$ inclusive E and σ_Y weak increase at ε_e growth is observed, moreover their absolute values for extruded and annealed specimens are close, but at $\varepsilon_e > 0.36$ the strongly pronounced antibatness of these parameters for the indicated specimen types is displayed. The cluster model of polymers amorphous state structure and developed within its frameworks polymers yielding treatment allows to explain such behavior of the studied samples [35, 65].

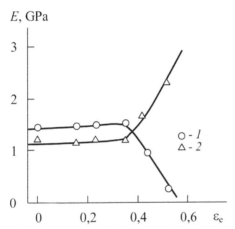

FIGURE 15.29 The dependences of elasticity modulus E_p on extrusion strain ε_e for extrudated (1) and annealed (2) REP [62].

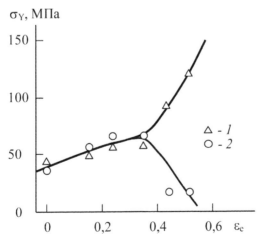

FIGURE 15.30 The dependences of yield stress σ_Y on extrusion strain ε_e for extrudated (1) and annealed (2) REP [62].

The cluster model supposes that polymers amorphous state structure represents the local order domains (nanoclusters), surrounded by loosely packed matrix. Nanoclusters consist of several collinear densely packed statistical segments of different macromolecules and in virtue of this they offer the analog of crystallite with stretched chains. There are two types of nanoclusters – stable, consisting of a relatively large segments number, and instable, consisting of a less number of such segments [65]. At temperature increase or mechanical stress application the instable nanoclusters disintegrate in the first place, that results to the two well-known effects. The first from them is known as two-stage glass transition process [11] and it supposes that at $T'_g = T_g - 50$ K disintegration of instable nanoclusters, restraining loosely packed matrix in glass state, occurs that defines devitrification of the latter [3, 5]. The well-known rapid polymers mechanical properties reduction at approaching to T_g [51] is the consequence of this. The second effect consists of instable nanoclusters decay at σ_Y under mechanical stress action, loosely packed matrix mechanical devitrification and, as consequence, glassy polymers rubber-like behavior on cold flow plateau [65]. The stress drop $\Delta\sigma_Y$ beyond yield stress is due to just instable nanoclusters decay and therefore, $\Delta\sigma_Y$ value serves as characteristic of these nanoclusters fraction[5]. Proceeding from this brief description, the experimental results, adduced in Figs. 15.28 ÷ 15.30, can be interpreted.

The rarely cross-linked epoxy polymer on the basis of resin UP5–181 has low glass transition temperature T_g, which can be estimated according to shrinkage measurements data as equal ≈ 333K. This means, that the testing temperature $T = 293$ K and T'_g for it are close, that is confirmed by small $\Delta\sigma_Y$ value for the initial REP. It assumes nanocluster (nanostructures) small relative fraction φ_{cl} [3–5] and, since these nanoclusters have arbitrary orientation, ε_e increase results rapidly enough to their decay, that induces loosely packed matrix mechanical devitrification at $\varepsilon_e > 0.36$. Devitrificated loosely packed matrix gives insignificant contribution to E_p [66, 67], equal practically to zero, that results to sharp (discrete) elasticity modulus decrease. Besides, at $T > T'_g$ φ_{cl} rapid decay is observed, that is, segments number decrease in both stable and instable nanocluster [5]. Since just these parameters (E and φ_{cl}) check σ_Y value, then their decrease defines yield stress sharp lessening. Now extruded at $\varepsilon_e > 0.36$ REP presents as matter of fact rubber with high cross-linking degree, that is reflected by its diagram $\sigma - \varepsilon$ (Fig. 15.28, curve 2).

The polymer oriented chains shrinkage occurs at the extruded REP annealing at temperature higher than T_g. Since this process is realized within a narrow temperature range and during a small time interval, then a large number of instable nanoclusters is formed. This effect is intensified by available molecular orientation, that is, by preliminary favorable segments arrangement, and it is reflected by $\Delta\sigma_Y$ strong increase (Fig. 15.28, curve 3).

The φ_{cl} enhancement results to E_p growth (Fig. 15.29) and φ_{cl} and E_p combined increase – to σ_Y considerable growth (Fig. 15.30).

The considered structural changes can be described quantitatively within the frameworks of the cluster model. The nanoclusters relative fraction φ_{cl} can be calculated according to the method, stated in Ref. [68].

The shown in Fig. 15.31 dependences $\varphi_{cl}(\varepsilon_e)$ have the character expected from the adduced above description and are its quantitative conformation. The adduced in Fig. 15.32 dependence of density ρ of REP extruded specimens on ε_e is similar to the dependence $\varphi_{cl}(\varepsilon_e)$, that was to be expected, since densely packed segments fraction decrease must be reflected in ρ reduction.

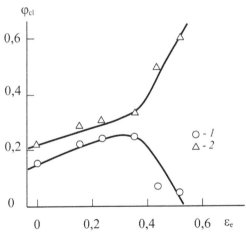

FIGURE 15.31   The dependences of nanoclusters relative fraction φ_{cl} on extrusion strain ε_e for extruded (1) and annealed (2) REP [62].

In Ref. [69], the supposition was made that ρ change can be conditioned to microcracks network formation in specimen that results to ρ reduction at large ε_e (0.43 and 0.52), which are close to the limiting ones. The ρ relative change ($\Delta\rho$) can be estimated according to the equation

$$\Delta\rho = \frac{\rho^{max} - \rho^{min}}{\rho^{max}}, \tag{15.33}$$

where ρ^{max} and ρ^{min} are the greatest and the smallest density values. This estimation gives $\Delta\rho \approx 0.01$. This value can be reasonable for free volume increase, which is necessary for loosely matrix devitrification (accounting for closeness of T and T'_g), but it is obviously small if to assume as real microcracks formation. As the experiments have shown, REP extrusion at $\varepsilon_e > 0.52$ is impossible owing to specimen cracking during extrusion process. This allows to suppose that value $\varepsilon_e = 0.52$ is close to the critical one. Therefore, the critical dilatation $\Delta\delta_{cr}$ value, which is necessary for microcracks cluster formation, can be estimated as follows [40]:

$$\Delta\delta_{cr} = \frac{2(1+v)(2-3v)}{11-19v}, \tag{15.34}$$

where v is Poisson's ratio.

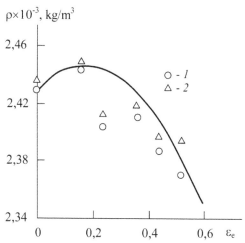

FIGURE 15.32 The dependence of specimens density ρ on extrusion strain ε_e for extruded (1) and annealed (2) REP [62].

Accepting the average value $v \approx 0.35$, we obtain $\Delta\delta_{cr} = 0.60$, that is essentially higher than the estimation $\Delta\rho$ made earlier. These calculations assume that ρ decrease at $\varepsilon_e = 0.43$ and 0.52 is due to instable nanoclusters decay and to corresponding REP structure loosening.

The stated above data give a clear example of large possibilities of polymer properties operation through its structure change. From the plots of Fig. 15.29 it follows that annealing of REP extruded up to $\varepsilon_e = 0.52$ results to elasticity modulus increase in more than 8 times and from the data of Fig. 15.30 yield stress increase in 6 times follows. From the practical point of view the extrusion and subsequent annealing of rarely cross-linked epoxy polymers allow to obtain materials, which are just as good by stiffness and strength as densely cross-linked epoxy polymers, but exceeding the latter by plasticity degree. Let us note, that besides extrusion and annealing other modes of polymers nanostructure operation exist: plasticization [70], filling [26, 71], films obtaining from different solvents [72] and so on.

Hence, the stated above results demonstrated that neither cross-linking degree nor molecular orientation level defined cross-linked polymers final properties. The factor, controlling properties is a state of suprasegmental (nanocluster) structure, which, in its turn, can be goal-directly regulated by molecular orientation and thermal treatment application [62].

In the stated above treatment not only nanostructure integral characteristics (macromolecular entanglements cluster network density v_{cl} or nanocluster relative fraction φ_{cl}), but also separate nanocluster parameters are important (see Section 15.1). In this case of particulate-filled polymer nanocomposites (artificial nanocomposites) it is well-known, that their elasticity modulus sharply increases at nanofiller particles size decrease [17]. The similar effect was noted above for REP, subjected to different kinds of processing (see Fig. 15.28). Therefore, the authors of Ref. [73] carried out the study of the dependence of elasticity modulus E on nanoclusters size for REP.

It has been shown earlier on the example of PC, that the value E_p is defined completely by natural nanocomposite (polymer) structure according to the Eq. (15.32) (see Fig. 15.26).

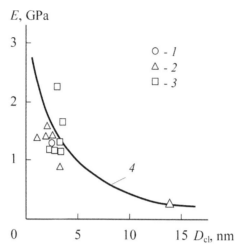

FIGURE 15.33 The dependence of elasticity modulus E_p on nanoclusters diameter D_{cl} for initial (1), extruded (2) and annealed (3) REP. 4 – calculation according to the Eq. (15.32) [73].

In Fig. 15.33, the dependence of E_p on nanoclusters diameter D_{cl}, determined according to the Eq. (4.64), for REP subjected to the indicated processing kinds at ε_e values within the range of 0.16÷0.52 is adduced. As one can see, like in the case of artificial nanocomposites, for REP strong (approximately of order of magnitude) growth is observed at nanoclusters size decrease from 3 up to 0.9 nm. This fact confirms again, that REP elasticity modulus is defined by neither cross-linking degree nor molecular orientation level, but it depends only on epoxy polymer nanocluster structure state, simulated as natural nanocomposite [73].

Another method of the theoretical dependence $E_p(D_{cl})$ calculation for natural nanocomposites (polymers) is given in Ref. [74]. The authors of Ref. [75] have shown, that the elasticity modulus E value for fractal objects, which are polymers [4], is given by the following percolation relationship:

$$K_T, G \sim (p - p_c)^{\eta}, \tag{15.35}$$

where K_T is bulk modulus, G is shear modulus, p is solid-state component volume fraction, p_c is percolation threshold, η is exponent.

The following equation for the exponent η was obtained at a fractal structure simulation as Serpinsky carpet [75]:

$$\frac{\eta}{v_p} = d - 1,$$ (15.36)

where v_p is correlation length index in percolation theory, d is dimension of Euclidean space, in which a fractal is considered.

As it is known [4], the polymers nanocluster structure represents itself the percolation system, for which $p = \varphi_{cl}$, $p_c = 0.34$ [35] and further it can be written:

$$\frac{R_{cl}}{l_{st}} \sim (f_{cl} - 0,34)^{v_p},$$ (15.37)

where R_{cl} is the distance between nanoclusters, determined according to the Eq. (4.63), l_{st} is statistical segment length, v_p is correlation length index, accepted equal to 0.8 [77].

Since in the considered case the change E_p at n_{cl} variation is interesting first of all, then the authors of Ref. [74] accepted $_{cl}$ = const = 2.5×10^{27} m^{-3}, l_{st} = const = 0.434 nm. The value E_p calculation according to the Eqs. (15.35) and (15.37) allows to determine this parameter according to the formula [74]:

$$E_p = 28,9(f_{cl} - 0,34)^{(d-1)v_p}, \text{ GPa.}$$ (15.38)

In Fig. 15.34 the theoretical dependence (a solid line) of E_p on nanoclusters size (diameter) D_{cl}, calculated according to the Eq. (15.38) is adduced. As one can see, the strong growth E_p at D_{cl} decreasing is observed, which is identical to the shown one in Fig. 15.33. The adduced in Fig. 15.34 experimental data for REP, subjected to hydrostatic extrusion and subsequent annealing, correspond well enough to calculation according to the Eq. (15.38). The decrease D_{cl} from 3.2 up to 0.7 nm results again to E_p growth on order of magnitude [74].

The similar effect can be obtained for linear amorphous polycarbonate (PC) as well. Calculation according to the Eq. (15.38) shows, n_{cl} reduction from 16 (the experimental value n_{cl} at $T = 293$K for PC [5]) up to 2 results to E_p growth from 1.5 up to 5.8 GPa and making of structureless ($n_{cl} = 1$) PC will allow to obtain $E_p \approx 9.2$ GPa, that is, comparable with obtained one for composites on the basis of PC.

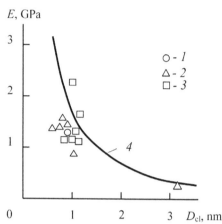

FIGURE 15.34 The dependence of elasticity modulus E_p on nanoclusters diameter D_{cl} for initial (1), extruded (2) and annealed (3) REP. 4 – calculation according to the Eq. (15.38) [74].

Hence, the stated in the present chapter results give purely practical aspect of such theoretical concepts as the cluster model of polymers amorphous state stricture and fractal analysis application for the description of structure and properties of polymers, treated as natural nanocomposites. The necessary nanostructure goal-directed making will allow to obtain polymers, not yielding (and even exceeding) by their properties to the composites, produced on their basis. Structureless (defect-free) polymers are imagined the most perspective in this respect. Such polymers can be natural replacement for a large number of elaborated at present polymer nanocomposites. The application of structureless polymers as artificial nanocomposites polymer matrix can give much larger effect. Such approach allows to obtain polymeric materials, comparable by their characteristics with metals (e.g., with aluminum).

KEYWORDS

- **amorphous polymer**
- **friction coefficient**
- **micromechanical model**
- **nanocluster**
- **natural nanocomposite**
- **reinforcement degree**

REFERENCES

1. Kardos, I. L., & Raisoni, I. (1975). The Potential Mechanical Response of Macromolecular Systems-F Composite Analogy. Polymer Engng. Sci., *15(N3)*, 183–189.
2. Ivanches, S. S., & Ozerin, A. N. (2006). A Nanostructures in Polymeric Systems. Vysokomolek Soed.B, *48(N8)*, 1531–1544.
3. Kozlov, G. V., & Novikov, V. U. (2001). The Cluster Model of Polymers Amorphous State. Uspekhi Fizicheskikh Nauk, *171(N7)*, 717–764.
4. Kozlov, G. V., & Zaikov, G. E. (2004). Structure of the Polymer Amorphous State. Utrecht, Boston, Brill Academic Publishers, 465 p.
5. Kozlov, G. V., Ovcharenko, E. N., & Mikitaev, A. K. (2009). Structure of the Polymer Amorphous State. Moscow, Publishers of the D.I. Mendeleev RKhTU, 392 p.
6. Kozlov, G. V., & Novikov, V. U. (1998). Synergetics and Fractal Analysis of Cross-Linked Polymers. Moscow, Klassika, 112 p.
7. Burya, A. I., Kozlov, G. V., Novikov, V. U., & Ivanova, V. S. (2003). Synergetics of Supersegmental Structure of Amorphous Glassy Polymers. Mater of 3rd Intern.Conf. "Research and Development in Mechanical Industry-RaDMI-03", September 19–23, Herceg Novi, Serbia and Montenegro, 645–647.
8. Bashorov, M. T., & Kozlov, G. V., & Mikitaev, A. K. (2010). Nanostructures and Properties of Amorphous Glassy Polymers. Moscow, Publishers of the D.I. Mendeleev RKhTU, 269 p.
9. Malamatov, A. Kh., Kozlov, G. V., & Mikitaev, M. A. (2006). Reinforcement Mechanisms of Polymer Nanocomposites. Moscow, Publishers of the D.I. Mendeleev RKhTU, 240 p.
10. Kozlov, G. V., Gazaev, M. A., Novikov, V. U., & Mikitaev, A. K. (1996). Simulation of Amorphous Polymers Structure as Percolation Cluster. Pis'ma v ZhTF, *22(N16)*, 31–38.
11. Belousov, V. N., Kotsev, B. Kh., & Mikitaev, A. K. (1983). Two-Step of Amorphous Polymers Glass Transition Doklady ANSSSR, *270(N5), 1145–1147.*
12. Ivanova, V. S., & Kuzeev, I. R., & Zakirnichnaya, M. M. (1998). Synergetics and Fractals. Universality of Metal Mechanical Behaviour. Ufa, Publishers of UGNTU, 366p.
13. Berstein, V. A., & Egorov, V. M. (1990). Differential scanning calorimetry in Physics-Chemistry of the Polymers. Leningrad, Khimiya, 256p.
14. Bashorov, M. T., Kozlov, G. V., & Mikitaev, A. K. (2009). A Nanoclusters Synergetics in Amorphous Glassy Polymers Structure. Inzhenernaya Fizika, N4, 39–42.
15. Bashorov, M. T., Kozlov, G. V., & Mikitaev, A. K. (2009). A Nanostructures in Polymers: Formation synergetics, Regulation Methods and Influence on the properties. Materialovedenie, N9, 39–51.
16. Shevchenko, V. Ya., & Bal'makov, M. D. (2002). A Particles-Centravs as Nanoworld objects. Fizika I Khimiya Stekla, *28(N6)*, 631–636.
17. Mikitaev, A. K., Kozlov, G. V., & Zaikov, G. E. (2008). Polymer Nanocomposites: Variety of Structural Forms and Applications. New York, Nova Science Publishers Inc., 318p.
18. Buchachenko, A. L. (2003). The Nanochemistry Direct Way to High Technologies of New Century. Uspekhi Khimii, *72(N5)*, 419–437.

19. Formanis, G. E. (2003). Self-Assembly of Nanoparticles is Nanoworld Special Properties Spite. Proceedings of Intern. Interdisciplinary Symposium "Fractals and Applied Synergetics", FiPS-03", Moscow, Publishers of MGOU, 303–308.

20. Bashorov, M. T., Kozlov, G. V., Shustov, G. B., & Mikitaev, A. K. (2009). The Estimation of Fractal Dimension of Nanoclusters Surface in Polymers. Izvestiya Vuzov, Severo-Kavkazsk region, estestv. Nauki, *(N6)*, 44–46.

21. Magomedov, G. M., & Kozlov G. V. (2010). Synthesis, Structure and Properties of Cross-Linked Polymers and Nanocomposites on its Basis. Moscow, Publishers of Natural Sciences Academy, 464p.

22. Kozlov, G. V. (2011). Polymers as Natural Nanocomposites: the Missing Opportunities. Recent Patents on Chemical Engineering, *4(N1)*, 53–77.

23. Bovenko, V. N., & Startsev, V. M. (1994). The Discretely-Wave Nature of Amorphous Poliimide Supramolecular Organization. Vysokomolek. Soed. B, *36(N6)*, 1004–1008.

24. Bashorov, M. T., Kozlov, G. V., & Mikitaev, A. K. (2009). Polymers as Natural Nanocomposites: An Interfacial Regions Identification. Proceedings of 12th Intern. Symposium "Order, Disorder and Oxides Properties". Rostov-na-Donu-Loo, (September 17–22), 280–282.

25. Magomedov, G. M., Kozlov, G. V., Amirshikhova, Z. M. (2009). Cross-Linked Polymers as Natural Nanocomposites: An Interfacial Region Identification. Izvestiya DGPU, estestv. I tochn. nauki, *(N4)*, 19–22.

26. Kozlov, G. V., Yanovskii, Yu. G., & Zaikov, G. E. (2010). Structure and Properties of Particulate-Filled Polymer Composites: the Fractal Analysis. New York, Nova Science Publishers Inc., 282p.

27. Bashorov, M. T., Kozlov, G. V., Shustov, G. B., & Mikitaev, A. K. (2009). Polymers as Natural Nanocomposites: the Filling Degree Estimations. Fundamental'nye Issledovaniya, N*4*, 15–18.

28. Vasserman, A. M., & Kovarskii, A. L. (1986). A Spin Probes and Labels in Physics-Chemistry of Polymers. Moscow, Nauka, 246p.

29. Korst, N. N., & Antsiferova, L. I. (1978). A Slow Molecular Motions Study by Stable Radicals EPR Method. Uspekhi Fizicheskikh Nauk, *126(N1)*, 67–99.

30. Yech, G. S. (1979). The General Notions on Amorphous Polymers Structure. Local Order and Chain Conformation Degrees. Vysokomolek.Soed.A, *21(N11)*, 2433–2446.

31. Perepechko, I. I. (1978). Introduction in Physics of Polymers. Moscow, Khimiya, 312p.

32. Kozlov, G. V., & Zaikov, G. E. (2001). The Generalized Description of Local Order in Polymers. In: Fractals and Local Order in Polymeric Materials. Kozlov, G. V., Zaikov, G. E., Ed., New York, Nova Science Publishers Inc. 55–63.

33. Tager, A. A. (1978). Physics-Chemistry of Polymers. Moscow, Khimiya, 416p.

34. Bashorov, M. T., Kozlov, G. V., Malamatov, A. Kh., & Mikitaev, A. K. (2008). Amorphous Glassy Polymers Reinforcement Mechanisms by Nanostructures. Mater of IV Intern.Sci.-Pract.Conf. "New Polymer Composite Materials". Nal'chik, KBSU, 47–51.

35. Bobryshev, A. N., Koromazov, V. N., Babin, L. O., & Solomatov, V. I. (1994). Synergetics of Composite Materials. Lipetsk, NPO ORIUS, 154p.

36. Aphashagova, Z. Kh., Kozlov, G. V., Burya, A. T., & Mikitaev, A. K. (2007). The Prediction of particulate-Filled Polymer Nanocomposites Reinforcement Degree. Materialovedenie, N*9*, 10–13.

37. Sheng, N., Boyce, M. C., Parks, D. M., Rutledge, G. C., Ales, J. I., & Cohen, R. E. (2004). Multiscale Micromechanical modeling of Polymer/Clay Nanocomposites and the Effective Clay Particle. Polymer, *45(N2)*, 487–506.

38. Dickie, R. A. (1980). The Mechanical Properties (Small Strains) of Multiphase Polymer Blends. In: Polymer Blends. Ed. Paul, D. R., Newman, S. New York, San Francisko, London, Academic Press, *1*, 397–437.

39. Ahmed, S., & Jones, F. R. (1990). A review of particulate Reinforcement Theories for Polymer Composites. J.Mater Sci., *25(N12)*, 4933–4942.

40. Balankin, A. S. (1991).Synergetics of Deformable Body. Moscow, Publishers of Ministry Defence SSSR, 404p.

41. Bashorov, M. T., Kozlov, G. V., & Mikitaev, A. K. (2010). Polymers as Natural Nanocomposites: Description of Elasticity Modulus within the Frameworks of Micromechanical Models. Plast. Massy, N*11*, 41–43.

42. Lipatov, Yu. S. (1980). Interfacial Phenomena in Polymers. Kiev, Naukova Dumka, 260p.

43. Yanovskii, Yu. G., Bashorov, M. T., Kozlov, G. V., & Karnet, Yu. N. (2012). Polymeric Mediums as Natural Nanocomposites: Intercomponont Interactions Geometry. Proceedings of All-Russian Conf. "Mechanics and Nanomechanics of Structurally-Complex and Heterogeneous Mediums Achievements, Problems, Perspectives". Moscow, IPROM, 110–117.

44. Tugov, I. I., & Shaulov, A. Yu. (1990). A Particulate-Filled Composites Elasticity Modulus. Vysokomolek. Soed. B, *32(N7)*, 527–529.

45. Piggott, M. R., & Leidner, Y. (1974). Microconceptions about Filled Polymers. Y. Appl. Polymer Sci., *18(N7)*, 1619–1623.

46. Chen, Z. Y., Deutch, Y. M., & Meakin, (1984). Translational Friction Coefficient of Diffusion Limited Aggregates. Y.Chem. Phys., *80(N6)*, 2982–2983.

47. Kozlov, G. V., Beloshenko, V. A., & Varyukhin. V. N. (1998). Simulation of Cross-Linked Polymers Structure as Diffusion-Limited Aggregate. Ukrainskii Fizicheskii Zhurnal, *43(N3)*, 322–323.

48. Novikov, V. U., Kozlov, G. V., & Burlyan, O. Y. (2000). The Fractal Approach to Interfacial Layer in Filled Polymers. Mekhanika Kompozitnykh Materialov, *36(N1)*, 3–32.

49. Stanley, E. H. (1986). A Fractal Surfaces and "Termite" Model for Two-Component Random Materials. In: Fractals in Physics. Ed. Pietronero L., Tosatti E. Amsterdam, Oxford, New York, Tokyo, North-Holland, 463–477.

50. Bashorov, M. T., Kozlov, G. V., Zaikov, G. E., & Mikitaev, A. K. (2009). Polymers as Natural Nanocomposites: Adhesion between Structural Components. Khimicheskaya Fizika i Mezoskopiya, *11(N2)*, 196–203.

51. Dibenedetto, A. T., & Trachte, K. L. (1970). The Brittle Fracture of Amorphous Thermoplastic Polymers. Y. Appl. Polymer Sci., *14(N11)*, 2249–2262.

52. Burya, A. I., Lipatov, Yu. S., Arlamova, N. T., & Kozlov, G. V. Patent by Useful Model N27 199. Polymer composition. It is registered in Ukraine Patents State Resister October 25 2007.

53. Novikov, V. U., & Kozlov, G. V. (1999). Fractal Parametrization of Filled Polymers structure. Mekhanika Kompozitnykh Materialov, 35(N3), 269–290.

54. Potapov, A. A. (2008). A Nanosystems Design Principles Nano- i Mikrosistemnaya Tekhnika, *3(N4)*, 277–280.

55. Bashorov, M. T., Kozlov, G. V., Zaikov, G. E., & Mikitaev, A. K. (2009). Polymers as Natural nanocomposites. 3. The Geometry of Intercomponent Interactions. Chemistry and Chemical Technology, *3(N4)*, 277–280.

56. Bashorov, M. T., Kozlov, G. V., Zaikov, G. E., & Mikitaev, A. K. (2009). Polymers as Natural Nanocomposites. 1. The Reinforcement Structural Model. Chemistry and Chemical Technology, *3(N2)*, 107–110.

57. Edwards, D. C. (1990). Polymer-Filler Interactions in Rubber Reinforcement. I. Mater. Sci., *25(N12)*, 4175–4185.

58. Bashorov, M. T., Kozlov, G. V., & Mikitaev, A. K. (2009). Polymers as Natural nanocomposites: the Comparative Analysis of Reinforcement Mechanism. Nanotekhnika, *(N4)*, 43–45.

59. Bashorov, M. T., Kozlov, G. V., Zaikov, G. E., & Mikitaev, A. K. (2009). Polymers as Natural Nanocomposites. 2. The Comparative Analysis of Reinforcement Mechanism. Chemistry and Chemical Technology, *3(N3)* 183–185.

60. Chen, Y. S., Poliks, M. D., Ober, C. K., Zhang, Y., Wiesner, U., & Giannelis, E. (2002). Study of the Interlayer Expansion Mechanism and Thermal-Mechanical Properties of Surface-Initiated Epoxy Nanocomposites. Polymer, *43(N17)*, 4895–4904.

61. Kozlov, G. V., Beloshenko, V. A., Varyukhin, V. N., & Lipatov, Yu. S. (1999). Application of Cluster Model for the Description of Epoxy Polymers Structure and Properties. Polymer, *40(N4)*, 1045–1051.

62. Bashorov, M. T., Kozlov, G. V., & Mikitaev, A. K. (2009). Nanostructures in Cross-Linked Epoxy Polymers and their Influence on Mechanical Properties. Fizika I Khimiya Obrabotki Materialov, *(N2)*, 76–80.

63. Beloshenko, V. A., Shustov, G. B., Slobodina, V. G., Kozlov, G. V., Varyukhin, V. N., Temiraev, K. B., & Gazaev, M. A. (February 27 1998). Patent on Invention "The Method of Rod-Like Articles Manufacture from Polymers". Clain for Invention Rights N95109832. Patent N2105670. Priority: 13 June 1995. It is registered in Inventions State Register of Russian Federation.

64. Aloev, V. Z., & Kozlov, G. V. (2002). Physics of Orientational Phenomena in Polymeric Materials. Nalchik, Polygraph-service and T, 288p.

65. Kozlov, G. V., Beloshenko, V. A., Garaev, M. A., & Novikov, V. U. (1996). Mechanisms of Yielding and Forced High-Elasticity of Cross-Linked Polymers. Mekhanika Kompozitnykh Materialov, *32(N2)*, 270–278.

66. Shogenov, V. N., Belousov, V. N., Potapov, V. V., Kozlov, G. V., & Prut, E. V. (1991). The Glassy Polyarylatesurfone Curves Stress-Strain Description within the Frameworks of High-Elasticity Concepts. Vysokomolek. Soed. F, *33(N1)*, 155–160.

67. Kozlov, G. V., Beloshenko, V. A., & Shogenov, V. N. (1999). The Amorphous Polymers Structural Relaxation Description within the Frameworks of the Cluster Model. Fiziko-Khimicheskaya Mekhanika Materialov, *35(N5)* 105–108.

68. Kozlov, G. V., Burya, A. I., & Shustov, G. B. (2005). The Influence of Rotating Electromagnetic Field on Glass Transition and Structure of Carbon Plastics on the Basis of lhenylone. Fizika I Khimiya Obrabotki Materialov, *(N5)*, 81–84.

69. Pakter, M. K., Beloshenko, V. A., Beresnev, B. I., Zaika, T. R., Abdrakhmanova, L. A., & Berai, N. I. (1990). Influence of Hydrostatic Processing on Densely Cross-Linked Epoxy Polymers Structural Organization Formation. Vysokomolek. Soed. F, *32(N10)*, 2039–2046.

70. Kozlov, G. V., Sanditov, D. S., & Lipatov, Yu. S. (2001). Structural and Mechanical Properties of Amorphous Polymers in Yielding Region. In: Fractals and Local Order in Polymeric Materials. Kozlov, G. V., Zaikov, G. E., Ed., New York, Nova Science Publishers Inc., 65–82.

71. Kozlov, G. V., Yanovskii, Yu. G., & Zaikov, G. E. (2011). Synergetics and Fractal Analysis of Polymer Composites Filled with Short Fibers. New York, Nova Science Publishers Inc., 223p.

72. Shogenov, V. N., & Kozlov, G. V. (2002). Fractal Clusters in Physics-Chemistry of Polymers. Nal'chik, Polygraphservice and T, 270p.

73. Kozlov, G. V., & Mikitaev, A. K. (2010). Polymers as Natural Nanocomposites: Unrealized Potential. Saarbrücken, Lambert Academic Publishing, 323p.

74. Magomedov, G. M., Kozlov, G. V., & Zaikov, G. E. (2011). Structure and Properties of Cross-Linked Polymers. Shawbury, A Smithers Group Company, 492p.

75. Bergman, D. Y., & Kantor, Y. (1984). Critical Properties of an Elastic Fractal. Phys. Rev. Lett., *53(N6)*, 511–514.

76. Malamatov, A. Kh., & Kozlov, G. V. (2005). The Fractal Model of Polymer-Polymeric Nanocomposites Elasticity. Proceedings of Fourth Intern. Interdisciplinary Symposium "Fractals and Applied Synergetics FaAS-*05*". Moscow, Interkontakt Nauka, 119–122.

77. Sokolov, I. M. (1986). Dimensions and Other Geometrical Critical Exponents in Percolation Theory. Uspekhi Fizicheskikh Nauk, *151(N2)*, 221–248.

APPENDIX

THE FRACTAL MODEL OF RUBBERS DEFORMATION

CONTENTS

The experimental data about rubbers deformation are usually interpreted within the frameworks of the high-elasticity entropic theory [1–3], elaborated on the basis of assumptions about high-elastic polymers incompressibility (Poisson's ratio $v = 0.5$) and polymer chains Gaussian statistics. As it is known [4], the Gaussian statistic is characteristic only for the networks, prepared by chains concentrated solution curing, in the case of their compression or weak (draw ratio $\lambda < 1.2$) tension. For such structures the fractal dimension $d_f = 2$ and in case of $v = 0.5$ the following classical expression was obtained [3]:

$$\sigma = \frac{E}{3}\left(\lambda - \lambda^{-2}\right), \tag{A.1}$$

where E is Young's modulus, proportional to temperature, the expressions of which through polymer structure parameters are made more exact repeatedly [1–3]. Calculations according to the Eq. (A.1) with fitted E value correspond quite well to the experiment in the relatively small strains only ($\lambda < 1.2$). At $1.2 \leq \lambda \leq 2$ the dependence (A.1) plot is disposed, as a rule, higher and at $\lambda > 2$ – essentially lower than an experimental curve $\sigma(\lambda)$, which at $\lambda > 4$ usually reaches asymptote $\sigma \sim \lambda^2$ [5].

The dependence (A.1) precision improvement was carried out traditionally by the usage of entropic theory phenomenological modifications, the main achievements of which are united in monography [3]. Moreover, the required exactness of calculation and experiment agreement is reached owing to the usage of agreement additional parameters, which are in reality fitted ones. Hence, the high-elasticity entropic theory loses the main advantage in comparison with the approach [6], based on the empirical dependences of elastic potential on strain and temperature invariants using, which at the agreement fitted parameters enough number ensures any beforehand given precision of an experimental data approximation. The main deficiency of both the entropic theory empirical modifications and elastic potential empirical model apart from a large number and not always clear physical significance of agreement parameters is the necessity of the same elasticity parameter, for example, E, different values using for both experimental data description, obtained at different loading conditions, and for the same data description, but within the frameworks of the entropic theory of various modifications or elastic potential [3, 5]. This gives vagueness of both elasticity parameters absolute value and relations between them [7].

Balankin [7–9] has shown that two main assumptions of elastomaterials entropic elasticity classical theory, mentioned above: (1) $d_f = 2$ and (2)

$v = 0.5$ [3] contradict each other (see the Eq. (1.9)). In Refs. [7–9], the relationship was obtained, coupling conditional stress σ and draw ration λ at uniaxial tension (compression) of elastomaterials within the frameworks of fractal analysis:

$$\sigma = \frac{E}{1 + 2v + 4v^3} \left[\lambda^{1+2v} - 2v^{-1-2v(1+v)} - (1 - 2v)^{-2v} \right], \tag{A.2}$$

which differs from the classical expression (A.1) even in the limit $v = 0.5$, when the Eq. (A.2) is transformed to the dependence (the Eq. (6.16)) [9].

The authors of Ref. [10] considered the elasticity and entropic high-elasticity fractal concept (the Eq. (A.2)) [7–9] application for elastomaterials deformation behavior description on the example of styrene-butadiene rubber (SBR) and nanocomposite on its basis with carbon soot content of 34 mas. % (SBR-S).

The experimental curves stress-draw ratio ($\sigma - \lambda$) for SBR and nanocomposite SBR-2 are presented in Fig. A.1 [11]. As one can see, the curves $\sigma - \lambda$ for these materials differ significantly even by exterior appearance: one can visually affirmed, that nanocomposite has higher Young's modulus E and much more strong strain hardening (i.e., modulus at large strains) than matrix rubber SBR. For these curves theoretical description the authors of Ref. [10] used two approaches: classical [3] and fractal [7] ones. The first from them supposes the relation between σ and λ, defined by the equation [3]:

$$\sigma = G\left(\lambda^2 - \lambda^{-1}\right), \tag{A.3}$$

where G is shear modulus, coupled with E according to the equation [4.58].

The adduced in Fig. A.1 experimental data and calculated according to the Eq. (A.3) dependences $\sigma(\lambda)$ for SBR are agreed well at $E = 1.82$ MPa. Within the frameworks of the classical high-elasticity theory the value G in the Eq. (A.3) is defined as follows [3]:

$$G = NkT, \tag{A.4}$$

where N is an active chains number per rubber volume unit, k is Boltzmann constant, T is testing temperature.

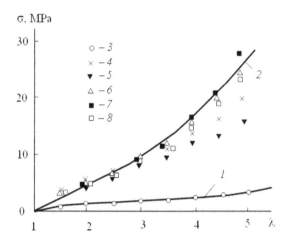

FIGURE A.1   The curves stress-draw ratio $\sigma - \lambda$ for styrene butadiene rubber (1, 3) and nanocomposite on its basis (2, 4 ÷ 8). 1, 2 – the experimental curve. Calculation: 3 – according to the Eq. (A.3) at $E = 1.82$ MPa; 4 – according to the Eq. (A.3) at nominal λ and $E = 3.29$ MPa; 5 – according to the Eq. (A.2) at $\lambda = \lambda_{mol}$ and $E = 6$ MPa; 6 – according to the Eq. (A.2) at $\lambda = \lambda_{mol}$, $E = 2.76$ MPa and $v = 0.433$; 7 – according to the Eq. (A.2) at $\lambda = \lambda_{mol}$, $E = 2.01$ MPa and $v = 0.5$, 8 – according to the Eq. (A.2) at nominal λ, $E = 2.56$ MPa and $v = 0.433$ [10].

According to the Eq. (A.4) cross-linking density for SBR $v_c = 1.52 \; 10^{26}$ m^{-3} can be obtained, that corresponds to molecular weight between cross-linking nodes $M_c = 4400$ g/mole. This value corresponds well to the known data ($M_c \approx 5000$ g/mole) for SBR [11].

However, the attempts to describe $\sigma - \lambda$ curve for nanocomposite SBR-S with the help of the Eq. (A.3) were unsuccessful (see Fig. A.1). Already at $\lambda = 3$ the essential discrepancy of theory and experiment (theoretical result are lower than experimental) is observed and this discrepancy grows at λ increasing. Let us note, that σ calculation according to the Eq. (A.3) for SBR-S was carried out in two variants. The first used the nominal value λ and the second used real (molecular) draw ratio λ_{mol}, determined according to the equation [12]:

$$\lambda = \lambda_p(1-\varphi_n) + \lambda_n\varphi_n, \tag{A.5}$$

where λ_p and λ_n are draw ratios for polymer matrix and nanofiller, accordingly, φ_n is nanofiller volume contents. It is obvious, that $\lambda_n = 0$ [13].

The usage of λ_{mol} values at σ calculation according to the Eq. (A.3) does not improve the conformity of theory and experiment (see Fig. A.1).

It is known through Ref. [14] that filler introduction in polymer matrix changes Poissonэs ratio and for composites the value ν can be estimated according to the equation:

$$\nu = \nu_p(1-\varphi_n) + \nu_n\varphi_n, \qquad\qquad (A.6)$$

where ν_p and ν_n are the Poisson's ratio values for polymer matrix and nanofiller accepted equal to 0.50 and 0.25 accordingly [13]. Then $\nu = 0.433$ for nanocomposite SBR-S.

One can see a good conformity of theory and experiment one can see (Fig. A.1) at the comparison of calculation according to the Eq. (A.2) results at $E = 2.76$ MPa u $\nu = 0.433$ and the experimental curve $\sigma(\lambda)$. The similar comparison, but at other initial conditions ($E = 2.01$ MPa, $\nu = 0.50$), has shown, that at $\lambda = 5$ faster growth of theoretical curve $\sigma(\lambda)$ in comparison with experiment begins. In other words, the incompressibility condition acceptance for nanocomposite (ν makes up 0.50) worsens the theory and experiment conformity. And at last, the calculation according to the Eq. (A.2) was carried out at nominal value λ, $E = 2.56$ MPa and $\nu = 0.433$. In this case the lower calculated values σ at $\lambda > 3$ in comparison with the experiment are obtained (Fig. A.1).

The parameter E increases from 1.82 MPa for SBR up to 2.76 MPa for SBR-S means according to the Eq. (A.4) ν_c enhancement from 1.52 10^{26} m^{-3} up to 2.52 10^{26} m^{-3} or M_c reduction from 4400 g/mole for SBR up to 3400 g/mole for SBR-S. It can be assumed, that this ν_c enhancement is the consequence of interfacial regions formation on the polymer – filler interfacial boundary, in which segments molecular mobility is frozen owing to filler particles fractal surface influence. This corresponds to the generalized definition of physical entanglements cluster network in amorphous state of polymers [15, 16], where each contact of two segments with freezing mobility is considered as the indicated network node. In this case relative volume fraction of such nodes (clusters) of entanglements network φ_{cl} is estimated according to the Eq. (1.11). For the constituent SBR polybutadiene (PB) and polystyrene (PS) the values C, l_0 and S are equal to 5.5 and 10 Å [17], 1.47 and 1.54 Å, 19.3 and 69.8 Å^2 [18], accordingly. In SBR case as C, l_0 and S the average values of these parameters for PB and PS were used. Then the value φ_{cl} will be equal to 0.121 and the possibility of independent calculation of nanocomposite structure fractal dimension d_f appears according to the Eq. (1.12), which proves to be equal to 2.888 [13].

The other variant of d_f calculation uses the Eq. (1.9) and, since $d = 3$, $v = 0.433$, then according to this equation $d_f = 2.866$ follows. Hence, d_f values according to the two adduced estimation methods are well agreed (the discrepancy makes up less than 1%). Such agreement indicates unequivocally on the definite intercommunication of parameters φ_n and φ_{cl} or interfacial regions fraction in nanocomposite [10].

Let us note in conclusion an interesting aspect of the Eq. (A.2) application for the description of curves $\sigma(\lambda)$, having strong strain hardening. The fractal dimension d_f reduction according to the Eq. (1.9) means v decrease and the member before square brackets in the Eq. (A.2) enhancement, that increases σ. But this decreases at the same time the first item in square brackets owing to the exponent $(1 + 2v)$ for λ reduction. Therefore, the strong strain hardening can be obtained at $\lambda_{mol} = \lambda_p$ usage, that compensates the indicated exponent decrease. Besides, the nanofiller volume contents φ_n increase results to λ_{mol} enhancement and, hence, to strain hardening intensification [10].

Hence, the stated above results show that the classical theory of entropic high-elasticity can be used for the description of stress-draw ratio curves for rubbers with weak strain hardening, but it is incorrect in case of nanocomposites with elastomeric matrix. The correct description of deformation behavior of the latter gives the high-elasticity fractal model that is due to fractal nature of filled rubbers structure [13].

KEYWORDS

- active chains
- deformation
- elastic potential
- fractal theory
- rubber
- strain hardening

REFERENCES

1. De Gennes, P. G. (1979). Scaling Concepts in Polymer Physics. Ithaca, NY, Cornell University Press, 352 p.
2. Grosberg, A. Y., & Khokhlov, A. R. (1989). Statistical Physics of Macromolecules. Moscow, Nauka, 344 p.
3. Bartenev, G. M., & Frenkel, S. Ya. (1990). Physics of Polymers. Leningrad, Khimiya, 432 p.
4. Panyukov, S. V. (1990). Scaling Theory of High-Elasticity Zhyrnal Eksperimental noi i Teoreticheskoi Fiziki, *98(2)*, 668–680.
5. Gul, V. E., & Kuleznev, V. N. (1979). Structure and Mechanical Properties of Polymers. Moscow, Vysshaya Shkola, 353 p.
6. Landau, L. D., & Lifshits, E. M. (1987). The Elasticity Theory. Moscow, Nauka, 248 p.
7. Balankin, A. S. (1991). The Fractals Elasticity and Entropic High-Elasticity Theory. Pis ma v ZhTF, *17(17)*, 68–72.
8. Balankin A. S. Fractal Dynamics of Deformable Solid. Metally, 1992, № 2, p. 41–51.
9. Balankin, A. S. (1992). Fractal Elastic Properties and Solids Brittle Fracture Dynamics Fizika Tverdogo Tela, *34(4)*, 1245–1258.
10. Kozlov, G. V., Aloev, V. Z., & Byrya, A. I. (2004). Simulation of Deformation Curves of Nanocomposites with Elastomeric Matrix: the Comparison of Classical and Fractal Theories. In: Selected Works of Republican Sci. Seminar "Mekhanika." Nal'chik, KBSAA, 58–61.
11. Edwards, D. C. (1990). Polymer-Filler Interactions in Rubber Reinforcement. J. Mater. Sci., *25(12)*, 4175–4185.
12. Buisson, G., & Ravi-Chandar, K. (1990). On the Constitutive Behavior of Polycarbonate under Large Deformation. Polymer, *31(11)*, 2071–2076.
13. Mikitaev, A. K., Kozlov, G. V., & Zaikov, G. E. (2008). Polymer Nanocomposites: Variety of Structural Forms and Applications. New York, Nova Science Publishers, Inc., 318 p.
14. Kubat, J., Rigdahl, M., & Welander, M. (1990). Characterization of Interfacial Interactions in High Density Polyethylene Filled with Glass Spheres Using Dynamic-Mechanical Analysis. J. Appl. Polymer. Sci., *39(5)*, 1527–1539.
15. Kozlov, G. V., & Novikov, V. U. The Cluster Model of Polymers Amorphous State. Uspekhi Fizicheskikh Nauk, 2001, *171(7)*, 717–764.
16. Kozlov, G. V., & Zaikov, G. E. (2004). Structure of the Polymer Amorphous State. Utrecht, Boston, Brill Academic Publishers, 465 p.
17. Aharoni, S. M. (1983). On Entanglements of Flexible and Rod-Like Polymers. Macromolecules, *16(9)*, 1722–1728.
18. Aharoni, S. M. (1985). Correlations between Chain Parameters and Failure Characteristics of Polymers below their Glass Transition Temperature. Macromolecules, *18(12)*, 2624–2630.

INDEX

N

Milton Keynes UK
Ingram Content Group UK Ltd.
UKHW031140141024
449569UK00024B/1172